Conservation Agriculture in India

This book examines the current situation, levels of adoption, management practices, and the future outlook of conservation agriculture in India, and also in other tropical and subtropical regions of the world.

While conservation agriculture is proposed as an important means to combat climate change, improve crop productivity and food affordability, and to protect the environment, the adoption of conservation agriculture in India, and south-east Asia more broadly, has been slow. This volume reflects on the current status of conservation agriculture in India, asking why adoption has been slow and putting forward strategies to improve its uptake. The chapters cover the various aspects of crop management such as soil, water, nutrients, weeds, crop residues, machinery, and energy, in a range of environments, including irrigated and rainfed regions. The impact of climate change and the economic considerations behind the adoption of conservation agriculture are also discussed. The volume concludes by discussing the future outlook for conservation agriculture in India, in particular drawing out parallels with other tropical and subtropical regions of the world.

This book will be of great interest to students and scholars of conservation agriculture, sustainable agriculture, crop and soil management, and environmental and natural resource management.

A.R. Sharma is Director of Research at Rani Lakshmi Bai Central Agricultural University, Jhansi, Uttar Pradesh, India. Previously, he served as Director, ICAR–Directorate of Weed Research, Jabalpur (2012–2017); and Professor and Head at the Division of Agronomy, ICAR–Indian Agricultural Research Institute, Pusa, New Delhi (2009–2012). He possesses an unparalleled record of academic achievements throughout, and has made outstanding contributions in the field of natural resource management over the last 35 years of his professional career including nutrient and weed management, and conservation agriculture.

Earthscan Food and Agriculture

For more information about this series, please visit: www.routledge.com/books/
series/ECEFA/

Conservation Agriculture in India

A Paradigm Shift for Sustainable Production

Edited by
A.R. Sharma

Routledge
Taylor & Francis Group
LONDON AND NEW YORK

earthscan
from Routledge

First published 2023
by Routledge
4 Park Square, Milton Park, Abingdon, Oxon OX14 4RN

and by Routledge
605 Third Avenue, New York, NY 10158

Routledge is an imprint of the Taylor & Francis Group, an informa business

British Library Cataloguing-in-Publication Data
A catalogue record for this book is available from the British Library

Library of Congress Cataloging-in-Publication Data
Names: Sharma, A. R., 1960– editor.
Title: Conservation agriculture in India : a paradigm shift for
sustainable production / Edited by A.R. Sharma
Description: First edition | New York, NY : Routledge, 2023. |
Series: Earthscan food and agriculture |
Includes bibliographical references and index.
Identifiers: LCCN 2022011231 (print) | LCCN 2022011232 (ebook) |
ISBN 9781032273877 (hardback) | ISBN 9781032273938 (paperback) |
ISBN 9781003292487 (ebook)
Subjects: LCSH: Sustainable agriculture–India. |
Vegetation and climate–India. | Alternative agriculture–India.
Classification: LCC S471.I4 C66 2023 (print) | LCC S471.I4 (ebook) |
DDC 338.10954–dc23/eng/20220330
LC record available at https://lccn.loc.gov/2022011231
LC ebook record available at https://lccn.loc.gov/2022011232

ISBN: 978-1-032-27387-7 (hbk)
ISBN: 978-1-032-27393-8 (pbk)
ISBN: 978-1-003-29248-7 (ebk)

DOI: 10.4324/9781003292487

Typeset in Goudy
by Newgen Publishing UK

Contents

Contributors

A. Arunachalam, ICAR–Central Agroforestry Research Institute, Jhansi, Uttar Pradesh, India

A.K. Biswas, ICAR–Indian Institute for Soil Science, Bhopal, Madhya Pradesh, India

A.R. Sharma, Rani Lakshmi Bai Central Agricultural University, Jhansi, Uttar Pradesh, India

A.R. Uthappa, ICAR–Central Agroforestry Research Institute, Jhansi, Uttar Pradesh, India

Amir Kassam, Global Conservation Agriculture Community of Practice (CA-CoP) and University of Reading, United Kingdom

Amit Kumar, Department of Botany, Dayalbagh Educational Institute, Dayalbagh, Agra, Uttar Pradesh, India

Arti Bhatia, ICAR–Indian Agricultural Research Institute, New Delhi, India

Asha Ram, ICAR–Central Agroforestry Research Institute, Jhansi, Uttar Pradesh, India

Avijit Ghosh, ICAR–Indian Agricultural Research Institute, New Delhi, India

C.M. Parihar, ICAR–Indian Agricultural Research Institute, Pusa, New Delhi

D.M. Mahala, ICAR–Indian Institute of Maize Research, New Delhi, India

Dharvinder Singh, Punjab Agricultural University, Ludhiana, Punjab, India

G. Pratibha, ICAR–Central Research Institute for Dryland Agriculture, Hyderabad, Telangana, India

Geeta Singh, ICAR–Indian Agricultural Research Institute, New Delhi, India

H. Arunakumari, ICAR–Central Research Institute for Dryland Agriculture, Hyderabad, Telangana, India

H.S. Dhaliwal, Punjab Agricultural University, Ludhiana, Punjab, India

H.S. Jat, ICAR–Central Soil Salinity Research Institute, Karnal, Haryana, India

H.S. Nayak, ICAR–Indian Agricultural Research Institute, New Delhi, India

H.S. Sidhu, Borlaug Institute for South Asia, CIMMYT, Ludhiana, Punjab, India

I. Srinivas, ICAR–Central Research Institute for Dryland Agriculture, Hyderabad, Telangana, India

I.P. Abrol, Centre for Advancement of Sustainable Agriculture, New Delhi, India

Inder Dev, ICAR–Central Agroforestry Research Institute, Jhansi, Uttar Pradesh, India

J. Somasundaram, ICAR–Indian Institute of Soil Science, Bhopal, Madhya Pardesh, India

K. Majumdar, International Plant Nutrition Institute, Gurgaon, Haryana, India

K. Patra, ICAR–Indian Agricultural Research Institute, New Delhi, India

K. Sammi Reddy, ICAR–Central Research Institute for Dryland Agriculture, Hyderabad, Telangana, India

K.K. Bandyopadhyay, ICAR–Indian Agricultural Research Institute, New Delhi, India

K.L. Sharma, ICAR–Central Research Institute for Dryland Agriculture, Hyderabad, Telangana, India

K.M. Hati, ICAR-Indian Institute of Soil Science, Bhopal, Madhya Pradesh, India

K.V. Rao, ICAR–Central Research Institute for Dryland Agriculture, Hyderabad, Telangana, India

M. Srinivas Rao, ICAR–Central Research Institute for Dryland Agriculture, Hyderabad, Telangana, India

M.D. Parihar, Chaudhary Charan Singh Haryana Agricultural University, Hisar, Haryana, India

M.L. Jat, International Maize and Wheat Improvement Center, Pusa, New Delhi, India

Manpreet Singh, Punjab Agricultural University, Ludhiana, Punjab, India

N.T. Yaduraju, ICAR–Directorate of Weed Research, Jabalpur, Madhya Pradesh, India

Naresh Kumar, ICAR–Central Agroforestry Research Institute, Jhansi, Uttar Pradesh, India

Naveen Gupta, Punjab Agricultural University, Ludhiana, Punjab, India

P.K. Sahoo, ICAR–Indian Agricultural Research Institute, New Delhi, India

Raj Gupta, Centre for Advancement of Sustainable Agriculture, New Delhi, India

Rajbir Singh, ICAR–Agricultural Technology Application Research Institute, Ludhiana, Punjab, India

Ranjan Bhattacharyya, ICAR–Indian Agricultural Research Institute, New Delhi, India

Shiv Kumar Lohan, Punjab Agricultural University, Ludhiana, Punjab, India

S.L. Jat, ICAR–Indian Institute of Maize Research, New Delhi, India

Sourav Ghosh, ICAR–Directorate of Onion and Garlic Research, Pune, Maharashtra, India

T. Satyanarayana, International Plant Nutrition Institute, Gurgaon, Haryana, India

T.K. Das, ICAR–Indian Agricultural Research Institute, New Delhi, India

U.K. Behera, College of Agriculture, Central Agricultural University, Kyrdemkulai, Meghalaya, India

U.K. Mandal, ICAR–Central Soil Salinity Research Institute, Regional Research Station, Canning Town, West Bengal, India

V.K. Singh, ICAR–Central Research Institute for Dryland Agriculture, Hyderabad, Telangana, India

Y.S. Saharawat, International Fertilizer Development Centre, New Delhi, India

Yadvinder-Singh, Borlaug Institute for South Asia, CIMMYT, Ludhiana, Punjab, India

Foreword I

The Green Revolution during the 1960s led to increased productivity and the elimination of acute foodgrain shortages in India. These technologies primarily involved growing of high-yielding dwarf varieties of rice and wheat, increased use of chemical fertilizers and other agrochemicals, and the expansion of irrigation facilities. This was also accompanied by the other so-called modern methods of cultivation, which included maximum tilling of land, virtually clean cultivation with complete removal of crop residues and in-field burning of biomass, fixed crop rotations mostly involving cereals, and the elimination of fertility-restoring pulses and oilseed crops in the highly productive north-western plain zone of the country.

Over the last 4–5 decades (since circa 1970), India has achieved not only self-sufficiency in foodgrain production but also the capability to export food commodities. This is cited as one of the greatest accomplishments of Indian agriculture in the post-independence era. However, the transformation from 'traditional animal-based subsistence farming' to 'intensive chemical- and tractor-based modern agriculture' has led to a multiplicity of issues associated with the sustainability of these production practices. The adoption of these technologies has led to declining factor productivity, degrading soil health, surface and groundwater eutrophication, air pollution, increasing cost of production, and lower profitability. Furthermore, soils are becoming impoverished due to the unbalanced use of fertilizers, and the discontinuation of traditional practices like mulching, intercropping, and inclusion of legumes in cropping systems. Consequently, the use of organic manures, compost, and green manure crops has also decreased considerably for various reasons. Similarly, water resources are under great stress due to their indiscriminate exploitation and also being polluted due to anthropogenic perturbations. Burning of fossil fuels and crop residues, and puddling for rice cultivation lead to the emission of greenhouse gases (GHGs), which are responsible for climate change and global warming. There is also a growing realization that productivity levels are stagnating, and the incomes of farmers are reducing due to the rising costs of the inputs and farm operations. Modern cultivation practices are not sustainable, and there is a need for a paradigm shift from the business as usual of crop production on arable lands.

Conservation agriculture (CA) is a new paradigm in resource management for alleviating the problems associated with the so-called modern cultivation practices. The three principles of CA are: (i) conversion to no-till or minimal mechanical soil disturbance, (ii) maintenance of permanent biomass mulch cover on the soil surface, and (iii) diversification of crop species and the use of cover crops and forages in the rotation cycle. These principles are universally applicable to all agricultural landscapes and land uses with locally fine-tuned and adapted practices. CA also requires suitable modifications in the use of supplemental irrigation, fertilizers, and weed and pest management practices as well as farm machinery compared with those for conventional tillage. It is a holistic approach to improving productivity and soil health.

Two major innovations in the latter half of the 20th century have led to a change in our thinking on crop production. These are the availability of: (i) new farm machinery, and (ii) effective herbicides, which suggest that ploughing of fields is no longer required for sowing, fertilizer placement, and weeding. A new generation of farm machinery can place the seed and fertilizer at an appropriate depth in the desired amounts. Furthermore, these machines can work in standing as well as slashed crop residues; thus, providing a very effective mulch cover for moisture and nutrient conservation, soil temperature moderation, and weed control. In addition to the availability of new herbicides, biotic systems of weed control (with mulching and cover cropping) have also necessitated a change in our thinking about weed management. Furthermore, other triggering factors for a shift towards CA are changing climate, increasingly scarce labour availability, degrading soil health, declining factor productivity, rising costs, and ever-decreasing farm incomes. Thus, CA systems help in overcoming the problems being experienced by ploughed systems.

CA-based technologies have been developed, and the benefits in terms of enhanced productivity, profitability, soil health, and climate resilience have been documented in most parts of the country. However, these technologies have been adopted on a limited scale due to the apparent apprehensions, lack of will, and some operational constraints. There exists an immense scope for bringing barren and fallow areas under profitable cropping systems with the adoption of CA-based technologies. Such a transformational change will require a coordinated effort involving multi-stakeholders to enhance farmers' awareness and demonstrate the effectiveness of these technologies on a large scale. Furthermore, necessary back-up in the form of suitable farm machinery (on a rental basis) is required to facilitate the adoption of these technologies.

CA is a challenging and exciting subject for researchers, requiring a multi-disciplinary approach to address all the related issues in a comprehensive manner. Resource management scientists should take the lead to finetune CA for the site-specific conditions and perfect the technologies in their respective domains. Focus must be on reducing the cost of production while enhancing productivity, and improving soil quality and soil health. CA-based on-farm research should be an integral part of on-station research, without which research findings cannot be transformed into technologies. CA has been adopted globally on about 200 M ha, including in North and South America, Australia, etc., and this transformation must also happen in India.

India has made commendable progress in developing CA-based technologies in different eco-regions. This book is an up-to-date synthesis of the available information on the potential and challenges of adopting CA in India. I am pleased that Dr. A.R. Sharma has taken this initiative to collate, organize, and present the available information, which is of interest to academicians, scientists, policy makers, research administrators, and field functionaries for enriching their knowledge on CA-based technologies. I commend the hard work and commitment of Dr. Sharma for preparing the book. It is indeed an important milestone in promoting the adoption of CA in India, South Asia, and elsewhere with similar biophysical and social environments.

(*Rattan Lal*)
(World Food Laureate 2020)
Distinguished University Professor of Soil Science
Director, CFAES Rattan Lal Center for Carbon Management
and Sequestration
IICA Chair in Soil Science and Goodwill Ambassador for
Sustainable Development Issues
Adjunct Professor, University of Iceland, & the Indian
Agricultural Research Institute
The Ohio State University

Foreword II

The invention of the plough for soil inversion and growing crops during the mid-19th century was one of the major milestones in the history of agriculture. For centuries, conventional agricultural systems have been characterized by intensive tillage and soil inversion operations, clean cultivation, extensive use of irrigation water, and chemical fertilisers. These have no doubt brought yield revolutions, but at the cost of over-exploitation of our natural resources. In South Asia, the adoption of intensive tillage practices since the Green Revolution in the mid-1960s has led to a factor productivity decline, deterioration in soil health, depletion of groundwater, increase in the cost of production, lower profitability, and environmental degradation. Therefore, it has been argued that the system of crop production should be suitably modified in accordance with the changing environment and deteriorating natural resources.

Also, the Sustainable Development Goals (SDGs) demand an emphasis on regenerative, climate-smart, and profitable farm innovations. Over recent years, attention has been given to conservation agriculture (CA) as a strategy towards 'sustainable intensification' and regenerative agriculture (RA), especially for smallholder farming in South Asia. Accordingly, CA has emerged as an alternative to inefficient tillage-based agriculture. CA is actually an ecosystem approach towards RA, mainly based on three interlinked principles: (i) no-till or minimum mechanical soil disturbance, (ii) permanent maintenance of soil mulch (crop biomass and cover crops), and (iii) diversification of the cropping system (economically, environmentally, and socially adapted rotations, including legumes and cover crops).

Conservation agriculture has been adopted so far on about 200 M ha in countries such as the United States, Canada, Brazil, Argentina, and Australia. In contrast, the uptake of CA has rather remained slow in Asian and African countries. In India, efforts to promote CA started in the mid-1990s, primarily in the rice–wheat cropping system in the north-western plains. The area of zero-till wheat reached about 3 M ha in the early part of the 21st century. In recent times, issues related to crop residue burning have assumed serious concern. In this context, the recycling of crop residues under the CA system is a rather practical, economically viable, and environmentally sound option to deal with the problem. It will help in managing crop residues in combine-harvested fields by avoiding their burning, reducing the cost of cultivation by eliminating elaborate tillage operations, and improving soil health due to residue recycling. Science-based evidence has demonstrated that the adoption of CA-based sustainable intensification has merits beyond resource conservation for minimizing climatic risks and in reducing environmental footprints due to reduced GHG emissions and sequestering C in the soil.

Conservation agriculture is a knowledge-intensive concept, which requires specialized expertise and location-specific adaptation. Despite its proven benefits, adoption of the CA system requires a paradigm shift from commodity- and technology-centric conventional agriculture to system-based management using portfolio approaches. The CA technologies are essentially herbicide-driven, machine-driven, and knowledge-driven. It, therefore, requires good expertise and resources for adoption on a large scale. For wider adoption, there is a need to change the mindset of policy makers, researchers, and farmers. Tremendous efforts will be needed to persuade farmers to adopt CA, especially in dryland areas. In addition to new science, knowledge, and capacity, the accelerated adoption of CA would need urgent policy changes, especially to incentivize farmers for the ecosystem services by adopting CA on their farms, thus helping the nation in carbon sequestration. We must aim to cover at least 20 M ha under CA in India by 2030.

A great deal of research has been carried out on CA across diverse cropping systems, soils, and ecological conditions in India. In most of these studies, CA has had either the same or higher crop yields, besides other direct and indirect benefits. However, there has not been a comprehensive synthesis on the work done on CA so far. I am pleased to note that Dr. A.R. Sharma has taken this initiative to compile the available information on CA for the benefit of researchers, extension agents, policy planners, and other stakeholders. I am sure this book will prove to be highly useful in appreciating the expected benefits and promote faster adoption of CA in India. Dr Sharma deserves full appreciation for the timely bringing out of this useful publication.

(R.S. Paroda)
(Padma Bhushan Awardee)
Founder Chairman, TAAS
Former Director General, ICAR and
Secretary, DARE, Government of India

Preface

Conservation agriculture is the fastest adopted technology globally, covering >200 M ha and showing a double-digit annual growth (>10%) over the last two decades. Currently, it occupies around 15% of the global cropped area, mainly in the countries of South America (>65% coverage), North America (>30%), and Australia and New Zealand (50%). Adoption in other parts of the globe, including Russia, Asia, Africa, and Europe, remains low (<5%) but it has been picking up in recent times. It is envisaged to cover 50% of the global cropped area under conservation agriculture by 2050.

In India, the Green Revolution during the mid-1960s resulted in a tremendous increase in foodgrain production, leading to food sufficiency over the next two decades by the mid-1980s. However, the productivity levels stagnated thereafter, and also there were emerging concerns about resource degradation due to excessive exploitation of natural resources in the highly productive zones of the country. Indiscriminate use of chemical fertilizers and other agro-chemicals, excessive exploitation of ground and surface water for flood irrigation, energy for intensive tillage operations, crop residue burning, and decreased use of organics are also responsible for deteriorating soil health. Furthermore, climate change has emerged as a major challenge, having adverse effects on agricultural productivity in conventional farming systems.

Research on conservation agriculture in India has been on-going since the mid-1990s, and picked up from the mid-2000s as seen from the number of research articles published in leading Indian journals and elsewhere. Most research on conservation agriculture in different regions of the country on crops like wheat, rice, maize, soybean, and other crops has shown that in >80% of cases, the yields are either greater than or equal to those with a conventional agriculture system but with less use of inputs, thus resulting in increased profitability and beneficial effects on soil health. However, the adoption levels on farmers' fields are low, except in some localized regions of north-western and central India for wheat cultivation, coastal areas of Andhra Pradesh for maize and sorghum cultivation, the north-eastern hill region for mustard cultivation, and the Konkan region of Maharashtra for rice and other crops. This is mostly due to a lack of awareness and expertise, farm machinery, incentives, and policy support.

This book is the first up-to-date compilation on the available information on various aspects of conservation agriculture in the Indian context. It presents the practical experiences of research workers associated with this subject for more than two decades. It covers the global scenario and status of conservation agriculture in India; management options; soil health and GHG emissions; economics, adoption and future of conservation agriculture in India, in four different sections and 17 chapters. In addition, internationally renowned scientists, Prof. Rattan Lal, World Food Laureate (2020) and Dr. R.S. Paroda, Padma Bhushan Awardee (2012) have forwarded their inputs for promotion of conservation agriculture in India.

I have acquired adequate knowledge and experience since the late 1990s on conservation agriculture based on my researches in the hilly regions of the western Himalayas, alluvial soils of the Indo-Gangetic plains, vertisols of central India, and now in the impoverished soils of Bundelkhand region. I am thankful to my mentors, Dr. J.S. Samra, Dr. C.L. Acharya, Dr. H.S. Gupta, Dr. Raj Gupta, and Dr. R.S. Paroda; my colleagues, Dr. U.K. Behera, Dr. T.K. Das, and Dr. M.L. Jat, and a large number of post-graduate students who guided and assisted me in my endeavours on this subject.

It is hoped that this book will provide insights and encouragement to my fellow agronomists and other resource management scientists to become not only preachers but practitioners of conservation agriculture to achieve the target of covering 20 M ha in the country by 2030. It will be immensely useful to post-graduate students, teachers and researchers, policy makers, extension personnel, and other stakeholders to enrich their knowledge and further refine this technology in their respective domains.

Jhansi, India (A.R. Sharma)
Editor

Acronyms and Abbreviations

ACRP	Agricultural Crop Residue Burning
AESR	Agroecological sub-region
AICRPDA	All India Coordinated Research Project on Dryland Agriculture
AMF	Arbuscular Mycorrhizal Fungi
APA	Alkaline phosphatase activity
AQI	Air Quality Index
ASS	Automatic Survey System
ATARI	Agricultural Technology Application Research Institute
AWRC	Available Water Retention Capacity
BBF	Broad-bed furrow
BD	Bulk density
BISA	Borlaug Institute for South Asia
BMP	Best management practice
BP	By-product
CA	Conservation agriculture
CAAAP	Conservation Agriculture Alliance for Asia and Pacific
CAAQMS	Continuous Ambient Air Quality Monitoring Station
CEC	Cation Exchange Capacity
CHC	Custom Hiring Centre
CIMMYT	International Centre for Maize and Wheat Improvement
CPE	Cumulative Pan Evaporation
CR	Crop Residues
CREAMS	Consortium for Research on Agroecosystem Monitoring and Modelling from Space
CRI	Crown root initiation
CRM	Crop residue management
CRP-CA	Consortium Research Platform on Conservation Agriculture
CSISA	Cereal Systems Initiative for South Asia
CT	Conventional tillage
CV	Coefficient of variation
CW	Cotton-wheat
DACFW	Department of Agriculture, Cooperation and Farmers' Welfare
DAP	Diammonium phosphate

DARE	Department of Agricultural Research and Education
DEW	Differential Earth Work
DHA	Dehydrogenase Activity
DI	Deficit irrigation
DLI	Differential Levelling Index
DLUC	Differential Land Uniformity Coefficient
DSH	Dry sub-humid
DSR	Direct-seeded rice
DSS	Decision support system
DT	Deep tillage
DTPA	Diethylene triamine penta acetic acid
ECAF	European Conservation Agriculture Federation
ET	Evapotranspiration
EUE	Energy-use efficiency
FAO	Food and Agriculture Organization
FB	Flat-bed
FDA	Fluorescein diacetate
FFP	Farmer fertilizer practice
FI	Flood irrigation
FIRB	Furrow-irrigated raised-bed
FLD	Frontline demonstrations
FP	Farmers' practice
FPO	Farmer producer organization
FYM	Farmyard manure
GDD	Growing degree days
GHG	Greenhouse gas
GNP	Gross national product
GoGHG	Global greenhouse gases
GR	Gross returns
GS	Green Seeker
GWP	Global warming potential
HS	Happy seeder
HST	Happy seeder Technology
HTC	Herbicide-tolerant crop
IARI	Indian Agricultural Research Institute
ICAR	Indian Council of Agricultural Research
ICT	Information Communication and Technology
IDM	Integrated disease management
IEC	Information Education and Communication
IFPRI	International Food Policy Research Institute
IGP	Indo-Gangetic plains
IISS	Indian Institute of Soil Science
INM	Integrated nutrient management
INR	Indian National Rupee
IPCC	Inter-Governmental Panel on Climate Change

IPM	Integrated pest management
IPNI	International Plant Nutrition Institute
IR	Infiltration rate
IRRI	International Rice Research Institute
ISAAA	International Service for the Acquisition of Agri-Biotech Applications
IW	Irrigation water
IWM	Integrated weed management
IWUE	Irrigation water-use efficiency
KAP	Knowledge attitude and practice
KVK	Krishi Vigyan Kendra
LCC	Leaf Colour chart
LED	Light-emitting diode
LFOM	Light fraction organic matter
LLL	Laser land levelling
LLWR	Least limiting water range
M ha	Million hectares
M t	Million tonnes
MB	Mouldboard
MBC	Microbial biomass carbon
MBN	Microbial biomass nitrogen
MCM	Million cubic meter
MDS	Minimum data set
MEA	Millennium Ecosystem Assessment
MER	Machinery energy ratio
MI	Mechanization index
MIT	Mineralization-immobilization turnover
MNRE	Ministry of New and Renewable Energy
MODIS	Moderate resolution imaging spectroradiometry
MOOC	Massive Open Online Course
MSH	Moist sub-humid
MSM	Manual survey method
MT	Minimum tillage
MW	Maize–wheat
MWD	Mean weight diameter
NAAS	National Academy of Agricultural Sciences
NARES	National Agricultural Research and Education System
NARP	National Agricultural Research Project
NASA	National Aeronautical Space Agency
NCT	National Capital Region
NDVI	Normalized difference vegetative index
NE	Nutrient expert
NRS	Nitrogen-rich strip
NT	No tillage
NUE	Nutrient-use efficiency

NW	North-western
OM	Organic matter
OPT	Optimum application rate
PAU	Punjab Agricultural University
PAWC	Plant-available soil water capacity
PB	Permanent bed
PBB	Permanent broad-bed
PM	Particulate matter
PMN	Potentially mineralizable nitrogen
PNB	Permanent narrow-bed
POM	Particulate organic matter
PPCB	Punjab Pollution Control Board
PRB	Permanent raised-bed
PTR	Puddle transplanted rice
RDF	Recommended dose of fertilizer
RDN	Recommended dose of nitrogen
RI	Response index
RM	Rice–maize
RS	Relay seeding/Remote sensing
RT	Reduced tillage
RWCS	Rice–wheat cropping system
RWS	Rice–wheat system
SDD	Stress degree days
SDI	Surface drip irrigation
SFD	Seed-cum-fertilizer drill
SHG	Self-help group
SMB	Soil microbial biomass
SMP	Soil matric potential
SMS	Straw management system
SOC	Soil organic carbon
SOM	Soil organic matter
SPAD	Soil plant analysis development
SPR	Soil penetration resistance
SQI	Soil quality index
SR	State recommendation/Strip rotor
SSDI	Sub-surface drip irrigation
SSNM	Site-specific nutrient management
SYI	Sustainability yield index
TBO	Tree bearing oilseed
TDR	Time domain reflectometry
THS	Turbo happy seeder
TOC	Total organic carbon
TP	Total product
TSN	Total soil nitrogen
WCCA	World Congress on Conservation Agriculture

WHO	World Health Organization
WP	Water productivity
WRG	Water Resources Group
WSA	Water-stable aggregate
WUE	Water-use efficiency
ZT	Zero tillage
ZTDD	Zero tillage double disc
ZTT	Zero tillage inclined T-type

Part I

Conservation Agriculture

Global Scenario and Status in India

1 Conservation Agriculture for Sustainable Intensification

Global Options and Opportunities

Amir Kassam, Y.S. Saharawat, and I.P. Abrol

Introduction

The term sustainable intensification has become popular in recent years. While its definition can vary, it can be considered in both a narrower and a broader sense as a process to optimize production system performance on farms and watersheds as well as at the sector level within the local and national economy so that it is optimal institutionally for society and the environment along the value chains serviced by the public, private, and civil sectors. The narrower ecological definition at the production level applies to the process of optimizing production systems performance. This involves maximizing yields (total output) and factor productivity (efficiency) of the whole production system in space and time with minimum negative consequences on the environment while maximizing system resilience to biotic and abiotic stresses and shocks, and building and sustaining ecosystem resources and functions and the flow of ecosystem societal services (Kassam 2013). This combines increasing biological outputs, productivity (efficiency), resilience, and ecosystem or environmental services through integrated production systems and landscape management in rainfed and irrigated conditions. Protection and management of all the ecosystem functions and services on agricultural and natural landscapes are considered.

In the broader context, the sustainable intensification definition applies at the sector level across value chains for society and the environment and would also encompass the existence of effective demand for biological products by consumers and industry, input and output supply chains for production inputs and biological outputs, as well as the existence of supporting social and economic organizations for social equity, employment, livelihoods, economic growth, and the environment. This implies improving the capacities of people and informal and formal institutions to deliver and utilize affordable inputs efficiently, distribute and utilize biological outputs efficiently so as to avoid excessive wastage, and harness ecosystem services at all landscape levels that benefit producers and consumers alike. This further implies the existence of a supportive knowledge and innovation system, and above all, of enabling policies for public, private, and civil sector engagement to sustain and keep improving the whole food and agriculture system for farmers, society, and the planet.

DOI: 10.4324/9781003292487-2

Whichever way sustainable intensification is defined, it is necessary to achieve desired yields of crops and livestock in ways that do not harm the natural resource base and the environment, and even improve them in terms of quality and functions. One of the common reasons promoted to justify the need for sustainable production intensification is to meet the increasing demand for food and raw materials for industry due to increases in population, income, and urbanization. The other reason is based on the fact that the land resource and environmental degradation caused by conventional tillage-based agricultural production systems, as well as by traditional grazing systems, has become unacceptable worldwide, including to producers themselves, to society in general, to governments, and to the development community globally.

Since the mid-1990s, much has been debated about the need for sustainable production intensification by the mainstream research and development community, and quite rightly so. Generally, at the ecological level, a sustainable production intensification approach would be adopted at the practical level by farmers only if it offers the following crop and land production performance: (i) maximum farm biological outputs with minimum inputs, i.e. maximum yields and total output with minimum costs, and therefore maximum profit; (ii) maximum efficiency of utilization of purchased inputs and natural resources, i.e. maximum individual factor productivities and maximum total productivity; (iii) maximum resilience to biotic and abiotic stresses, including those arising from climate change requiring climate change adaptability and mitigation; and (iv) best quality ecosystem services at all levels from production fields to landscapes and territorial, to sustain maximum output and productivity and meet societal and environmental needs. Additionally, a suitable sustainable production approach must also have the ability to help mitigate potential damage to agricultural land or restore any damage or loss that may occur in agro-ecological land potentials and in ecosystem services during production or rehabilitate agricultural land that has been abandoned due to degradation. Consequently, terms such as regenerative agriculture, climate-smart agriculture, or ecological agriculture have recently become fashionable in describing the kind of agriculture needed to achieve sustainable intensification.

Some 85% of our crop lands globally are under tillage systems (Kassam et al. 2021), and therefore unsustainable in every sense. These systems rely on regular tillage of different types for land preparation, seeding and crop establishment, and for weeding. In addition, farmers managing much of the tilled cropland, especially mechanized tillage farmers, also rely on the use of herbicides to control weeds, making tillage-based production systems unsustainable due to continuous soil degradation and erosion, inefficient use of all purchased inputs such as seeds, nutrients, pesticides, fuel, and labour. We need a production paradigm which not only enables farmers to build and manage production in systems that are sustainable over the longer term but that also is able to restore the total agronomically attainable agro-ecological land potentials (of annual and perennial crops including grasses, pastures, shrubs, trees, and livestock, and of ecosystem services) that have been lost, and at the same time is capable of self-recuperation to cope with normal wear and tear. Indeed, what we need now is a paradigm which has all the above key characteristics, upon which to build sustainable food and agriculture systems, and its principles can be put into practice gainfully by any serious land-based farmer, small or large, rich or poor, men or women, mechanized or not, and by any government and by any public, private, and civil sector institution which is willing to service and support the farmers and food and agriculture system (Kassam 2016a; Paroda 2016).

These characteristics thus allow the conservation or preservation of the natural resource base, as well as the restoration or rehabilitation of any lost agro-ecological functions and productivity potential due to land degradation and erosion, and lead to the enhancement of the ecological functions and productivity potential of the resource base and the environment. Thus, it can be argued that only when all these features are operating satisfactorily in a production system, does the system qualify to be labelled as being 'smart' or 'regenerative'.

These multiple outcomes from a production system, and the fact that they all can be harnessed simultaneously, imply that the terms agro-ecology, regenerative, sustainable, sustainable intensification, and smart or climate-smart agriculture can have real meanings and can be made to work at the practical level by farmers and supported by their service providers and supported by relevant institutions and policies.

The ability of agriculture to meet future demand placed upon it by society is generally analysed by mainstream scientists and policy analysts in terms of available resources and production inputs to supply the required level of agricultural products. Similarly, production systems are commonly assessed on the efficiency and effectiveness of different combinations of inputs, technologies, and/or practices to produce certain agricultural outputs. It is only relatively recently that analyses have begun to address externalities of production systems, such as environmental damage, the associated input factor inefficiencies, and poor resilience against major external biotic and abiotic challenges. However, relatively rarely do mainstream researchers question the actual agricultural paradigm (characterized here as conventional tillage-based agriculture) itself in terms of its continuing appropriateness for the sustainable development agenda and for the environmental challenges faced by agriculture around the world. Equally, the delivery of supportive, regulatory, provisioning, and cultural ecosystem services to society by conventional tillage agriculture has not been an area of serious mainstream research concern.

In general, mainstream approaches to agricultural assessments are simplistic and limited in scope. As a result, they are unable to identify and address the root causes of the damage caused to land resources, the environment, and human health by the current agricultural paradigm. Such assessments are also decoupled from the human and ethical consequences of the demands and pressures placed upon agricultural production by the food and agriculture system as a whole, including consumer demand, diets, human health, industry, government, and the economy. These aspects are also important causes of unsustainability when it is considered that the world already produces more than twice the amount of food needed to feed its total population while wasting 30% of it, and yet mainstream scientists, global models, and multi-national corporations keep arguing the need for even greater production to meet the current and future demand. Ultimately, even ecologically sustainable production systems cannot continue to remain sustainable if they are driven by demand levels and lifestyles that are excessive, wasteful, unhealthy, and unjust, supported by national and global food distribution systems that are discriminatory and not accessible to all (Kassam and Kassam 2020). These aspects and issues of food system which are a major driver for agricultural intensification are outside the scope of this chapter which focuses mainly on the ecological sustainability of agricultural production.

Extent and Seriousness of Land Degradation

The seriousness of agricultural land degradation, and its end result of desertification has been receiving considerable attention from the international community for decades. A major reason for the slow progress in reversing the land degradation trends is the general lack of understanding and awareness about the root causes of land degradation and abandonment. Worldwide empirical and scientific evidence clearly shows that the root causes of soil degradation in agricultural land use and decreasing productivity – as seen in terms of loss of soil health and eventual abandonment of land – are closely related to the soil life-disrupting agricultural paradigm based on mechanical soil tillage, the agricultural methods of using mouldboard ploughs, disc harrows, tine, rotavators, hoes, and other mechanical tools to prepare the fields for crop establishment and weed control. This mechanical disturbance leads to losses in soil organic matter, soil structure, and soil health, and debilitates many important soil- and landscape-mediated ecosystem processes and functions.

For the most part, agricultural soils worldwide have been mechanically destructured, agricultural landscapes are kept exposed and unprotected, and soil life is starved of organic matter, thus being reduced in biological activity and deprived of habitat. The loss of soil biodiversity, damaged soil structure and its self-recuperating capacity or resilience, increased compaction of topsoil and subsoil, poor infiltration and increased water runoff and wind and water erosion, and greater infestation by insect pests, pathogens, and weeds indicate the current poor state of the health of most agricultural soils.

In the developing regions, a combination of all these elements is a major cause of low and stagnant/declining agricultural productivity and inadequate food and nutrition security, poor adaptation of agriculture to climate change, and a general lack of pro-poor development opportunities for smallholder farmers.

In industrialized countries, the poor condition of soils and sub-optimal yield ceilings due to excessive soil disturbance through mechanical tillage is being exacerbated by: (a) the over-reliance on the application of mineral fertilizers as the main source of plant nutrients and (b) reducing or doing away with crop diversity and rotations, including legumes. The situation is now leading to further problems of increased threats from insect-pests, pathogens, and weeds, against which farmers are forced to apply ever more pesticides including herbicides, and which further damages biodiversity and pollutes the environment.

It is reported that we have lost some 400 M ha of agricultural land from degradation since World War II (Montgomery 2007). This abandonment is due to the severe degradation and erosion arising from tillage-based agriculture systems in both industrialized and less industrialized countries. A recent study puts the annual global cost of land degradation due to land use and cover change at 300 billion USD, of which sub-Saharan Africa accounted for some 26%, Latin America and the Caribbean some 23%, and North America some 12% (Khonya et al. 2016). Other reports indicate much higher costs, and in cases where priceless ecosystem services are lost, it is argued that it is not possible to put a cost value. This shows that our agro-ecosystems globally are facing a serious challenge of reversing the trends and of rehabilitating abandoned lands into productive and regenerative agriculture. However, solutions for sustainable soil management in farming have been known for a long time, at least since the mid-1930s when the mid-west of the USA suffered massive dust storms and soil degradation due to a combination of intensive inversion ploughing of the prairies and multi-year drought.

The main purpose of tillage throughout ages has been two-fold, namely: to mechanically break and loosen the soil and to bury weeds in order to prepare a clean-looking seed-bed for sowing and crop establishment. Subsequently, during the season, the tillage operation is often used to control weeds. It is a commonly held belief by conventional non-organic and organic tillage farmers that the main benefit from tillage is to control or even eradicate weeds. However, in reality, tillage has been shown to increase weed infestation, and it has never been able to eliminate weed infestation.

In 1943, Edward H. Faulkner wrote a book 'Plowman's Folly' in which he provocatively stated that it can be said with considerable truth that the use of the plough has actually destroyed the productiveness of our soils. More recently, David Montgomery in his well-researched book 'Dirt: The Erosion of Civilizations' shows that in general, with any form of tillage, including non-inversion tillage, the rate of soil degradation (and loss of soil health) and soil erosion is generally by orders of magnitude greater than the rate of soil formation, rendering agro-ecosystems unsustainable (Montgomery 2007). Similar to Faulkner, Montgomery concluded that tillage has caused the destruction of the agricultural resource base and of its productive capacity nearly everywhere in the world and continues to do so.

Tillage-based production systems everywhere have converted our agricultural soils and landscapes into – for lack of a better term – 'dirt' and even worse in terms of excessive use of agrochemicals, seeds, water, and energy, whilst increasing production costs, decreasing factor productivity, and reducing overall resilience. These have led to degraded agro-ecosystems and dysfunctional societal ecosystem services, including poor water quality and quantity, disrupted water, nutrient, and carbon cycles, suboptimal water, nutrient, and carbon provisioning and regulatory water services, and loss of soil and landscape biodiversity.

Conservation Agriculture: A New Paradigm for Sustainable Production

In light of the above, the need for a new paradigm of agriculture has become increasingly clear. In recent decades, the situation of system unsustainability has begun to change, at least on the agricultural production side, as CA systems in rainfed and irrigated agriculture based on annual and perennial cropping systems, backed up by some 45–50 years of research and practical experience, have spread in all continents and in most agro-ecologies around the world.

CA has been defined as a production system based on the application of three interlinked principles, namely: (1) continuous no or minimum mechanical soil disturbance; (2) permanent maintenance of biomass mulch soil cover; and (3) diversification of crop species (Kassam et al. 2020). In a CA system, the core practices that would correspond to these principles would be complemented by practices related to integrated crop, soil, nutrient, pest, water, and energy management (Lal 2015).

However, even production systems such as CA ultimately also have their limits when they are linked to unconstrained food systems in which food demand seems to be growing at a rate far in excess to what is needed to achieve food security for the total population. Indeed, while farmers of the world already produce enough food to meet the global food needs of more than twice the global population, some 2 billion people remain poor and food insecure, and some 2 billion people are obese and overfed. In addition to the significant amount of food that is reported to go to waste, a significant portion of the food produced is fed to livestock to meet the increasing demand for livestock products by the population with higher incomes and urban lifestyles (Kassam and Kassam 2020). The inequity in the food distribution system due to income differentials cannot be addressed simply by adopting new production paradigm such as CA. It would also require structural changes in the political economy of the global food and agriculture system to make food more affordable by poorer sections of the population, the notion of food and land justice and food sovereignty to become more generally acceptable, a greater proportion of food from primary production being used for human consumption than used as livestock feed for secondary production, and a substantial move towards more healthy plant-based diets. These issues, although being beyond the scope of this chapter, are nonetheless of great importance if we are to minimize the unreasonable pressures currently being put on the natural resources and the environment as well as on the production systems and farmers, and on the food systems.

Making Sustainable Production Real Through CA

It would appear from the current scientific literature on sustainable production that we seem to have rediscovered the power of agro-ecology as a central element in helping us to understand what may constitute ecological sustainability of production systems and how it can be harnessed by all land-based agricultural producers globally. This time round we seem to have discovered which principles of agro-ecology constitute the foundation of ecological sustainability in production systems upon which to build overall economic, social, and environmental sustainability, and how to turn these principles into adapted CA systems and practices that are locally formulated and contextually robust or resilient. Unlike the struggling and vulnerable tillage production systems and food and agriculture systems built upon these, we appear to have come to realize that with CA production systems, we can achieve sustainable production intensification and potentially build sustainable food and agriculture systems provided some of the core elements of the unsustainable tillage agricultural production systems indicated above can be managed differently (Kassam 2016b).

The innovations represented by the core practices of CA and how these integrate with conventionally defined good agricultural practices are of great importance to the future of agriculture and food systems globally. They can appear counter-intuitive, and they seem to question the very foundation of conventional tillage-based agriculture and the underlying assumptions of the industrial monolithic mind-set of the 'Green Revolution' agriculture. CA principles and systems have set in motion a global paradigm change in agriculture and food systems, benefitting all farmers and their rural communities who have adopted them seriously, as well as the greater society of which they are a part and the environment in general. They are providing solutions to global challenges such as local and global food insecurity, poverty alleviation, agricultural land degradation and soil erosion, loss of efficiency and resilience in conventional agriculture, loss of ecosystem services and biodiversity, stagnating yields and land productivity, and climate change adaptability and mitigation, as outlined below.

Since the mid-1990s, it has become apparent that climate change was upon us, and that it was necessary to develop climate-smart agriculture, defined as agriculture which is adapted to climate change and is able to mitigate climate change by decreasing the emissions of CO_2, CH_4, and N_2O. In recent years, CA has become accepted as being climate-smart and able to serve as the core of climate-smart agriculture.

In CA systems, all parts of the system function better and plants are larger and stronger, efficient, and resilient. Sometimes, CA has been referred to as being made up of three 'Rs' – Roots, Residues, and Rotations. The new knowledge being generated about CA systems is of global significance and provides evidence of the need to transform all land-based agriculture to CA so that they can more fully contribute to the future needs of the global society and of the planet. Conventional tillage-based agriculture is considered 'bad business as usual' and can no longer be relied upon to meet future needs.

Principles of CA described above, upon which to build sustainable production systems (Kassam, 2021), first and foremost, are functionally biological and ecological in nature, meaning that when put into practice, using locally formulated adapted practices, they provide a biologically active ecological foundation for sustainable production by maintaining as many of the ecosystem functions below and above the ground in space and time that are present in natural ecosystems.

The actual forms of core CA practices generally establish many interactive physical, biological, chemical, and hydrological processes that restore and maintain soil health and functions (Anderson 2015), and which together with other best agricultural practices, such as integrated crop, soil, nutrient, water, pest, energy, labour, and farm power management, lead to improved performance in terms of biological output, ecosystem services, efficiency, and resilience from all land-based production systems. The interlinked core practices provide the following ecological improvements that are regenerative and enhance land productivity potentials and enable the best phenotypic performance from any adapted traditional or modern genotype:

- *Continuous no or minimum mechanical soil disturbance.* Sow seed or plant crops directly into untilled soil and no-till weeding in order to maintain soil organic matter; promote soil biological processes; protect soil structure and porosity and overall soil health; and enhance productivity, system efficiency, resilience, and ecosystem services.
- *Permanent maintenance of biomass mulch soil cover.* Use crop biomass (including stubble) and cover crops to protect the soil surface; conserve water and nutrients; supply organic matter and carbon to the soil system; and promote soil biological activity to enhance and maintain soil health (including structure and aggregate stability), contribute to integrated weed, pest, and nutrient management, and enhance productivity, system efficiency, resilience, and ecosystem services.
- *Diversification of crop species.* Use diversified cropping systems with crops in associations, sequences, or rotations that will contribute to enhanced crop nutrition; crop protection; soil organic matter build-up; and productivity, system efficiency, resilience, and ecosystem services. Crops can include annuals, trees, shrubs, nitrogen-fixing legumes, and pasture, as appropriate, including cover crops.

Global empirical and scientific evidence in support of the above contributions to the key elements of sustainable production intensification is now overwhelming and continues to accumulate in all continents and agro-ecologies. One feature of soil health improvement is the enhancement of soil life and land's productivity potential due to increased soil organic matter. Soil life, comprising of all forms of microorganisms and mesofauna, generates advantages leading to a better soil environment that improves soil structure and functions, root growth, and the relationship with soil microorganisms, as well as improving above-ground crop growth and yield. A number of adaptability advantages begin to operate in CA systems, improving over time, leading to improved farm output, efficiency in input use, enhanced resilience to extreme events, and to harnessing of a range of ecosystem services (Kassam 2020a,b). Greater and more stable yields, minimum use of purchased inputs, and minimization of soil degradation and erosion, and increased livestock carrying capacity are some of the major benefits in CA systems that are not available under conventional tillage production systems.

Thus, unlike conventional tillage agriculture where the focus is on intensive tillage and high application of mineral fertilizers and pesticides, the focus in CA is on soil health and system health aimed at obtaining more output from less input, and with stability and least environmental damage. The key to soil health has been the central role played by soil organic matter in building soil life and biodiversity, leading to improved soil structure and pore volume, aggregate stability, and establishing a symbiotic relationship between crop root systems and soil microorganisms, all leading to improved phenotypic expression in terms of growth, yield, efficiency, resilience, ecosystem services, and minimizing all inputs and maximizing outputs. The counter-intuitive element in CA is the fact that it calls for no mechanical tillage and establishing soil and cropping system health based on soil biology, crop diversity, and biological pest control.

Global Adoption and Spread of CA

The good news is that we now have a new agricultural paradigm staring us in the face. It is called Conservation Agriculture, and it offers research, education, and development opportunities to all stakeholders – public, private, and civil sectors – in the national and international food and agriculture systems to help accelerate the ongoing agricultural transformation. Since 2008–9, the annual rate of expansion of CA cropland area has been some 10 M ha. In 2008–9, CA covered some 107 M ha of annual rainfed and irrigated cropland, corresponding to 7.4% of global annual cropland, and in 2013–14 it covered some 160 M ha of annual cropland, corresponding to about 11% of global annual cropland. In 2015–16, CA covered more than 180 M ha of annual cropland, corresponding to 12.5% of global annual cropland (Kassam et al. 2019), and in 2018–19 CA covered more than 205 M ha, corresponding to 14.7% of global annual cropland (Kassam et al. 2021). Some 50% of CA land is in low-income countries, particularly in Latin America and Asia, and during the last decade it has begun to spread in west and central Asia and Africa as farmers and their communities learn how to overcome the constraints to adoption of CA. CA principles are also being applied to perennial crops in orchard systems involving olives, vines, and fruit trees, in plantation systems such as oil palm, cocoa, coffee, rubber, and coconut, and in agroforestry systems where CA systems with trees are being referred to as being part of 'evergreen agriculture'. This ongoing transformation is an illustration that farmers are willing to take greater control of their future by experimenting with radically new and innovative no-till CA principles and related practices in building sustainable and regenerative farming systems. Some 45–50 years of CA research in different parts of the world has shown that CA principles are universally applicable and that sustainable production and land management are possible for all farmers, small-scale and large-scale, rich and poor, men and women.

Thus, globally, agriculture is undergoing a fundamental change, a process that began in earnest around 1990 in response to unacceptable degradation that was being caused by all forms of tillage agriculture. The realization that tillage agriculture is inherently unsustainable came about as a result of the 'dust bowls' in the mid-West United States caused by the use of mouldboard ploughs to open up the prairies, accompanied by multi-year droughts. This disaster led to successful trials with no-till farming initially in the United States, United Kingdom, and parts of Africa, but it then took until around the end of the 1980s for no-till agriculture to take off, initially in the United States, Brazil, Argentina, Canada, and Australia, and later more generally in all continents, including in the last ten years in Asia, Europe, and Africa. The modern form of no-till agriculture has been referred to, since the mid-1990s, as Conservation Agriculture with its three interlinked principles, described above, of no or minimum mechanical soil disturbance (no-till seeding/crop establishment and weeding); maintenance of soil mulch cover with crop biomass, stubbles, and cover crops; and diversified cropping systems with annuals and perennials through rotations, sequences, and associations.

The no-till champions of the 1970s and 1980s, comprising of farmers and supported by a small number of research and extension agronomists and engineers, and no-till soil and water conservationists, led to globalization of the awareness of the severe soil and land degradation being caused by tillage agriculture and its inherent long-term ecological unsustainability. It also led to the awareness of the consequent loss of farm output and profit, of production efficiency, of resilience, and of ecosystem services (and total abandonment of agricultural land due to natural resources degradation including biodiversity and the environment). During this period, the transformation of tillage agriculture to CA was not driven by mainstream research, education, extension, and development systems, in fact, quite the contrary. Some national research systems openly opposed the spread of CA, and to some extent this is the case even today. As a result, the adoption and spread of CA has been very much a farmer-led phenomenon.

The globalization of CA began in earnest with the establishment, in 2001, of the process of holding a World Congress on Conservation Agriculture every two to three years, sponsored by the FAO and agricultural development bodies in Europe, Africa, Latin America, Australia, and North America along with some multi-national corporations involved in providing agricultural inputs to farmers. The first Congress was held in Spain (2001), the second in Brazil (2003), the third in Kenya (2006), the fourth in India (2009), the fifth in Australia (2011), the sixth in Canada (2014), the seventh in Argentina (2017), and the eighth in Switzerland (2021).

This ongoing transformation is an illustration that farmers are willing to take greater control of their future by experimenting with radically new and innovative no-till CA principles and related practices in building sustainable and regenerative farming systems. What is more exciting is the fact that CA principles are actually being applied to all land-based farming systems – rainfed systems, irrigated and partially irrigated systems, non-organic and organic systems, annual cropping systems including rice-based systems, perennial cropping systems including mixed-systems, horticultural systems, plantation and orchard systems, agroforestry systems, and crop–livestock systems, including with trees. Where purchased inputs are not available, the CA approach has shown that it is possible to intensify sustainably using local resources, including adapted traditional cultivars. Thus, there are many farmers, especially in Asia and Africa, who are practicing uncertified organic CA systems. No wonder, the certified organic tillage-based farming sector is taking a closer and serious look at CA and certified conventional tillage organic systems are also being converted to certified organic CA systems.

Despite the significant progress in CA adoption, and the fact that the world already produces far more food than is required to feed the global population, the world is being frightened by some of the multi-nationals and mainstream international research and development agencies who continue to push for the destructive industrial version of the Green Revolution agriculture. Many of our universities too, in all continents, have been far too slow when it comes to offering solutions in terms of sustainable production systems. One only has to watch the Massive Open Online Courses (MOOCs) being offered by some of the universities in Europe and America to appreciate how far behind the global education system is in preparing and equipping new graduates with the knowledge and management skills required to implement and sustain the required policy and institutional support to mainstream CA systems. Business as usual still prevails in nearly all our agricultural institutions, but we now have a fantastic opportunity to promote a radical transformation of agricultural land use systems worldwide.

Other Innovations

There are other innovations that are coming on stream through precision farming which is opening up possibilities for more efficient nutrient management, e.g., the 4R nutrient management approaches – right fertilizer, right rate, right place, and right time – as well through better understanding of the co-evolved relationship between the plant rhizosphere and soil microorganisms that can influence nutrient availability and uptake. Similarly, more efficient pest (weeds, insect pests, pathogens) management is becoming possible which brings together the power of CA along with precision farming and with co-evolved relationships between plants and natural enemies of pests such as those in push–pull systems, and also with a co-evolved relationship between the plant rhizosphere and soil microorganisms that can influence gene functions that can influence whether the plant is going to be attacked by pathogens or not. Allelopathy is another area which is offering opportunities for innovations that can help manage weeds without relying on herbicides.

Needs and Opportunities

It is clear that there is an urgent need and opportunities to facilitate and support the spread of CA. The following needs and opportunities exist globally, but particularly for the South Asia region, to accelerate the uptake and spread of CA:

- The science underlying the paradigm of CA needs to be established as a regular part of concern for managing a vibrant and innovative knowledge system that can help to generate new knowledge, new technologies, and associated practice, and new enabling social and institutional arrangements to support and sustain the spread and widest benefit sharing from the application of all rainfed and irrigated CA systems. This must include the urgent realignment of research, education, and development institutions, public and private, towards the adoption and spread of CA.

- The damage that has been caused by modern and traditional tillage-based agriculture worldwide must not be underestimated. The rate of land degradation and abandonment must be stopped and reversed, so that they are restored based on the application of the principles of CA.
- There are exciting opportunities for the agriculture sector as a whole to deliver its full range of ecosystem services – supporting, regulating, provisioning, and cultural – through CA-based agricultural land and landscape management. Ecosystem services are also societal services, and farmers, land managers, food and agricultural service sectors, and policy and supporting institutions, all have a duty and a responsibility to minimize or avoid the degradation of agro-ecosystems and ecosystem services and to restore them to their ecologically desirable state.
- There is an opportunity for everyone, but particularly the youth, to feel excited and confident about the future of food and agriculture, and land use management in general, and their role and responsibility in particular, in creating a rewarding and ethically responsible post-modern food and agriculture sector.
- There is a need and to promote incentive-based agricultural development and stop subsidy-based agricultural development strategies. Farmers should be awarded and rewarded for adopting systems and practices of CA that provide ecosystem services to society.
- There is a need to promote the formation of national CA associations to promote and establish farmer-driven processes of adoption and spread of CA, and to access and attract the institutional support required to maintain a competitive and innovative CA-based food and agriculture sector.
- There is a need and an opportunity to create multi-stakeholder national platforms for CA development and dissemination in the Asia region and to facilitate information sharing and communication, and to monitor CA adoption, its contribution to agriculture and rural development, and to promote the mainstreaming of CA on farms and in all the supporting public, private, and civil sectors.

For Whom Are We Working? Whose Voices Do We Represent and Amplify? Is There a Future for the Youth in Agriculture?

The question must be asked: For whom are we working? Clearly, the answer is first and foremost, we are speaking for sustainable development of the national, regional, and international food and agriculture system. This means we are working for all the farmers, men and women, small-scale and large-scale, and rich and poor. However, in the development context, we must represent the voices and the needs of the poor smallholder farmers, their households and livelihoods, and their well-being. We must also represent the voices of the urban poor who cannot afford to buy food because they are not gainfully employed. Additionally, we must represent the global agricultural land use community in order that all agricultural land use worldwide is eventually brought under sustainable use and management.

CA systems have shown that the ecological foundation of sustainability in agricultural land use can be provided by CA systems for all farmers, in all land-based agro-ecologies to improve their biological, economic, and environmental performance. For poor farmers and their families in particular, CA systems offer something uniquely special – the ability to achieve sustainable production intensification with minimum or no purchased inputs to build local food security, strengthen livelihoods, and restore the flow of ecosystem services. They make their agriculture climate-smart and offer confidence and hope for the future for themselves and for their children, many of whom may not remain in farming once they have had an opportunity to educate themselves.

Ultimately, no amount of ecological sustainability in a production system such as CA can withstand the unlimited demand for food and non-food commodities placed upon the global land resource base. The world already produces enough food to feed more than its current population, but a significant proportion of it is wasted or fed to livestock. Thus, in general, mainstream approaches to agricultural assessments regarding national and international food security appear to be simplistic and limited in scope. As a result, they are unable to identify and address the root causes of the damage caused to land resources, the environment, and human health by the current agricultural paradigm. Such assessments are also decoupled from the human and ethical consequences of the demands and pressures placed upon agricultural production and farmers by the food and agriculture system as a whole which has continuously failed to achieve global food security for all. A number of recent analyses show that the global food system is broken, along with a number of other global chronic crisis. Increasingly, it appears that the overly dominant global capitalist economic system and the associated multi-national corporate sector that is driven by the goals of infinite growth for profit in a finite planet is not capable of solving the global burden of crises. It is not capable of protecting and meeting the food, seed, and land sovereignty needs of the smallholder farmers worldwide, nor is it capable of meeting the human nutrition and health needs of society and the planet (Kassam and Kassam 2020).

Prospects for a Brighter Future

In light of the above, we take the view that agriculture and all the related sectors have a bright future. The global transformation of agricultural land use towards CA gives us hope and confidence that we can establish and maintain development pathways for sustainable production intensification and sustainable land management. The really exciting part is that we have rediscovered that this can be done provided we respect and understand how nature works in its ecosystems and how we can build upon this understanding of sustainable land-based agro-ecosystems everywhere for all. Therefore, there is much to be excited about the for future of agriculture and land management broadly defined, and this is particularly relevant for youths, males and females, who can and must consider farming, agricultural sciences, and all the agricultural and land management professions to be respectable areas of opportunity to serve personal, national, and international ambitions.

It is particularly encouraging that the CA Global Community at the 8th World Congress on Conservation Agriculture set a notional goal of transforming 50% of the global cropland area into CA by 2050 (WCCA 2021). They have suggested a global plan of action to be developed by the CA Community at the national and regional level calling for all stakeholders to become involved. We therefore commend everyone, irrespective of whether they are in the public sector, private sector, or civil sector, whether they are plant breeders, agronomists, or plant protection professionals, agricultural engineers or crop physiologists, microbiologists or economists, rural sociologists or anthropologists, or irrigation and water management experts or those keen on gender, communication, ecosystem services, or climate change, or any of the food and agricultural and land use management disciplines, to respond fully and with confidence to the potentials and opportunities offered by the new agricultural paradigm of CA. Farmers, large and small, rich and poor, men and women, throughout the world are seizing this opportunity. They, more than anyone else, deserve your help, support, and trust because the future of our planet and of humankind is in their hands. How well they perform their custodian role and feed the population of the future will depend on how much we all care for them and how much we do to give them an effective voice in sustainable development.

There is a hopeful and exciting future emerging in post-modern agriculture based on the CA paradigm. Solutions are now available to manage agriculture sustainably and anyone who wishes to serve the farmers and agriculture development must understand and protect its ecological foundations. This also applies to farmers who wish to farm sustainably, and to those who wish to take care of the agricultural land resources and the environment on behalf of society and nature. This is the grand challenge to the education and research system globally, to ensure that new knowledge for innovative technologies is generated and transmitted appropriately to future generations. Equally important will be the need to maintain an enabling policy environment that will help to transform and build institutions and human capacity for innovations to serve the food and agriculture sector broadly defined.

Where to Look for More Information?

A general source of information on CA is the FAO website (www.fao.org/conservation-agriculture) and also the websites of the European Conservation Agriculture Federation (ECAF) (https://ecaf.org/), the Africa Conservation Tillage Network

(ACTN) (www.act-africa.org/) and the Conservation Agriculture Alliance for Asia and Pacific (CAAAP) (www.caa-ap.org/). More information regarding the development of CA systems globally can be found in books and journals, and on websites of national and regional CA organizations.

Books include Goddard et al. (2006), Baker et al. (2007), FAO (2011, 2016), Jat et al. (2014), Farooq and Siddique (2014), Chan and Fantle-Lepczyk (2015), and Dang et al. (2020). Nationally oriented information on CA development is available for several countries including Australia (Crabtree 2010), Canada (Lindwall and Sonntag 2010), Brazil (Junior et al. 2012; de Freitas and Landers 2014), Argentina (Peiretti and Dumanski 2014), and United States (Lessiter 2018).The three-volume book on Advances in Conservation Agriculture (Kassam 2020a,b, 2021) is a global source of information on Systems and Science (Volume 1), Practice and Benefits (Volume 2), and Adoption and Spread (Volume 3).

Lessiter (2011) provides a narrative on 40 legends of the past in no-till farming in North America. The International Soil and Water Conservation Research published a special issue on Pioneers in soil conservation and Conservation Agriculture which provides good information on the adoption of CA in several countries (Dumanski et al. 2014).

The Proceedings of World Congresses on CA are a good source of historical and current information on CA research and adoption. Similarly, proceedings of Africa Congresses on CA provide good information on research and development work in Africa (Kassam et al. 2017; Mkomwa and Kassam 2021). Websites of national and regional CA associations are a good source of CA information on adoption and spread as well as on research.

Conclusion and Future Outlook

CA systems are leading to a paradigm change in the food and agriculture system globally. The resulting impact is the opening up of new and more profitable ways of managing agricultural lands and improving livelihoods, investing in agricultural land for commercial purposes, and enhancing and being rewarded for ecosystem services. Agriculture is no longer the sector to employ the poor and the uneducated. It is a place where greater technical and managerial skills are going to be demanded in order to save the human race and the planet. Agriculture has become a calling for many, especially the youth, to reengage and double their efforts to achieve and sustain food security, address agricultural land degradation, achieve more from less, and respond to climate change. We must concentrate on promoting all aspects of CA for the benefit of farmers, wherever he or she may be farming, however poor or rich, small or large, as well for society and the planet. All disciplines and people have a role to play because the option and opportunity, which we all must seize, is at the level of a paradigm change – like moving from a flat Earth mindset to a round reaching mindset. All aspects of the food and agriculture systems must be realigned to the new paradigm over the coming decades across the world to achieve an attainable target of 700 M ha of cropland under CA (50% of cropped area) by 2050.

References

Anderson, R.L. 2015. Integrating a complex rotation with no-till improves weed management in organic farming: A review. *Agronomy for Sustainable Development* 35: 967–974.

Baker, C.J., Saxton, K.E., Ritchie, W.R., et al. (Eds.). 2007. *No-Tillage Seeding in Conservation Agriculture*, 2nd Edn. CABI Publishing, Wallingford, UK.

Chan, C. and Fantle-Lepczyk, J. 2015. Conservation Agriculture in Subsistence Farming: Case Studies from South-Asia and Beyond, 278 p. CABI, Wallingford.

Crabtree, W.L. 2010. Search for sustainability with no-till bill in dryland agriculture. Crabtree Agricultural Consulting, Western Australia, 204 p. Available at: www.no-till.com.au.

Dang, Y., Dalal, R. and Menzies, N. 2020. No-till Farming Systems for Sustainable Agriculture: Challenges and Opportunities. Springer Nature Switzerland, AG.

de Freitas, P.L. and Lander, J.N. 2014. The transformation of agriculture in Brazil through development and adoption of zero tillage conservation agriculture. *International Soil and Water Conservation Research* 1(2): 35–46.

Dumanski, J., Reicosky, D. C. and Peiretti, R.A. 2014. Pioneers in soil conservation and conservation agriculture. Special Issue *International Soil and Water Conservation Research* 2(1): 1–4.

FAO. 2011. *Save and Grow. A Policymakers Guide to the Sustainable Intensification of Smallholder Crop Production*. FAO, Rome. Available at: www.fao.org/ag/save-and-grow/.

FAO. 2016. *Save and Grow in Practice: Maize, Rice, Wheat – A Guide to Sustainable Production*. FAO, Rome.

Farooq, M. and Siddique, K.H.M. (Eds.). 2014. *Conservation Agriculture*. Springer International, Switzerland.

Goddard, T., Zoebisch, M.A., Gan, Y.T., et al. (Eds). 2006. *No-till Farming Systems*. Special Publication No. 3. World Association of Soil and Water Association, Bangkok, Thailand.

Jat, R.A., Sahrawat, K.L. and Kassam, A.H. (Eds.). 2014. *Conservation Agriculture: Global Prospects and Challenges*. CABI Publishing, Wallingford, UK.

Junior, R.C., de Araujo, A.G. and Llanillo, R.F. 2012. No-till Agriculture in Southern Brazil: factors that facilitated the evolution of the system and the development of the mechanization of conservation farming. FAO and IAPAR.

Kassam, A. 2013. Sustainable intensification and Conservation Agriculture. Technical Centre for Agricultural and Rural Cooperation (CTA), *Knowledge for Development* website (Knowledge for development (cta.int) (http://knowledge.cta.int/).

Kassam, A. 2016a. Integrated innovative production technologies in cereal systems: Global perspectives. Keynote address, Special Session II: Scaling Agronomic Innovations in Cereal-based Systems of South Asia. Fourth International Agronomy Congress, New Delhi, India 22–26 November 2016.

Kassam, A. 2016b. Conservation agriculture for Sustainable Intensification: Global Options and Opportunities. Keynote Address, Symposium VIII: Conservation Agriculture and Smart Mechanization: Fourth International Agronomy Congress, New Delhi, India 22–26 November 2016.

Kassam, A. (Ed.) 2020a. *Advances in Conservation Agriculture, Volume 1: Systems and Science*. Burleigh Dodds, Cambridge, UK.

Kassam, A. (Ed.) 2020b. *Advances in Conservation Agriculture, Volume 2: Practice and Benefits*. Burleigh Dodds, Cambridge, UK.

Kassam, A. (Ed.) 2021. *Advances in Conservation Agriculture, Volume 3: Adoption and Spread*. Burleigh Dodds, Cambridge, UK.

Kassam, A. and Kassam, L. (Eds.) 2020. *Rethinking Food and Agriculture: New Ways Forward*. Woodhead Publishing, Cambridge, UK.

Kassam, A., Derpsch, R. and Friedrich, T. 2020. Development of conservation agriculture systems globally, pp. 31–86. In: *Advances in Conservation Agriculture*, Ed. Kassam, A., Vol. 1. Burleigh Dodds, Cambridge, UK.

Kassam, A., Friedrich, T. and Derpsch, R. 2019. Global spread of conservation agriculture. *International Journal of Environmental Studies* 76(1): 29–51.

Kassam, A., Friedrich, T. and Derpsch, R. 2021. Successful experiences and learnings from conservation agriculture worldwide. Keynote address, sub-theme 1. 8th World Congress on Conservation Agriculture, 21–23 June 2021, Bern, Switzerland.

Kassam, A. H., Mkomwa, S. and Friedrich, T. 2017. *Conservation Agriculture for Africa*. CABI Publishing, Wallingford, UK.

Khonya, E., Mirzabaev, A. and von Braun, J. 2016. *Economics of Land Degradation and Improvement: A Global Assessment for Sustainable Development*. IFPRI and ZEF. Springer Open.

Lal, R. 2015. A system approach to conservation agriculture. *Journal of Soil and Water Conservation* 70(4): 82A–88A.

Lessiter, F. 2011. 40 legends of the past. *40[th]Anniversary Issue of No-till Farmer*, November 2011.

Lessiter, F. 2018. *From Maverick to Mainstream. A History of No-Till Farming*. No-till Farmer, 416 p.

Lindwall, C.W. and Sonntag, B. 2010. *Landscape Transformed: The History of Conservation Tillage and Direct Seeding*. Knowledge Impact in Society, Saskatoon, University of Saskatchewan, Canada.

Mkomwa, S. and Kassam, A. 2021. *Conservation Agriculture in Africa: Climate Smart Agricultural Development*. CAB International, Wallingford, UK.

Montgomery, D.R. 2007. *Dirt: The Erosion of Civilizations*. University of California Press, Berkeley, CA.

Paroda, R.S. 2016. Scaling innovations in natural resource management for sustainable agriculture. Plenary lecture. Fourth International Agronomy Congress, New Delhi, India, 22–26 November, 2016.

Peiretti, R. and Dumanski, J. 2014. The transformation of agriculture in Argentina through soil conservation. *International Soil and Water Conservation Research* 1(2): 14–20.

WCCA. 2021. Declaration. The 8th World Congress on Conservation Agriculture. 21–23 June, Bern, Switzerland.

2 History and Current Scenario of Conservation Agriculture

Benefits and Limitations

A.R. Sharma and A.K. Biswas

Introduction

Soil is the basic natural resource that sustains various types of ecosystems on Earth and fulfils all the ambient facilities required for growing plants, such as anchorage, nutrients, water, etc. It is the basis of crop production and all life processes in the world. It is the greatest sink of natural resources and also plays an important role in sustaining a favourable environment and climate regulation. As a consequence of intensive agriculture to meet the growing food requirements of an ever-increasing population, severe degradation of soil quality, soil erosion, desertification, salinization, and also compromised soil health have resulted during the last few decades. Decreasing factor productivity and profitability of agricultural systems poses serious challenges to meeting the future demand for food, feed, fodder, and fibre. This has become a major threat to modern agriculture as a consequence of the rapid degradation of natural resources, climate change, and residue management problems with increased mechanization. The significant impacts of these soil-degrading factors have already started making their influence on the productivity of land, and is estimated that it will be much more devastating in the next few decades if necessary corrective measures are not initiated. Conventional farming practices, particularly tillage and crop residue burning, have substantially degraded the soil resource base (Montgomery 2007a; Farooq et al. 2011a), with a concomitant reduction in crop production capacity (WRI 2000), and widespread soil erosion, nutrient mining, depleting water table, and eroding biodiversity being global concerns which are threatening the food security and livelihood opportunities of farmers, especially the poor and under-privileged. Under conventional farming practices, continued loss of soil is expected to become critical for global agricultural production (Farooq et al. 2011a). New and improved best management practices are required to ensure sustainable use of these resources and to optimally utilize the advantages of its associated ecosystem services. Also, new and better tools for the assessment and interpretation of soil quality indicators are crucial for soil diagnosis and to help farmers decide on the best management practices to adopt on specific pedo-climatic situations.

DOI: 10.4324/9781003292487-3

Estimates reveal that an annual loss of 75 billion tonnes of soil translates into US$400 billion per year; about US$70 per person per year (Vlek 2008). Almost, 10 Million ha of good-quality agricultural land is lost globally per annum due to soil degradation processes, which adversely affect the agricultural production and profitability of various production systems. Degradation of natural resources is adversely influencing the livelihood opportunities of poor farmers and dragging them into the vicious trap of debt-driven poverty. According to the FAO, 1.5 billion people depend directly on land that is degrading. Another study indicated that land degradation is worsening rather than improving in most regions, with declining trends only on some 24% of the global land area. According to this study, the main driver of degradation is poor land management (Paroda 2009). Soil carbon is an important key resource for crop production and is considered to be 'black gold'. However, an enormous quantity of soil carbon is lost due to inefficient production methods. Mechanization has led to carbon loss in the form of CO_2 to the extent of 78 billion metric tonnes (Lal 2004).

Conservation Agriculture – A Paradigm Shift

Conservation Agriculture (CA) is an approach of managing agro-ecosystems for improved and sustained productivity, and increased profits and food security, while preserving and enhancing the resource base and the environment. The adoption of CA practices has helped millions of farmers worldwide in increasing productivity and profitability through arresting land degradation, improving input-use efficiency, adapting and mitigating climatic extremes, and improving farm profitability in diverse ecologies across the world. Realizing the potential impacts of CA since the mid-1990s, significant efforts have been made in the direction of adoption and popularization of conservation agriculture covering more than 10.3 M ha in Asia. Conservation agriculture and its fundamental principles: minimum (or no) soil disturbance, permanent soil organic cover, and crop rotation/intercropping certainly figure among the possibilities that contribute for a sustainable soil management. CA is characterized by three linked principles, namely (i) continuous minimum mechanical soil disturbance, (ii) permanent organic soil cover, and (iii) diversification of crop species grown in rotations, sequences, or associations. Conservation agriculture involves a total paradigm shift from conventional agriculture systems. Some of the distinguishing features of conventional and conservation agriculture systems are presented in Table 2.1.

Historical Aspects of CA

Agriculture is believed to have started since the dawn of human civilization. It was about 8000–10000 BC when settled farming started, which was virtually a sort of zero-till (ZT) cultivation of crops. When some iron-made hand tools became available, the ancient dwellers started clearing the land using draft animals and opening the soil for sowing seeds. Subsequently, mould-board

Table 2.1 Distinguishing features of conventional vs. conservation agriculture systems

Conventional agriculture	Conservation agriculture
• Crop cultivation through the use of science and technology to dominate nature	• Crop cultivation through the least interference with natural processes
• Excessive soil erosion due to intensive mechanical tillage	• No-till or drastically reduced tillage (zero/minimal tillage), less erosive
• Residue burning or removal (bare surface)	• Surface retention of residues (permanently covered)
• Low rate of water infiltration	• Infiltration rate of water is high
• Use of ex situ FYM/composts as organic matter	• In situ retention of crop residues organics/composts
• Green manure crops are incorporated	• Brown manuring/cover crops (surface retention)
• Weed control due to tillage (established weeds) but also stimulates more weed seeds to germinate	• Weeds are a problem in the early stages of adoption but decrease with time (needs proper management through application of suitable herbicides)
• Weed control during critical period of crop–weed competition	• Season-long weed control, with focus on minimization of weed seed bank
• Free-wheeling of farm machinery, increased soil compaction	• Controlled traffic, less compaction in crop area
• Monocropping/culture, less efficient rotations	• Diversified and more efficient rotations
• More labour intensive, uncertainty of timely operations	• More mechanized operations, ensure timeliness of operations
• More prone to stresses, yield losses more under stress conditions	• More resilience to stresses, yield losses are less under stress conditions
• Productivity of gains in long run are in declining order (less sustainable)	• Productivity gains in long run are in incremental order (more sustainable)

ploughs were invented which turned the soil and incorporated weeds and residues into the soil. Tractors made an entry in the early 1900s, which could pull multiple ploughs and carry out other farm operations. The discovery of 2,4 D in the mid-1940s led to a revolution in the chemical control of broad-leaved weeds in cereal crops. The development of ZT seeders in the 1960s, followed by new-generation seed-cum-fertilizer drills and new herbicide molecules for weed control in the 1980s and 1990s revolutionized agriculture in many countries of North and South America (Moyer et al. 1994). Over the past 2–3 decades, advances in machinery and herbicides have made ZT practical on a commercial scale globally, including some parts of South Asia.

Agricultural milestones showing the major inventions and transformation leading to adoption of CA globally are given in Table 2.2.

Tull (1829), who is regarded as the 'Father of Tillage' carried out numerous experiments dealing with cultural practices and published a book 'Horse Hoeing Husbandry'. This was regarded as the most authoritative book on tillage in

Table 2.2 Agriculture milestones leading to transformation from conventional agriculture to CA

Period	Major invention and transformation
8000–10000 BC	Planting stick – the earliest version of zero-till, enabled the planting of seeds without cultivation
6000 BC	Draft animals replaced humans in powering the plough
3500 BC	Plough share – a wedge-shaped implement tipped with an iron blade was used to loosen the top layer of soil
1100 AD	Mouldboard plough – having a curved blade inverted the soil, buried weeds and residues
Mid-1800s	Steel mouldboard plough – invented by John Deere in 1837 was able to break up prairie sod
Early 1900s	Tractors – could pull multiple ploughs and did other farm operations
1940s–1950s	Discovery of 2,4 D in 1945 revolutionized chemical weed control, and other herbicides like atrazine and paraquat enabled management of weeds with less tillage
1960s	Zero-till seeders – opened a slice/small groove for placing seeds, keeping soil disturbance to a minimum
1970s–1980s	Discovery of glyphosate in early 1970s for non-selective weed control, new-generation farm machinery for placing seed and fertilizers in standing/loose residues
1990s	Introduction of herbicide-tolerant GM crops and development of low-dose high-potency post-emergence selective herbicides in most crops
Early 2000s	Zero-till revolution in Brazil and most other countries of North and South America
2010–2020	Conservation agriculture became the fastest adopted technology globally

Source: Adapted from Bolliger et al. (2006); Huggins and Reganold (2008).

English circles for many decades. He believed that soil should be finely pulverized to provide proper tilth for the growing plants. He propounded a theory that '*Soil particulars are ingested through openings in plant roots due to the processes caused by the swelling of growing roots*'. This was later found not to be true, but the adoption of this kind of agriculture for decades and centuries led to adverse effects on soil by causing erosion and loss of top fertile soil.

Tillage as a soil management concept was questioned for the first time in the 1930s. The 'Dust Bowl era' between 1931 and 1939 exposed the vulnerability of plough-based agriculture, as wind blew away precious topsoil from the drought-ravaged southern plains of the US, leaving behind failed crops and farms. It was realized that tillage is the root cause of agricultural land degradation – one of the most serious environmental problems worldwide –posing a threat to crop production and rural livelihoods, particularly in poor and densely populated areas of developing countries.

Ideas for reducing tillage and keeping soil covered with crop biomass followed, and the term 'conservation tillage' was introduced for practices aimed at erosion control (Derpsch 2003). Seeding machinery developments followed to seed directly without any soil tillage. At the same time, theoretical concepts resembling today's CA principles were elaborated by Faulkner (1945) in his book 'Plowman's Folly'. Nature magazine described an "agricultural bombshell" when Faulkner blamed the then-universally used mould-board plough for disastrous tillage of the soil. He questioned the use of the plough for cultivation of crops and showed that all standard wisdom used as a rationale for ploughing and working the soil was invalid. His ideas were considered 'mad' and 'without merit' while he was alive, but later, he was regarded as the first true conservationist.

Fukuoka (1975) worked for more than 65 years at his farm in Japan and developed a system of natural farming. He did not plough his fields, used no agricultural chemicals nor prepared fertilizers, did not flood his rice fields as farmers have done in Asia for centuries, and yet his yields equalled or even surpassed the most productive farms in Japan. His book 'One Straw Revolution' became one of the best-selling books in agriculture for the innovative system of cultivating the crops.

In the recent times, Montgomery (2007b) wrote the award-winning book 'Dirt – The Erosion of Civilizations', which showed that with any form of tillage, including non-inversion tillage, the rate of soil degradation and soil erosion is greater than the rate of soil formation. According to his research, tillage has caused the destruction of the agricultural base and of its productive capacity nearly everywhere, and that it continues to do so. The slow pace at which soil rebuilds makes its conservation essential.

Zero tillage entered farming practice in the USA in the 1960s (Islam and Reeder 2004). US scientists also introduced ZT into Brazil simultaneously with the local scientists and farmers. Later, ZT combined with permanent ground cover and rotations was termed as conservation agriculture (FAO 1987). It took around 20 years before CA reached significant adoption levels. During this time, farm equipment and agronomic practices in ZT systems improved and developed to optimize the performance of crops, machinery, and field operations. This process continues till today; the creativity of farmers and researchers is still producing improvements to the benefits of the production system, the soil, and the farmer. While tillage-based agriculture has been researched for several centuries, CA is only about half a century old, and the foundations of CA systems can only be understood as the agro-ecosystems evolve under the new production management.

Two decades of extensive research and experimentation with ZT methods allowed these systems to emerge in Brazil, involving no-soil turning, maintenance of a permanent vegetative cover, and rotations of both cash and cover crops (Bolliger et al. 2006). From the early 1990s, the adoption of CA started growing exponentially, leading to a revolution in agriculture in North and South America (Kassam et al. 2019). Realizing that the age-old practice of turning the soil before planting a new crop is a leading cause of farmland degradation, many farmers across the globe are now looking to make ploughing a thing of the past and turning to a more sustainable approach, known as CA.

Global Spread of CA

Farmer-led transformation of agricultural production systems based on CA is progressing globally. About 69.9 M ha (38.7%) of the total global area under CA is in South America, corresponding to some 63.2% of the cropland of the region, and some 63.2 M ha (35.0%) is in North America, mainly in the USA and Canada, corresponding to 28.1% of the crop land of the region. Some 22.7 M ha (12.6%) is in Australia and New Zealand, corresponding to 45.5% of the cropland, and some 13.9 M ha (7.7%) is in Asia, corresponding to 4.1% of the cropland of the region. Approximately 10.8 M ha of the total global CA area is in the rest of the world, comprising 5.7 M ha in Russia and Ukraine, 3.6 M ha in Europe, and 1.5 M ha in Africa, corresponding to 3.6, 5.0, and 1.1% of their total cropland area, respectively. In terms of CA adoption and uptake, Europe and Africa are the developing continents (Table 2.3).

It has been estimated that the global extent of CA cropland in 2008–9 covered about 106 M ha (7.5% of the global cropland) (Table 2.4). Since 2008–9, the adoption has increased exponentially with the impulse of the need for a new paradigm for sustainable intensification of crop production including the delivery of ecosystem services and as a base for 'climate-resilient agriculture'. In 2015–16, CA cropland was about 180 M ha (12.5% of the global cropland) representing a difference of some 74 M ha (69%) over the seven-year period since 2008–9. Since 2008–9, the annual rate of change has been 10.5 M ha from 106–180 M ha, showing the increased interest of farmers in the CA farming system approach to sustainable production and agricultural land management. The number of countries where CA adoption and uptake has occurred increased from 36 to at least 78 in 2015–16. The growth of area under CA has been especially significant in South America where Argentina, Brazil, Paraguay, and Uruguay are using the system on >70% of their total cropped area.

Table 2.3 Global spread of cropland area under CA by region in 2015–16

Region	CA cropland area (M ha)	% of global CA cropland area	% of cropland area in the region
South America	69.9	38.7	63.2
North America	63.2	35.0	28.1
Australia and New Zealand	22.7	12.6	45.5
Asia	13.9	7.7	4.1
Russia and Ukraine	5.7	3.2	3.6
Europe	3.6	2.0	5.0
Africa	1.5	0.8	1.1
Global total	**180.4**	**100.0**	**12.5**

Source: Kassam et al. (2019).

Table 2.4 Increase in cropland area (M ha) under CA in different countries of the world

Country	2008–09	2015–16	% increase
USA	26.5	43.2	63.0
Brazil	25.5	32.0	25.5
Argentina	19.7	31.0	57.4
Canada	13.4	19.9	48.5
Australia	12.0	22.3	85.8
Paraguay	2.4	3.0	25.0
Kazakhstan	1.3	2.5	92.3
China	1.3	9.0	592.3
India	-	1.5	-
Others	4.4	16.1	265.9
Total	**106.5**	**180.5**	**69.5**

Source: Kassam et al. (2019).

In the past decade, CA has become the fastest growing production system for many reasons, including: (i) greater farm productivity and farm output, (ii) reducing cost of production and improving profitability, (iii) greater resilience to biotic and abiotic stresses, (iv) minimizing soil erosion and degradation, (v) building soil health, and (vi) adapting and mitigating climate change. Whereas in 1973–74, CA was applied only on about 2.8 M ha worldwide (Figure 2.1), the area had grown to 6.2 M ha in 1983–84 and to 38 M ha in 1996–97. In 1999, worldwide adoption was 45 M ha, and by 2003, the area had grown to 72 M ha. During the period from 1999 to 2013, CA cropped land expanded at an average rate of 8.3 M ha per year from 72 to 157 M ha. Since 2008–9, the spread of CA worldwide appears to have been expanding at the rate of 10.5 M ha per annum and reached 180.5 M ha by 2015–16. It is estimated that, at this rate, the global spread of CA area might have exceeded 225 M ha by now, which is >15% of total cropland area.

Since the mid-1990s, the adoption of CA has been triggered with the introduction of genetically modified crops, 47% of the area of which is under herbicide-resistant and 41% with stacked traits (herbicide + insect resistant) (ISAAA 2019). The biotech crops are also considered as the fastest adopted crop technology in the history of modern agriculture. These two technologies, CA and GM crops, have revolutionized world agriculture and shown double-digit growth over the past two decades, and covered 180.5 and 190 M ha in 2015–16 and 2019, respectively (Figure 2.1).

CA in Asia

The concept of CA is relatively new in Asia, however Asian countries have adopted CA in many areas since 2008–9. The area under CA has increased more than 4-fold from 2.6 M ha in 2008–9 to some 13.9 M ha in 2015–16. In 2008–9, the CA area was reported in only two countries of the Asia region, but

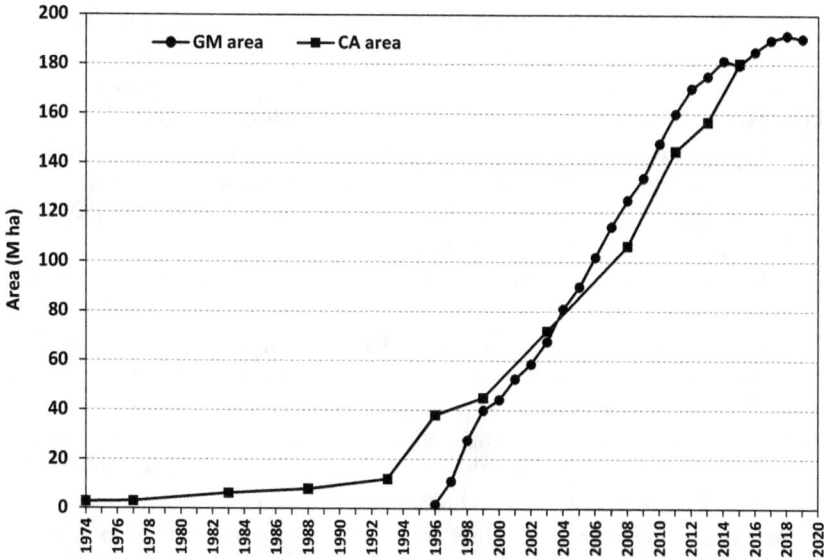

Figure 2.1 Global area under genetically modified (GM) crops and conservation agriculture (CA).

Sources: Kassam et al. (2019); ISAAA (2019).

in 2015–16, it was reported in 18 countries. China occupies the dominant position in terms of area (9.0 M ha), followed by Kazakhstan (2.5 M ha) (Kassam et al. 2019). The history of CA in South Asia is inextricably linked with wheat production constraints in the rice–wheat system, which is one of the dominant cropping systems. Wheat, the staple cereal crop of South Asia, showed declining yield trends in the 1980s due to late planting, poor plant stand, less seed replacement, inefficient fertilizer and irrigation system, weed competition, deterioration in soil health, and the menace of littleseed canary grass (*Phalaris minor*) (Malik et al. 2005). The productivity and sustainability threat to the rice–wheat system further intensified because of the inefficient production practices, excessive and imbalanced fertilizer application, over-exploitation of resources, especially water, energy, labour, climate change, and socio-economic changes. The rice–wheat cropping system in South Asia, therefore, suffers from conflicts in economic, social, climatic, ecological, and production-related issues. Zero tillage technology in wheat was introduced in the early 1980s in South Asia by using a seed drill imported from New Zealand for the first time in Punjab, Pakistan. In 1996, ZT was again attempted and evaluated in farmers' fields.

The International Maize and Wheat Improvement Center (CIMMYT) through the Rice-Wheat Consortium (RWC) and collaboration from the Australian Centre of International Agriculture Research (ACIAR), International Rice Research Institute (IRRI), and National Agriculture Research Institutes helped in planning and executing resource-conserving technologies (RCTs). The RWC, through its collaborators, planned for the introduction of second-generation machinery from 2002 onwards. The change towards CA perceived a fundamental shift from the age-old practice of excessive ploughing to a new paradigm shift, whose best exponent is RCTs, based on no-till (minimal soil disturbance and compaction), innovative cropping systems, and management of crop residues rather burning. The development, testing, and refinement of RCT practices were based on an innovative low-cost seed-cum-fertilizer drill (costing US$400–500), which can plant seeds with minimal disturbance to soil. With the commissioning of the RWC in 1994, ZT and reduced-till practices spread very rapidly, increasing from just a few thousand ha in 1997–98 to more than 2.18 M ha in 2004–5 and 2.6 M ha in 2010–11. With the advancement of CA, second-generation drills or planters were developed for seeding in the presence of anchored and loose residue. In the past few years, a major emphasis has been put towards developing CA-based small machinery, keeping in view the small farmer's needs. The success of CA's spread is attributed to a multi-stakeholder approach and low-priced drills, which increased the demand in other countries and across regions too. Based on the success and formative nature of the framework for promotion of RCTs to both small- and large-scale farmers from service providers, use of a farmer-participatory approach with on-farm demonstration with involvement of local manufacturers and their national partners has made available cheap, affordable, and effective ZT drills. Nonetheless, the alarming threat of climate change and reduced water availability in the region indicate the existence of potential for a further increase in cereal system profitability if CA-based RCTs in the IGP of South Asia are promoted. Widespread adoption of CA practices at a rapid pace in different countries is testimony to the higher benefits to farmers in comparison with conventional agriculture with respect to reducing costs, enhancing profits, and conserving precious resources.

Rice–wheat is a major cropping sequence in the Indo-Gangetic Plains (IGP) of South Asia; covering over 13.5 M ha in Bangladesh, India, Nepal, and Pakistan, and a source of livelihood to millions of people. The problems of the post Green Revolution due to intensive farming, imbalanced use of fertilizers, and faulty irrigation practices caused soil degradation and depletion of soil organic carbon (SOC), water resources, and environment pollution leading to stagnation or a decline in yields of the rice–wheat cropping system (RWCS). In this system, large-scale adoption of ZT wheat with some 5 M ha was reported but only modest adoption of permanent ZT and full CA (Farooq and Siddique 2014). The exception appears to be India and Pakistan, where significant adoption (1.5 and 0.6 M ha, respectively) of ZT practices by farmers has occurred in recent years in the rice–wheat double-cropping system. Bangladesh has begun to report some CA areas with a rice-based cropping system, particularly on permanent beds. This is expected to expand because farmers now have access to ZT seeding machines from service providers where locally produced CA equipment is available (Saharawat et al. 2022).

CA in India

Green revolution technologies paved the way for increased productivity and elimination of acute food grain shortages in India during the 1960s, primarily by cultivation of high-yielding dwarf varieties of rice and wheat, increased use of chemical fertilizers and other agrochemicals, and the development of irrigation facilities accompanied by the so-called modern methods of cultivation, which included maximum tilling of land, virtually clean cultivation with complete removal of crop residues and other biomass from the field, fixed crop rotations mostly involving cereals, and oilseed crops in the highly productive north-western plain zone of the country. Transformation of 'traditional animal-based subsistence farming' to 'intensive chemical- and tractor-based conventional agri-culture' has led to a multiplicity of issues associated with the sustainability of these production practices. There have been emerging concerns about natural resource degradation due to the adoption of these technologies. Burning of fossil fuels, crop residues, and excessive tillage including puddling for rice cultivation, are leading to the emissions of greenhouse gases, which are responsible for cli-mate change and global warming. It has been realized that the productivity levels of predominant crops and the cropping systems are stagnating and the incomes of farmers are reducing due to the rising costs of the inputs and farm operations.

Indian agriculture produces about 500–550 M t of crop residues annually. These crop residues are used as animal feed, soil mulch, manure, thatching for rural homes, and fuel for domestic and industrial purposes, and thus are of tremendous value to farmers. However, a large portion of these crop residues, about 90–140 M t is burnt on-farm primarily to clear the fields to facilitate timely sowing of succeeding crops. The problem of on-farm burning of crop residues has intensified in recent years due to use of combines for harvesting and the high cost of labour in removing the crop residues by conventional methods. The residues of rice, wheat, cotton, maize, millet, sugarcane, jute, rapeseed-mustard, and groundnut crops are typically burnt on-farm across the country, particularly in north-west India where the rice–wheat system is mechanized. Burning of crop residues leads to a series of complex problems such as the release of soot particles and smoke, causing human health problems; emission of greenhouse gases such as carbon dioxide, methane, and nitrous oxide adding to global warming; loss of plant nutrients such as N, P, K, and S; adverse impacts on soil properties; and wastage of valuable crop residues mainly due to non-availability and easy access of the quality crop planters which can sow the succeeding crop into loose or anchored residues. Use of crop residues as soil organic amendment in the system of agriculture is a viable and valuable option. The adoption of CA is the most viable, profitable, and eco-friendly option for conservation of soil, water, nutrient, and energy, and from a climate change point of view. In India alone, this has led to an overall saving of USD 164 million, with an investment of only USD 3.5 million on ZT technology with an internal rate of return of 66%. In addition to the saving on inputs, CA-based crop management practices have other potential benefits such as natural resource conservation, reduced emissions of greenhouse gases, and better resilience to climatic extremes.

An experiment conducted in the IGP of India by Ram et al. (2013) revealed that soybean and wheat planted on raised-beds and straw application recorded ~17% and 23% higher water-use efficiency (WUE), respectively, compared with flat-bed planting. Similarly, Das et al. (2016) reported that CA had significantly higher WUE as compared to conventional tillage in a pigeonpea–wheat system. Ghosh et al. (2015) also reported that the mean wheat equivalent yield was 47% higher with CA as compared to conventional agriculture in maize–wheat crop rotation. The results across the IGP in India suggest that double ZT with retention of crop residues resulted in higher system productivity over conventional and ZT without residues (Jat et al. 2011). A comparison of different CT practices in a rice–wheat cropping system on crop productivity over six years revealed that rotary, strip, and ZT drilling, and bed planting of rice and wheat provided higher yields (2–8%), reduced cost (9–27%), and energy efficiency (21–32%). In situ recycling of wheat straw provided rice yields (6.3 t/ha) that were 11% and 7% higher than residue retrieval and burning, respectively. In another study in New Delhi, direct-seeded rice (DSR) alone gave about 0.5 t/ha lower yield than transplanted rice in a rice–wheat cropping system (Sharma et al. 2012). However, the loss was compensated when brown manuring with *Sesbania* or greengram residue incorporation in the previous summer season. The highest productivity was recorded under DSR followed by a ZT wheat and greengram cropping system.

A large share of the CA in India is confined to the Indo-Gangetic plains. Attempts were made as early as 1980s for ZT (dry seeding) as a part of 'vertisol technology' launched by the International Crops Research Institute for the Semi-Arid Tropics (ICRISAT) in India for standardization and popularization of this resource-conserving technology. Dry seeding was one of the key components of vertisol technology in rainfed agriculture. The purpose was to increase crop yields and expand cropping intensities in rainfed areas of semi-arid tropics. The technology was partly adopted and was not pursued in rainfed areas. The concept of ZT (and also CA) has been well tested, perfected, and widely adopted in irrigated areas of the Indo-Gangetic plain.

In India, CA adoption remains at a nascent stage despite considerable research and efforts to promote it over the last 2 decades. The potential for area expansion of this technology is estimated to be 20 M ha by 2030, however only about 10% of that has been achieved as of now. CA-based management practices have shown the benefits in enhancing the natural resource base, improving input use efficiency, soil aggregation, soil health, farm productivity, and mitigation of climate change. ZT has been a success in the rice–wheat cropping system due to a reduction in the cost of production by Rs 2000–3000 per ha, which is the main driver behind its spread (Malik et al. 2005), and improved soil health (Jat et al. 2009; Gathala et al. 2011). The potential of C sequestration in C-depleted soils in India is high with the adoption of ZT. Long-term C sequestration and build-up of soil organic matter constituted a practical strategy to mitigate GHG emissions and impart greater resilience to production systems to climate change (Saharawat et al. 2012) and improved environmental quality (Pathak et al. 2011). Crop residue management provides an opportunity to protect the topsoil, enriched with organic matter, moderate soil temperature, improve soil biological activities (Gathala et al. 2011), and also enhanced the water and nutrient use efficiency (Jat et al. 2012). The development of non-selective contact and post-emergence herbicide for weed control provided a base for recommendation, allowing consistent yield and creditable performance, particularly reducing the incidence of *Phalaris minor* in wheat (Malik et al. 2005).

North-Western India

In India, the rice–wheat cropping system is predominant in the north-western part of the Indo-Gangetic Plains (IGP), and is the major source of calories and protein requirements. However, over the years, this cropping system has led to over-exploitation of natural resources, degradation of soil health, and enhanced global warming (Saharawat et al. 2010). In the rice–wheat system, wheat straw is mostly removed from the fields as a dry fodder for livestock, whereas rice straw is burnt in the fields due to the short time lapse between rice harvest and wheat seeding (Gupta and Seth 2007). In the IGP of NW India, nearly 44.5 M t of rice residues and 24.5 M t of wheat straw are burned annually. Burning of crop residues is a serious concern in NW India as major N, S, and C fractions in the residue are lost during burning. This accelerates the losses of organic matter, increases CO_2 emissions, and reduces soil microbial activity. Crop residues retention at the soil surface conserves soil and water for sustaining crop production and increases SOC, thereby improving soil properties such as soil structure, cation exchange capacity, water-holding capacity, and lower bulk density. In this aspect, CA could be a better alternative which not only utilizes crop residues but at the same time recycles plant nutrients in soil, improves soil properties, and provides environmental benefits by avoiding in situ burning. Wheat cultivation through conservation tillage is being promoted in the rice–wheat cropping system on a large area and promoted in the north-western states such as Punjab, Haryana, Bihar, and Uttar Pradesh. It is rapidly gaining popularity because farmers have realized that CT practices are time consuming, resulting in delayed seeding and yield losses in wheat after rice crop.

Cultivation of conventional puddled rice has led to over-exploitation of groundwater, leading to an alarming drop of the water table in many parts of north-western India. Furthermore, conventional puddled transplanted rice (PTR) requires large amounts of energy and labour, as well as consuming larger quantities of irrigation water, and affecting physical and chemical soil properties, thereby adversely influencing productivity of the succeeding upland crop (e.g. wheat) (Gathala et al. 2011). This calls for immediate action through the adoption of best management practices for improving soil and environment quality, and maintaining ecosystem services. CA has shown its effectiveness in sustaining and improving the productivity of the RWCS, while at the same time preserving scarce natural resources such as energy, labour, time, water, and environment quality. The CA systems are efficient in slowing down the degradation of the physical, chemical, and biological quality of soil while reducing the cost of production. Sustaining productivity of the RWCS cannot be maintained unless the declining trend in soil fertility resulting from the nutrient mining by these crops is replenished.

The development of a suitable crop production technology is the need of the hour that avoids puddling, requires less water, saves labour for transplanting, maintains rice yield potential, and is environmentally friendly. Direct dry seeding of rice (DSR) into soil has proved to be an appropriate alternative to manually transplanted PTR (Prasad 2011). Maize, due to its higher water-use efficiency, can be an excellent alternative to PTR in NW India, where lowering of the ground water level is a major concern. Mungbean cultivated between wheat and rice is beneficial for enhancing the carbon and nitrogen concentration in soil, thereby improving the overall soil quality (Singh et al. 2015). Several studies conducted across the production systems under varied ecologies have revealed the potential benefits of CA-based crop management technologies in resource conservation, use efficiency of external inputs, yield enhancement, soil health improvement, and adaptation to changing climates (Table 2.5).

There is an improvement in various parameters of soil health (physical, chemical, and biological quality) under conservation tillage by increasing the carbon and nutrients concentrations at the surface soil (Singh et al. 2015). The

Table 2.5 Effect of different CA-based crop management technologies on crop yields, water savings, and water productivity

Technology	Location	Crop/cropping system	Yield gain over conventional practices (kg/ha)	Water saving over conventional practices (ha-cm)	Increase in water productivity (kg/m³)
Laser levelling	Meerut	Rice–wheat	750	26.5	0.06
	Karnal	Rice–wheat	810	24.5	–
	Ludhiana	Rice	750	22.0	–
ZT	Karnal	Wheat	150–400	2–4	0.10–0.21
	Meerut	Wheat	610	2.2	0.28
	Delhi	Corn	150	8.0	0.21
ZT with surface residue	Karnal	Rice–wheat	500	61	0.24
	Meerut	Wheat	410	10	0.13
Direct-seeded rice	Ghaziabad	Rice	120	25	0.08
	Kamal	Rice	62	18	0.10
Raised-bed planting	Meerut	Corn	324	12	0.80
	Meerut	Wheat	310	16	0.58

Source: Malik et al. (2005), Gupta and Seth (2007).

advantages of ZT after burning or removal of crop residues in the RWCS are reported in the IGP, particularly in NW India (Erenstein and Laxmi 2008). In most cases, ZT is practiced in wheat for the timely sowing of wheat, control of *Phalaris minor*, reducing the cost of cultivation and saving water, without taking into account improvements in soil properties and nutrient availability. Studies on changes in macro- and micro-nutrient availability as well as soil properties under different CA-based practices in RWCS are very limited. However, associated with yield potential, the rice–wheat system in the IGP is defined into two broad categories, i.e. an irrigated environment favourable for rice and wheat exists in the western part of the IGP (Punjab and Haryana states, with western Uttar Pradesh), and the eastern part of the IGP which includes the districts favourable for rainfed rice or irrigated or rainfed wheat, consisting of West Bengal, the northern parts of Bihar, and eastern Uttar Pradesh. Maize is a high input-responsive crop with higher yield potential. It requires much less irrigation water than both direct-seeded and puddled transplanted rice. Puddling in rice and entire biomass removal with intensive tillage causes a loss of carbon and other nutrients (Beri et al. 2003) and development of water repellency in soil (Singh et al. 2005). Moreover, continuous pumping of groundwater over the years to meet the high water requirement of rice has resulted in a drastic decline in groundwater tables (Sharma et al. 2012), leading to a potential reduction in water availability. CA practices in the maize–wheat cropping system, where all three principles of CA (ZT, residue retention, and crop rotation) along with raised-bed and flat planting are carried out, have shown promising results.

Eastern India

Cereal production systems in eastern India are traditional, with low yield and low farm income, and have largely missed out on the benefits of the Green Revolution. To enhance productivity and farmers' incomes, and alleviate environmental and management constraints in the rice–wheat cropping system, new approaches that are more productive and sustainable methods are required to be developed. CA together with best management practices (BMPs) used in other parts of the IGP offer potential to be extended in the eastern IGP. A study conducted during 2009–11 with the objective of evaluating a range of approaches for enhancing the productivity and economic returns of the rice–wheat cropping system revealed that avoiding tillage in wheat and including mungbean in the cropping system increased the yields of wheat and the succeeding rice crop by 21–31% and 5–10%, respectively (Laik et al. 2014). The yields of wheat and rice increased further by 46–54% and by 10–24%, respectively, with the inclusion of more CA components. In another experiment aimed to include higher cropping intensity and diversification (potato and maize–rice–cowpea rotation) with CA components, 144–163% higher rice equivalent system productivity was attained. Higher irrigation water productivity in the winter season and net returns ($2855–4193 per ha) were recorded under this system. The results showed that there is enormous untapped potential to improve overall system performance through the adoption of CA in integration with BMP in the E-IGP of India. Similarly, full implementation of CA practices in Odisha, i.e. minimum tillage, maize–cowpea intercropping, and mustard residue retention led to significantly higher system productivity and net benefits than traditional farmer practices, i.e. conventional tillage, sole maize cropping, and no mustard residue retention (Pradhan et al. 2016). The dominance analysis demonstrated increasing benefits of combining conservation practices that exceeded thresholds for farmer adoption. It is true that yield is primarily a time phenomenon, i.e. it is a function of the time for which the crop remains in the field. That means that the longer the crop remains in the field for its growth and development, the higher will be the grain yield. In the E-IGP, sowings are delayed up to December. This is where maximum gains in productivity will occur, especially from areas where sowing of wheat is delayed beyond December.

Central India

Limited water resources, with uncertainty of rainfall and poor nutrient input, soil-related constraints like low water infiltration, high incidence of inundation, accelerated runoff, and soil erosion are major constraints to sustainable productivity in vertisols of Central India. These soils occupy a total area of 70.3 M ha, constituting 22% of the total geographical area of the country, of which 34.3% and 30.2% are in Maharashtra and Madhya Pradesh, respectively (Kushwah et al. 2016). In vertisols, the production systems are quite heterogeneous in terms of land and water management and cropping systems. These include the core rainfed areas which cover up to 60–70% of the net sown area and the remaining irrigated production systems. The rainfed cropping systems are mostly single cropped in the alfisols, while in vertisols, a second crop is generally taken on the residual moisture. In black soils, the farmers keep lands fallow during the rainy season and grow winter crops on conserved moisture. Sealing, crusting, sub-surface hard pans, and cracking are the key constraints which cause high erosion and impede infiltration of rainfall. The choice and type of tillage largely depend on the soil type and rainfall. Leaving crop residues on the surface in CA is a major concern in these rainfed areas due to its competing use as fodder, leaving very little or no residues available for surface application. Agroforestry and alley cropping systems are other options for CA practices. This indicates that the concept of CA has to be adopted in a broader perspective in arid and semi-arid areas. Experience has shown that a reduced tillage soybean–wheat system is a suitable option for growing soybean and wheat crops in vertisols while saving energy and labour. This also improves soil organic carbon, and physical and biological properties.

The results of a long-term tillage experiment at Bhopal revealed that yield levels of soybean–wheat system under conservation tillage (i.e. no-tillage and reduced tillage) were on a par with conventional tillage, besides saving of energy and labour (Hati et al. 2015). Conservation tillage practices were as effective as conventional tillage in terms of crop productivity of soybean and wheat. However, improvements in selected soil physical properties, like soil water storage, bulk density, aggregate stability, penetration resistance, and saturated hydraulic conductivity (Ks), were recorded in ZT and reduced tillage (RT) compared with conventional tillage (CT). Soil organic carbon (SOC) and also the aggregate associated carbon content (0–15 cm depth) were significantly higher in ZT and RT where wheat residues were left after harvest than that in CT system after ten years of a cropping system. It is concluded that ZT and RT systems with management of residues and recommended rate of N for a soybean–wheat system would be a suitable practice for sustainable production of a soybean–wheat cropping system in vertisols of central India.

Wheat residue incorporation or retention coupled with application of 28 kg N/ha through fertilizer or organic manures is more beneficial than burning in terms of enhanced crop productivity and soil fertility in vertisol (Kushwah et al. 2016). A five-year field study conducted with wheat residues on a vertisol in the soybean–wheat system at Bhopal demonstrated that in contrast to residue burning, the soil incorporation or retention of wheat residue when coupled with supplemental N supply through organic or inorganic sources resulted in increased yield and nutrient uptake by crops, P recovery efficiency, and improvements in soil organic carbon and soil fertility (Reddy and Srivastava 2005). Similarly, long-term incorporation or retention of crop residues coupled with appropriate fertilization have been shown to enhance soil quality and productivity. Adopting ZT for wheat had a positive effect on soil quality regardless of the treatments used for rice under rice–wheat cropping systems on a vertisol (Mohanty et al. 2007).

Southern India

Much of the research done on CA in southern India is related to the conservation of soil and rainwater and drought proofing, which is an ideal strategy for adaptation to climate change. Important technologies include *in situ* moisture conservation, rainwater harvesting and recycling, efficient use of irrigation water, energy efficiency in agriculture, and use of poor-quality water. Watershed management is an accepted strategy for the development of rainfed agriculture through soil and water conservation by moderating the runoff and minimizing flooding during high-intensity rainfall. There is scope for the integration of watershed management and CA towards achieving the task of sustainability. While reduced tillage is possible in some production systems in high-rainfall regions in southern India, non-availability of crop residue for surface application is a major constraint. In dryland ecosystems, it is possible to raise a second crop with residual soil moisture by covering the soil with crop residues. Rainfall and soil type have a strong influence on the performance of reduced tillage. In arid regions (<500 mm rainfall), reduced tillage was found to be on a par with conventional tillage and the weed problem was controllable in arid inceptisols and aridisols. In semi-arid (500–1000 mm) regions, conventional tillage was superior. However, low tillage

+ interculture was superior in semi-arid vertisols and low tillage + herbicide was superior in aridisols. In sub-humid (>1000 mm) regions, the weed problem was severe due to rainfall and thus, there is a possibility of reducing the weed population by using herbicides in a reduced tillage condition.

Significant effects of tillage as well as conjunctive nutrient–use treatments were observed on sorghum and mungbean grain yields at Hyderabad (Sharma et al. 2009). Conventional tillage up to the 8th year of the study maintained 12.8% and 11.2% higher sorghum and mungbean grain yields, respectively, compared to reduced tillage. After eight years, reduced tillage tended to be equal or better than conventional tillage in improving crop yields. Overall, CA has been reported to improve crop productivity and resource-use efficiency, and reduce global warming potential more effectively than CT.

Benefits of CA

Several benefits of CA have been demonstrated based on years of research and adoption on farmers' fields, which can be seen at the farm, regional, and national levels. These can be classified into four broad categories: (i) resource saving, (ii) soil health improvement, (iii) environmental, and (iv) productivity and economics. Some of these benefits are listed in Table 2.6.

Table 2.6 Proven benefits of research and adoption of CA at the regional and national level

Parameter	Benefit accrued
I. Resource saving	
Fuel consumption	Up to 80% fuel used is conserved by converting to CA
Time conservation	Only 1–3 trips are required over a field with CA compared with 5–10 trips in tillage-based farming
Labour consumption	Up to 60% fewer mandays are needed compared with conventional tillage
Time flexibility	CA allows late decisions to be made about growing crops
Reduced irrigation requirements	Improved water-holding capacity and reduced evaporation lessen the need for irrigation
Reduced germination of weeds	Absence of physical soil disturbance and retention of surface residues under CA reduce new weed seed germination
Reduced fertilizer requirements	Adding crop residues and inclusion of cover crops over a period improve soil fertility
II. Soil health improvement	
Physical properties	
Preservation of soil structure	Tillage destroys natural soil structure, while CA minimizes structural breakdown
Improved aeration	Improvement in earthworm numbers, organic matter and soil structure result in improved soil aeration and porosity over time
Improved infiltration	Residues reduce surface sealing by raindrop impact and slows down velocity of runoff water

Table 2.6 Cont.

Parameter	Benefit accrued
Preventing soil erosion	Preserving soil structure, earthworms, and organic matter, together with surface residues reduce wind and water erosion
Soil moisture conservation	Physical disturbance of soil exposes it to drying, whereas CA along with surface residues reduces drying
Moderating soil temperatures	Under CA soil temperature in summer is lower than under tillage, winter temperatures are usually higher when the crop residues are retained on soil surface
Chemical properties	
Increased soil organic matter	By leaving the previous crop residues on the soil surface to decay, soil organic matter near the surface is increased
Increased soil nitrogen	Tillage mineralizes soil N and provides a short-term boost to plant growth, but such N is mined from the soil organic matter, further reducing total soil organic matter
Natural mixing of soil P and K	CA increases earthworms, which mix large quantities of soil P and K in the root zone
Biological properties	
Preservation of earthworms	Tillage destroys human's most valuable soilborne ally, earthworms, while CA encourages their multiplication
Microbial biomass and soil enzymes	CA favours microbial biomass growth and increased FDA, DHA, alkaline phosphatase

III. Environmental benefits

Reduced air pollution	Reduction in diesel consumption results in lower emissions of CO_2 into the atmosphere
Reduced pollution of waterways	Decreased runoff of water from soil and chemicals it transports reduces pollution of streams and rivers
Moderating canopy temperatures	Canopy temperature under ZT + crop residues has been observed to be lower than under CT. This avoids terminal heat stress in crops like wheat
Reduced emission of GHGs	CA involves a set of climate-resilient technologies, which cause lower emission of GHGs like CH_4, CO_2, N_2O, etc.

IV. Higher productivity and economics

Increased crop yields	All the above factors are capable of improving crop yields to levels well above those attained by tillage, but only if the CA system and processes are fully practiced
Lower costs	Total capital and/or operating costs required to establish tillage crops are reduced by up to 50% when CA substitutes for tillage
Longer replacement intervals for machinery	Because of reduced hours per ha, tractors and advanced no-till drills are replaced less often and reduce capital costs over time
Future improvements expected over time	Modern advanced CA systems and equipment have removed earlier expectations of depressed crop yields in the short term to gain long-term benefits of CA

Source: Adapted from Baker and Saxton (2007).

Table 2.7 Disadvantages and limitations of CA

Parameter	Limitation
Risk of crop failure	Where inappropriate ZT tools, and weed or pest control measures are used, there may be greater risk of yield reductions or failure than for tillage
Larger tractors required	Although the total energy input is reduced under CA, most of that input is applied in a single operation, which may require a large tractor
New machinery required	Because no-tillage is a relatively new technique, new and different equipment has to be purchased or hired
New pest and disease problems	Absence of physical disturbance and retention of surface residues may encourage some pests and diseases
Fields are not smoothed	Absence of physical disturbance prevents soil movement by machines for smoothing and levelling purposes
Fertilizers/pesticides are not incorporated	General incorporation of fertilizers/pesticides is difficult in the absence of physical burial by machines
Use of agricultural chemicals	The reliance of no-tillage on herbicides for weed control is a cost and environmental negative
Shift in dominant weed species	Chemical weed control tends to be selective and may lead to weed shift and resistant weeds
New skills are required	CA is a more exacting farming method, requiring the learning and implementation of new skills
Non-availability of expertise	Until the specific requirements of successful no-tillage are fully understood by experts, the quality of advice to practitioners from consultants will remain variable
Untidy field appearance	Farmers who have become used to the appearance of neat, clean tilled seedbeds often find the retention of surface residues (trash) untidy

Source: Adapted from Baker and Saxton (2007).

Disadvantages and Limitations of CA

There are a number of problems that may be encountered with the adoption of CA. One of the major constraints is the mindset of the farming community. In the past, farmers have realized huge economic benefits through intensive agriculture practices. A complete shift from conventional intensive tillage to ZT or CA needs an extensive educational programme by demonstrating the benefits accrued by CA on a large scale. The disadvantages and limitations of CA are presented in Table 2.7.

Conclusion and Future Outlook

CA technologies are the future of sustainable agriculture. There are potential benefits of CA across different agro-ecoregions and farmer groups. The benefits range from the nano-level (improving soil properties) to the micro-level (saving inputs, reducing cost of production, increasing farm income), and macro-level (reducing poverty, improving food security, alleviating global warming). Despite several advantages, there has been relatively slow adoption of this sustainable agricultural system on smallholder farms under Indian conditions. Adoption of CA practices not only results in yield levels on a par with conventional tillage systems but also leads to a 20–25% reduction in energy consumption and a reduction in the cost of land preparation. An overall saving to the extent of Rs 5000 per ha can be accrued as compared to farmers' current practice, besides other benefits such as natural resource conservation, reduced emissions of greenhouse gases, and better resilience to climatic extremes. However, moving from conventional to CA-based technologies involves a paradigm shift in key elements including approaches to develop component technologies of cultivar choices, nutrient, water, weed, and pest management, while optimizing cropping systems.

The adoption of CA may not result solely from the biophysical feasibility and benefits of the technology alone but needs support and incentives from institutional arrangements (mechanization, markets, credit, policies, etc.) that enable its adoption by farmers with different socio-economic circumstances. Policy support at all levels is needed to promote the adoption of innovative sustainable agricultural system by all farmers. There is a need for aggressive demonstration and information dissemination programmes and for these to be well complemented by skill development of the farmers. There is a need for a national movement for promoting CA. Appropriate institutional arrangements need to be evolved for small and marginal farmers who may not be able to afford to maintain the machines and other equipment for practicing CA. In addition, a massive training programme for capacity development of farmers needs to be developed. *Krishi Vigyan Kendras* in partnerships with research institutions engaged in CA may take the lead in this endeavour.

References

Baker, C.B. and Saxton, K.E. 2007. The what and why of no-tillage farming. In: *No-tillage seeding in Conservation Agriculture*, 2nd Edn. (Eds.) Baker, C.J. and Saxton, K.E.. FAO and CAB International.

Beri, V., Sidhu, B.S., Gupta, A.P., et al. 2003. Organic Resources of a Part of Indo-Gangetic Plains and Their Utilization, 93 p. Department of Soils, Punjab Agricultural University, Ludhiana, India.

Bolliger, A., Magid, J., Amado, J.C.T., et al. 2006. Taking stock of the Brazilian "Zero-Till Revolution": A review of landmark research and farmers' practice. *Advances in Agronomy* 91: 47–110.

Das, T.K., Bandyopadhyay, K.K., Bhattacharyyaa, R., et al. 2016. Effects of conservation agriculture on crop productivity and water-use efficiency under an irrigated pigeonpea-wheat cropping system in the western Indo-Gangetic Plains. *Journal of Agricultural Science*, Cambridge 154: 1327–1342.

Derpsch, R. 2003. Conservation tillage, no-tillage and related technologies, pp. 181–190. In: García-Torres, L., Benites, J., Martínez-Vilela, et al. (Eds.) Conservation Agriculture. Springer, Dordrecht.

Erenstein, O. and Laxmi, V. 2008. Zero tillage impacts in India's rice-wheat systems: a review. *Soil and Tillage Research* 100: 1–14.

FAO. 1987. Conservation agriculture website. www.fao.org.

Farooq, M., Flower, K.C., Jabran, K., et al. 2011. Crop yield and weed management in rainfed conservation agriculture, *Soil and Tillage Research* 117: 172–183.

Faulkner, E.H. 1945. *Plowman's Folly.* Michael Joseph, London.

Fukuoka, M. 1975. *One Straw Revolution.* Rodale Press. English Translation of Shizen Noho Wara Ippeon No Kakumei. Hakujusha Co., Tokyo, Japan.

Gathala, M.K., Ladha, J.K., Saharawat, Y.S., et al. 2011. Effect of tillage and crop establishment methods on physical properties of a medium-textured soil under a seven-year rice–wheat rotation. *Soil Science Society of America Journal* 75: 1851–1862.

Ghosh, B.N., Dogra, Pradeep, Sharma, N.K., et al. 2015. Conservation agriculture impact for soil conservation in maize–wheat cropping system in the Indian sub-Himalayas. *International Soil and Water Conservation Research* 3: 112–118.

Gupta, R.K. and Seth, A. 2007. A review of resource conserving technologies for sustainable management of the rice-wheat cropping systems of the Indo-Gangetic Plains (IGP). *Crop Protection* 26: 436–447.

Hati, K.M., Chaudhary, R.S., Mandal, K.G., et al. 2015. Effects of tillage, residue and fertilizer nitrogen on crop yields, and soil physical properties under soybean–wheat rotation in Vertisols of Central India. *Agricultural Research* 4: 48–56.

Huggins, D.R. and Reganold, J.P. 2008. No-till: The quiet revolution. *Scientific American Inc.* (July 2008): 71–77.

ISAAA. 2019. *Global Status of Commercialized Biotech / GM Crops.* International Service for the Acquisition of Agri-biotech Applications (ISAAA) Brief No. 55. Ithaca, New York.

Islam, R. and Reeder, R. 2014. No-till and conservation agriculture in the United States: An example from the David Brandt farm, Carroll, Ohio. *International Soil and Water Conservation Research* 2(1): 97–107.

Jat, M.L., Gathala, M.K., Ladha, J.K., et al. 2009. Evaluation of precision land leveling and double zero-till systems in rice-wheat rotation: water use, productivity, profitability and soil physical properties. *Soil and Tillage Research* 105: 112–121.

Jat, M.L., Malik, R.K., Saharawat, Y.S., et al. 2012. Proceedings of Regional Dialogue on Conservation Agricultural in South Asia, p 32. APAARI, CIMMYT, ICAR, New Delhi, India.

Jat, M.L., Saharawat, Y.S. and Gupta, Raj. 2011. Conservation agriculture in cereal system of South Asia: Nutrient management perspectives. *Karnataka Journal of Agricultural Science* 24(1): 100–105.

Kassam, A., Friedrich, T. and Derpsch, R. 2019. Global spread of conservation agriculture. *International Journal of Environmental Studies* 76(1): 29–51.

Kushwah, S.S., Reddy, D. Damodar, Somasundaram, J., et al. 2016. Crop residue retention and nutrient management practices on stratification of phosphorus and soil organic carbon in the soybean–wheat system in vertisols of Central India. *Communications in Soil Science and Plant Analysis* 47: 2387–2395.

Laik, R., Sharma, Sheetal, Idris, M., et al. 2014. Integration of conservation agriculture with best management practices for improving system performance of the rice-wheat rotation in the Eastern Indo-Gangetic Plains of India. *Agriculture, Ecosystems and Environment* 195: 68–82.

Lal, R. 2004. Soil carbon sequestration impacts on global climate change and food security. *Science* 34: 1623–1627.

Malik, R.K., Gupta, R.K., Singh, C.M., et al. 2005. Accelerating the adoption of resource conservation technologies in rice-wheat system of the Indo-Gangetic Plains. Proceedings of Project Workshop, Directorate of Extension Education, Chaudhary Charan Singh Haryana Agricultural University, June 1–2, 2005. Hisar, India.

Mohanty, M., Painuli, D.K., Misra, A.K. et al. 2007. Soil quality effects of tillage and residue under rice–wheat cropping on a Vertisol in India. *Soil and Tillage Research* 92: 243–250.

Montgomery, D.R. 2007a. Soil erosion and agricultural sustainability. *Proceedings National Academy of Sciences* USA 104: 13268–13272.

Montgomery, D. 2007b. *Dirt: The Erosion of Civilizations*. California University Press, Berkeley.

Moyer, J.R., Roman, E.S., Lindwal, C.W., et al. 1994. Weed management in conservation tillage systems for wheat production in North and South America. *Crop Protection* 13(4): 243–259.

Paroda, R.S. 2009. Global Conventions and Partnerships and their Relevance to Conservation Agriculture, pp. 30–35. In: 4th World Congress on Conservation Agriculture held in New Delhi on 4–7 February 2009.

Pathak, H., Saharawat, Y.S., Gathala, M., et al. 2011. Impact of resource-conserving technologies on productivity and greenhouse gas emission in rice-wheat system. *Greenhouse Gases: Science and Technology* 1: 261–277.

Pradhan, A., Idol, T. and Roul, P.K. 2016. Conservation agriculture practices in rainfed uplands of India improve maize-based system productivity and profitability. *Frontiers in Plant Science* 7: 1008.

Prasad, R. 2011. Aerobic rice systems. *Advances in Agronomy* 111: 207–236.

Ram, H., Yadvinder-Singh, Saini, K.S., et al. 2013. Tillage and planting methods effects on yield, water use efficiency and profitability of soybean–wheat system on a loamy sand soil. *Experimental Agriculture* 49(4): 524–542.

Reddy, D. and Srivastava, S. 2005. Management of mechanical harvest-borne wheat residue under soybean-wheat system on vertisols. Indian Institute of Soil Science, Bhopal, Madhya Pradesh.

Saharawat, Y.S., Gill, M., Gathala, M., et al. 2022. Conservation agriculture in South Asia. In: *Advances in Conservation Agriculture*, Volume 3 – Adoption and Spread. Ed. Kassam, A. Burleigh Dodds. Science Publishing.

Saharawat, Y.S., Ladha, J.K., Pathak, H., et al. 2012. Simulation of resource-conserving technologies on productivity, income and greenhouse gas emission in rice-wheat system. *Journal of Soil Science and Environmental Management* 3(1): 9–22.

Saharawat, Y.S., Singh, B., Malik, R.K., et al. 2010. Evaluation of alternative tillage and crop establishment methods in a rice–wheat rotation in north-western IGPs. *Field Crops Research* 116: 260–267.

Sharma, K.L., Grace, J.K., Srinivas, K., et al. 2009. Influence of tillage and nutrient sources on yield sustainability and soil quality under sorghum–mungbean system in rainfed semi-arid tropics. *Communications in Soil Science and Plant Analysis* 40: 2579–2602.

Sharma, A.R., Jat, M.L., Saharawat, Y.S., et al. 2012. Conservation agriculture for improving productivity and resource-use efficiency: Prospects and research needs in Indian context. *Indian Journal of Agronomy* 57(3rd IAC Special Issue): 131–140.

Singh, V.K., Dwivedi, B.S., Shukla, A.K., et al. 2005. Diversification of rice with pigeonpea in a rice–wheat cropping system on a Typic Ustochrept: effect on soil fertility, yield and nutrient-use efficiency. *Field Crops Research* 92: 85–105.

Singh, G., Kumar, D. and Sharma, P. 2015. Effect of organics, biofertilizers and crop residue application on soil microbial activity in rice-wheat and rice-wheat-mungbean cropping systems in the Indo-Gangetic plains. *Cogent Geoscience* 1:1.

Tull, J. 1829. *Horse-Hoeing Husbandry*, 465 p. William Cobbett, Fleet Street, London.

Vlek, P.L.G. 2008. The incipient threat of land degradation. *Journal of Indian Society of Soil Science* 56(1): 1–13.

WRI. 2000. *People and ecosystems, the frayling web of life*, 36 p. World Resource Institute, United Nations Development Programme, World Bank, Washington, USA.

3 Conservation Agriculture in India

Progress in Research, Adoption, and the Way Forward

A.R. Sharma

Introduction

The adoption of green revolution technologies during the 1960s led to increased productivity and the elimination of acute food grain shortages in India. These technologies primarily involved growing of high-yielding dwarf varieties of rice and wheat, increased use of chemical fertilizers and other agrochemicals, and the expansion of irrigation facilities. This was accompanied by the other so-called modern methods of cultivation, which included maximum tilling of land, virtually clean cultivation with complete removal of crop residues and other biomass from the field, fixed crop rotations mostly involving cereals, and the elimination of fertility-restoring pulses and oilseed crops in the highly productive north-western plain zone of the country. Also, there was greater dependence on chemical fertilizers and reduced application of organic additions like compost and manures.

Over the last 4–5 decades, India has not only achieved self-sufficiency in food grain production but also the capability to export food commodities. This is cited as one of the greatest accomplishments of Indian agriculture in the post-independence era. However, the transformation from 'traditional animal-based subsistence farming' to 'intensive chemical- and tractor-based modern agriculture' has led to a multiplicity of issues associated with the sustainability of these production practices. Conventional crop production technologies are characterized by: (i) intensive tillage to prepare fine seed- and root-bed for sowing to ensure proper germination and initial vigour, faster absorption of moisture, control of weeds and other pests, mixing of fertilizers and organic manures; (ii) fixed crop rotations mostly involving cereal crops and excluding legumes; (iii) clean cultivation involving the removal or burning of all residues after harvesting leading to continuous mining of nutrients and moisture from the soil profile, and bare soil with no cover; (iv) indiscriminate use of pesticides, and excessive and imbalanced use of chemical fertilizers leading to a decline in input-use efficiency and factor productivity, and an increase in pollution of the environment, ground water, streams, rivers, and oceans; and (v) energy-intensive farming systems.

DOI: 10.4324/9781003292487-4

Conventional Versus Conservation Agriculture Systems

Conventional agriculture systems are characterized by intensive tillage operations, clean cultivation, fixed cropping, indiscriminate use of irrigation water, and chemical fertilizers. The adoption of these systems has led to declining factor productivity, deteriorating soil health, surface and groundwater pollution, increasing cost of production, and lower profitability. It is realized that soils are becoming impoverished due to imbalanced use of fertilizers, and the discontinuation of traditional practices like mulching, intercropping, and inclusion of legumes in cropping systems. Further, the use of organic manures, composts, and green manure crops has also decreased considerably for various reasons. Similarly, water resources are under great stress due to their indiscriminate exploitation and they are getting polluted due to various human interferences. Burning of fossil fuels and crop residues, and puddling for rice cultivation are leading to the emissions of greenhouse gases, which are responsible for climate change and global warming. Further, there is now a growing realization that productivity levels are stagnating, and the incomes of farmers are reducing due to the rising costs of the inputs and farm operations. It is feared that modern cultivation practices are not sustainable in the long-run, and there is a need to change the way we produce crops on arable lands.

Conservation agriculture (CA) is considered as a new paradigm in resource management for alleviating the problems associated with the so-called modern cultivation practices. Three principles of CA are: (i) continuous no or minimal mechanical soil disturbance that is just enough to get the seed into the ground for good germination, (ii) maintenance of permanent biomass mulch on the ground surface, and (iii) diversification or rotation of crop species. These principles are universally applicable to all agricultural landscapes and land uses with locally formulated and adapted practices. Strictly speaking, if the three principles are applied separately, they do not constitute a CA system. For example, the use of zero tillage (ZT) practice on its own does not qualify the production system to be CA-based, unless it is linked to the application of the other two practices of soil mulch cover and diversified cropping. Nonetheless, individual adoption of each of these three pillars is still better than the conventional system. For full adoption of CA, a stepwise system may be advocated since most farmers may be reluctant to adopt all three pillars at the same time.

CA also requires suitable modifications in water, fertilizer, weed, and pest management practices as well as farm machinery compared with conventional tillage (CT). It is a holistic approach towards increased productivity and improved soil health. The advantages of CA over CT have been reviewed by several workers (Sharma et al. 2012; Jat et al. 2020; Saharawat et al. 2021). In recent times, harvesting of most crops with combines and burning of residues in situ in India has become a major national issue, despite restrictions imposed and incentives offered by the government. This is an unhealthy practice as it leads to the loss of precious plant nutrients and environmental pollution. In view of this, innovative approaches are needed to deal with emerging challenges under the changed circumstances.

Two major innovations in the latter half of the 20th century have led to a change in our thinking on crop production. These are: (i) the availability of new farm machinery, and (ii) effective herbicides, which suggest that ploughing of fields is no longer required for sowing, fertilizer placement, and weeding. The new-generation farm machinery can place the seed and fertilizer at an appropriate depth in the desired amount. Further, these machines can work in standing (anchored) as well as loose crop residues; thus providing a very effective mulch cover for moisture and nutrient conservation, temperature moderation, and weed control. There is a need for farmer training and the use of proper sprayers to improve the effectiveness of herbicide use. Integrated weed management, including hand weeding, along with integrated pest and disease management are needed. The availability of new herbicide molecules has also necessitated a change in our thinking about weed management. Further, other triggering factors for a shift towards CA are labour scarcity, deteriorating soil health, declining factor productivity, rising costs, low water productivity, and low income. Thus, CA systems can help in overcoming the problems being experienced in conventional farming systems.

CA Research in India

In India, efforts have been made to advance CA research since the mid-1990s through eco-regional programmes such as the Rice–Wheat Consortium (RWC), national initiatives like the National Agricultural Technology Project (NATP), National Agricultural Innovation Project (NAIP), ICAR platform on CA, National Innovation on Climate Resilient Agriculture (NICRA), regional bilateral collaborative programs like Cereal Systems Initiative for South Asia (CSISA), Sustainable and Resilient Farming Systems Initiative for Eastern Gangetic Plains (SRFSI), CGIAR Research Programs (CRPs) on Climate Change, Agriculture and Food Security (CCAFS), Wheat-Agri-Food System (WHEAT), Maize Agri-Food System (MAIZE), and regional platforms like Borlaug Institute for South Asia (BISA), by involving a large number of institutions and organizations. The uptake of CA in India has been slow but, with science-based evidence on multiple benefits in addressing the growing complexity of challenges and to deliver to several the Sustainable Development Goals, CA has emerged as one of the major frontiers of future farming. However, scaling CA-based management practices in a diversity of farm typologies and production ecologies for impact at scale needs a collaborative approach of a consortium of projects/programmes/institutions involved in CA research for development (CAR4D) in India.

An analysis of the research papers published on topics related to CA was undertaken in the three major Indian journals, namely *Indian Journal of Agronomy*, *Indian Journal of Weed Science*, and *Indian Journal of Agricultural Sciences*. The articles on ZT alone (partial CA) or in combination with crop residue, weed control, fertilizer ,and water management in a single crop or cropping system were considered. During the period from 1995 to 2000, there

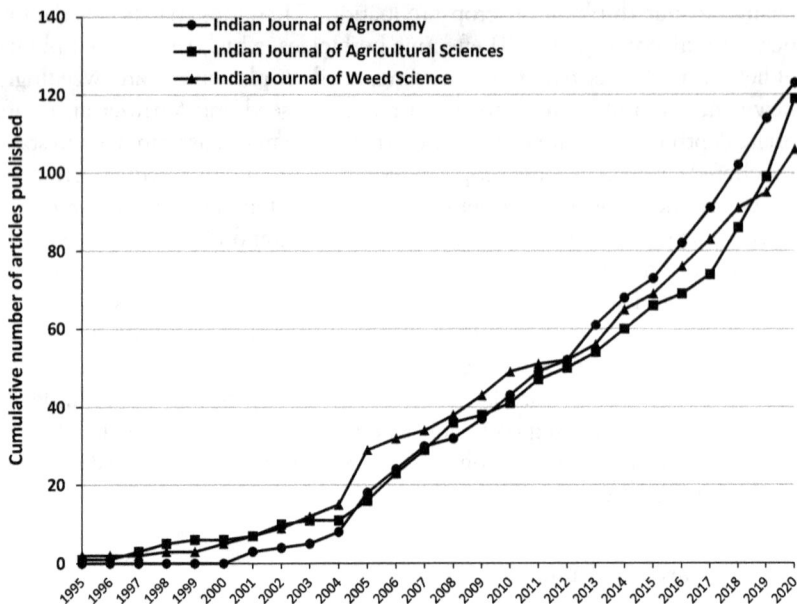

Figure 3.1 Research articles related to CA published in the three leading Indian journals.

were no articles published on CA in the *Indian Journal of Agronomy*, while there were only six and five articles published in the *Indian Journal of Agricultural Sciences* and *Indian Journal of Weed Science*, respectively. From 2000 to 2003 also, there were only a few sporadic articles published in these journals. However, it was from 2004 onwards that the research work on CA picked up and progressed almost linearly in all three journals. After 2010, there was a steep rise in the number of articles, and recent years have witnessed a jump in CA-based articles (Figure 3.1). This indicates that research work on CA has now become one of the major areas of investigation by resource management scientists in India.

An analysis was done on the comparison of grain yields obtained under CT and CA (ZT without or with crop residue) (Table 3.1). The yields varying under CA by a margin of ±5% were considered as higher or lower than CT. Most of the research on CA in India has been published on wheat (58.3%), which is a dominant crop of the winter season in north-western and central parts of the country. It was found that in 46% of the articles, the yields of wheat under CA were higher than under CT, while in 42% of the articles, the yields were the same. The yields under CA were lower than CT in 12% of articles. This suggests that yields of wheat under CA were either the same or higher than for CT in 88% of the articles. This was because wheat was sown closer to the optimum time under CA, while the lower yields in 12% cases were due to delayed sowing and lack of application of full CA protocols. On the other hand, in the case of rice,

Table 3.1 Effect of CA (zero-tillage with or without residue) vs. conventional tillage (CT) on crop productivity in India (no. of articles)

Crop	Yields equal under CA and CT	Yield higher under CA than CT	Yields lower under CA than CT	Total
Wheat	57	63	17	137
Rice	9	13	12	34
Maize	7	14	13	34
Soybean	1	5	4	10
Mustard	3	3	1	7
Chickpea	3	2	1	6
Cotton	2	1	1	4
Greengram	2	1	0	3
Total	**84**	**102**	**49**	**235**

only 27% of the articles reported equal yields and 38% reported higher yields under CA, while 35% reported that yields were lower under CA than CT. In the case of maize, the number of articles reporting lower yields was 38%, while 62% reported yields under CA which were either equal to or higher than CT. The number of articles in other crops, viz. cotton, chickpea, soybean, mustard, and greengram was relatively small, but a greater percentage of the articles reported yields either equal to or more under CA than CT. This suggests that most trials on crops grown in the winter season (wheat, chickpea, mustard) showed either equal or better performance under CA than CT, while a greater percentage of articles on rainy season crops (rice, maize, soybean) showed relatively lower yields under CA than CT. This may be because weeds, which are considered a major handicap under CA, showed greater infestation in the rainy season crops due to warm and humid conditions, while the weed problem is much less and easier to manage in the winter season crops. Weed infestation, especially of the troublesome weed, *Phalaris minor*, in wheat has been observed to be much less under CA than CT (Om et al. 2004; Malik et al. 2014). CA-based technologies are herbicide-driven, and efficient weed management is one of the major considerations for the success of CA (Sharma and Singh 2014; Bhullar et al. 2016). It has been conclusively proved based on long-term trials that weed problems are reduced in successive cropping cycles by following all the CA protocols along with integrated weed management (Baker and Saxton 2007; Chauhan et al. 2012; Fonteyne et al. 2020).

Further scrutiny of the articles published state-wise revealed that CA has been the major area of research in the north-western and central parts of the country (Table 3.2). The highest number of articles were published from Delhi, followed by Uttar Pradesh, Madhya Pradesh, Punjab, Uttarakhand, and Haryana. These are also the states where there has been sizable coverage of CA-based technologies. This was followed by the eastern states of Bihar and West Bengal. The work done on CA in the southern and western regions of the country has been relatively less. Evidently, the number of articles also depended on the size of the state

Table 3.2 State-wise number of research papers published in national journals (1995–2020)

State	IJA	IJWS	IJAS	Total
Delhi	13	2	47	62
Uttar Pradesh	19	12	15	46
Madhya Pradesh	8	21	11	40
Punjab	10	15	4	29
Uttarakhand	9	15	2	26
Haryana	10	6	7	23
Bihar	8	1	5	14
West Bengal	6	5	2	13
Jammu-Kashmir	6	1	4	11
Himachal Pradesh	5	1	4	10
Karnataka	2	2	5	9
Chhattisgarh	2	6	0	8
Rajasthan	4	3	1	8
Andhra Pradesh	0	4	2	6
Odisha	3	3	1	7
Telangana	4	0	2	6
NEH	3	0	3	6
Assam	4	0	1	5
Tamil Nadu	2	3	0	5
Jharkhand	2	3	0	5
Maharashtra	1	1	1	3
Kerala	2	0	1	3
Gujarat	0	2	0	2
Andaman & Nicobar Islands	0	0	1	1
Total	**123**	**106**	**119**	**348**

IJA – *Indian Journal of Agronomy*
IJWS – *Indian Journal of Weed Science*
IJAS – *Indian Journal of Agricultural Sciences*

and the number of researchers, but a greater focus of research on CA in north-western India suggested greater adoption of these technologies, particularly for the growing of wheat in the predominantly followed rice–wheat cropping system.

Adoption of CA in India

Since the mid-1990s, there has been greater focus on the development and promotion of CA-based technologies, primarily for growing wheat under ZT in the predominantly followed rice–wheat cropping systems of the Indo-Gangetic plains. However, efforts have also been made over the past decade for the promotion of ZT technologies in other crops including maize, sorghum, mustard, and chickpea in the non-Indo-Gangetic plains including central India, coastal Andhra Pradesh, NEH region, and Konkan region of Maharashtra (Sharma et al. 2018). Some successful examples of the adoption of these technologies are given below.

North-Western India

Rice–wheat is the major cropping system in the Indo-Gangetic plains (10.5 M ha) comprising Punjab, Haryana, Delhi, Uttar Pradesh, Bihar, and parts of Uttarakhand and Rajasthan. In this system, wheat is normally sown in a fine seedbed prepared with 4–5 tillage operations, which takes 10–15 days and Rs. 3000–3500 per ha for land preparation. The tillage operations increase the cost of production, but they have hardly any benefit for increasing the grain yield. Further, lower yields of wheat are often due to delayed sowings as a considerable amount of time is lost during land preparation and pre-sowing irrigation. There is also great concern about the reduction in soil fertility, scarcity of farm labour, declining water table, and high cost of production under conventional agriculture. In order to mitigate these problems, it was considered essential to adopt technically feasible, economically viable, and ecologically permissible technology to ameliorate late sowing, minimize weed infestation, lower cost of production, improve fertilizer/water-use efficiency, and improve soil fertility.

Initiatives have been taken since the early 1990s to promote the adoption of CA-based technologies in the alluvial soils of the Indo-Gangetic plains. These programmes were initially implemented through the CGIAR-sponsored CIMMYT/IRRI/NARS management of the Rice–Wheat Consortium, which led to the adoption of ZT cultivation of wheat on nearly 3 M ha by 2000. However, the area under ZT wheat declined subsequently as the harvesting of wheat progressed with combine harvesters leaving the residue on soil surface. The conventional ZT seed drill did not work in the loose residue conditions after combine harvesting. Thus, many farmers started burning the rice residue *in situ* so as clear the fields for timely sowing of the next crop of wheat. Subsequently, improved versions of seeding machines, known as happy seeders and super seeders, were developed. The happy seeder cuts and lifts the residue, and sows the seeds directly into the soil, depositing the cut straw as mulch over the sown area. This machine can work efficiently in anchored or loose residue conditions and has been reported to achieve the same or higher yields compared with tilled systems (Keil et al. 2021). Higher profits from the happy seeder system stem from slightly higher yields and lower input costs for land preparation. This technology has the potential to eradicate the practice of rice residue burning due to its ability to sow wheat directly into large amounts of anchored and loose residues. This technology has spread to several thousand ha in the last few years due to the easy availability of happy seeders at subsidized cost and restrictions imposed by the government on the burning of rice residues. It has been estimated that 20% of the area under direct-seeded rice and about 30% of the area under ZT wheat was covered in the state of Punjab during 2020–21, which is likely to increase further in the coming years as farmers gain confidence and perfect the technology to meet their requirements. Custom-hiring centres like those of combine harvesting service providers are now available in most parts of the region where farm machinery especially for residue management is available at an affordable price to small-holder farmers.

Central India

Crop production in the central plateau region of India is dominated by rice, soybean, maize, and sugarcane in the rainy season, followed by wheat, chickpea, lentil, peas, and mustard in the winter season. Soils are deep black vertisols in most of these areas, which have the property of swelling and shrinkage depending on the moisture content, and these are considered as difficult soils from the point of view of tillage and seed-bed preparation. Farmers follow conventional practices like intensive ploughing of the land, clean cultivation (removal or burning of all crop residues and stubble), fixed crop rotations, ad little use of organic manure, and moderate use of chemical fertilizers and pesticides, including herbicides. Combine harvesting of major crops is followed predominantly and the crop residues are invariably burnt in most areas. There is only a small area under greengram/blackgram during summer due to social problems like open cattle grazing. Due to the rising costs of cultivation, the profitability margins are generally low. Keeping this in mind, it is necessary to promote the adoption of resource conservation technologies to reduce the cost of cultivation and improve soil health, besides other benefits. Having the proper equipment for ZT planting and using mulch to keep the surface soil moist are important for the success of CA in these soils (Mohanty et al. 2007; Patil et al. 2016).

ICAR-Directorate of Weed Research, Jabalpur, took a major initiative and launched a flagship research programme in 2012 to develop and promote technologies related to weed management in CA. After laser levelling of the fields, trials were initiated on ZT sowing of wheat, chickpea, mustard, and maize (winter), followed by greengram (summer), while retaining the residues of the previous crop *in situ*. Sowing of seed and the placement of basal fertilizer was done with a happy seeder immediately after the harvesting of the previous crop with a combine harvester. Following the success of these crops, the technology was extended to rainy season crops including rice, soybean, maize, and pigeonpea. All crops showed either the same yield or a 5–10% higher yield under CA than CT when sowing was done at the same time (Table 3.3).The increase in yield is expected to be much higher because CA allows crops to be sown at the optimum time, which is at least 1–2 weeks before conventional sowing.

Simultaneously with the on-station trials, on-farm trials in participatory mode were also undertaken in farmers' fields of Jabalpur district in 2012–13. Initially, the farmers expressed serious doubt about growing a crop without ploughing. However, the farmers were persuaded to provide their lands for demonstrating the potential of CA technology with an assurance that they would be compensated if the technology failed to perform. Wheat was sown using a happy seeder for the first time in the fields, without tilling in the existing rice stubbles. Contrary to the general belief of the farmers, the crop showed good emergence and stand establishment. The weed population in these CA-based trials was lower than in the crop in which the land was prepared by a conventional cultivator and disc harrow. Herbicide application controlled the weed flora effectively and increased

Table 3.3 Yield performance of different crops under conventional and conservation agriculture systems at the DWR during 2012–2016 (figures in parentheses indicate the number of trials)

Season/crop	Grain yield (t/ha)		% increase under CA over CT
	*Conventional agriculture (CT)	**CA-based practices	
Rainy season			
Rice (10)	4.02	4.06	0.0
Maize (12)	4.03	4.39	8.9
Soybean (4)	0.94	1.06	11.3
Pigeonpea (5)	2.15	2.23	3.7
Winter season			
Wheat (10)	4.19	4.56	8.8
Chickpea (8)	1.89	2.01	6.3
Mustard (4)	1.83	1.94	6.0
Gobhisarson (4)	3.97	3.72	6.3
Field pea (4)	1.56	1.67	7.2
Summer season			
Greengram (9)	1.05	1.04	0.0

Notes: *Conventional agriculture involved 3–4 ploughings with a disc harrow, cultivator, and rotavator before sowing. All crop residues were removed from the field.
**CA-based practices involved zero-till sowing in full residue of the previous crop.

Note: Crop under CA and CT was sown at the same time.

Source: Sharma and Singh (2018).

the wheat yield by almost two times as compared to that cultivated by conventional practice.

Following the success of wheat, greengram was grown during the summer (mid-April to mid-June) under ZT using a happy seeder while retaining the residues of wheat on the soil surface. A spray of glyphosate was used before sowing where there were previously growing weeds in the field. Subsequently, imazethapyr was applied at 15–20 days of growth. Residue mulch helped to check the soil moisture loss through evaporation, moderated soil temperature, prevented emergence of weeds, and improved soil health. The crop under CA yielded 1.0–1.4 t/ha compared with 0.7–0.8 t/ha under the conventional farmers' practice.

After obtaining very encouraging results in 2012–13, many on-farm research (OFR) trials were undertaken in a participatory mode from 2013 to 2014 in different locations around Jabalpur (Gosalpur, Shahpura, Kundam, Bankhedi, and Majholi). Sowing of rice in the rainy season, wheat in winter, and greengram in summer was done on a 1-acre area (0.40 ha) in each farmer's field in participatory mode, as a field under conventional practice was also maintained nearby. From 2014 onwards, the OFR trials on CA-based technology were extended under the *Mera Gaon Mera Gaurav* (My Village My Pride) programme in the adjoining districts of Mandla, Seoni, Narsinghpur, and Katni. In each district,

Table 3.4 Comparative performance of ZT-sown crops and FP in OFR trials in different districts of Madhya Pradesh during 2015–16 (figures in parentheses indicate the number of trials)

District/villages/crops	Grain yield (t/ha)		% increase under CA over FP
	*CA-based practice	**Farmers' practice (FP)	
Jabalpur (villages – Bharda, Nipaniya, Nibhora, Neelkheda, Repura, Katangi, Chanti)			
Rice (8)	3.7–4.0	3.2–3.4	16–18
Wheat (16)	3.8–4.2	3.0–3.2	27–34
Greengram (9)	0.9–1.3	0.6–0.8	50–62
Mandla (villages – Bhawal, Bijegaon, Gojarsani, Lalipur, Harratikur)			
Rice (10)	3.9–4.3	3.0–3.2	30–34
Wheat (17)	4.2–4.4	3.2–3.4	29–31
Greengram (12)	1.0–1.2	0.7–0.9	33–42
Narsinghpur (villages – Baglai, Khamariya, Simariya, Kushiwara)			
Rice (7)	3.5–4.2	2.6–3.5	20–34
Wheat (8)	4.0–4.4	3.7–4.0	8–10
Katni (villages – Banda, Bichhiya, Chhitwara, Ghughra, Lakhapateri)			
Rice (8)	4.5–4.7	4.1–4.3	9–10
Wheat (10)	4.8–5.1	3.4–3.7	38–42
Greengram (8)	1.1–1.3	0.8–0.9	38–44
Seoni (villages – Ghughri Nagar, KhootKhamariya, NaganDeori, Dongargaon, Salaiya)			
Rice (7)	4.1–4.4	3.4–3.6	20–22
Wheat (12)	4.2–5.0	2.8–3.7	35–50
Greengram (5)	1.2–1.7	0.8–0.9	50–80

Source: Sharma and Singh (2018).

Notes: *CA-based practice involved ZT sowing in standing residue of the previous crop, 10–15 days ahead of farmers' usual practice.
**Farmers' practice involved burning or removal of the previous crop residues, followed by 3–4 ploughings before sowing.

five villages and 8–10 farmers from each locality were identified. Resource conservation technologies such as direct-seeding of rice, brown manuring with *Sesbania*, ZT sowing of crops, residue retention on soil surface, growing of summer legumes like greengram or *Sesbania* in the crop rotation, and integrated weed management were demonstrated in diversified cropping systems. About 100 OFR trials were laid out in 25 villages of the five districts of Madhya Pradesh during 2015–16. The performance of all crops sown at the optimum time under CA was far better than the delayed-sown crops under conventional farmers' practice (Table 3.4). The increase in grain yield ranged from 9–36% in rice, 27–50% in wheat, and

33–80% in greengram. In addition to the indirect benefits on soil health and the environment at large, there was an increase in net profit because of reduced costs of ploughing, and inputs such as fertilizer and water.

CA-based technology like zero-till sowing of crops in the presence of residue of the previous crop with improved weed management practices has proved to be the most promising technology in the vertisols of Madhya Pradesh. Adoption of CA technologies at the DWR, Jabalpur, ensured timely sowing of crops (by the end of June for rainy-season crops, the end of October for mustard and chickpea, mid-November for wheat, and the end of March for greengram); increase in cropping intensity from <150% in 2012 to 300% in 2016; large savings in diesel cost, machinery repair, and irrigation water; increased productivity (>10 t/ha/year) and profitability; and apparent improvement in soil health. This has proved to be a climate-resilient technology as it avoided burning of crop residues, puddling for rice transplanting, and ensured C-sequestration through residue recycling and zero-till cultivation. Contrary to the general belief, weed infestations were reduced considerably under CA compared with the conventional practices. Adopting integrated weed management throughout the crop season in combination with preventive measures led to decreased herbicide applications over time and thereby safer environment. This technology has found rapid acceptance among the farmers of Jabalpur, Katni, Seoni, Narsinghpur, and Mandla districts of Madhya Pradesh. It has spread to several thousand ha and the demand for happy seeder machines has increased in the region. Farmers have been highly convinced with this technology as it saved time, provided good weed control, maintained soil moisture status, and improved soil fertility, besides being environment friendly.

Coastal Andhra Pradesh

Rice is predominantly grown in the eastern and coastal areas of India, generally followed by fallow. Relay planting with short-duration pulses/oilseeds was practiced previously but the yields were low and highly variable due to poor crop stand and weed growth. Blackgram was popular in coastal Andhra Pradesh, but was affected by yellow vein mosaic virus and the parasitic weed *Cuscuta campestris*.

Cultivation of maize under ZT conditions immediately after the harvest of wetland rice has gained popularity in areas of Andhra Pradesh and Telangana (Jat et al. 2009a; Sreelatha et al. 2015). A large number of farmers are practicing ZT planting and getting higher profits in these states under the rice–maize system. Similarly, ZT sorghum in less irrigated areas has gained popularity among farmers and its cultivation has been virtually revolutionized in rice fallows (Chapke et al. 2011; Chapke and Babu 2016; Chapke et al. 2017). Sowing is done manually in wet soil in holes after harvesting of the preceding rice crop during mid-December, and fertilizers are applied after about one month, and two or three irrigations. Weeds are controlled by a tank-mix application of atrazine + paraquat (0.75 kg + 0.50 kg/ha) just after sowing but before crop emergence. It has been reported that greater grain yields of maize (8–10 t/ha) and sorghum (6–8 t/ha) are obtained

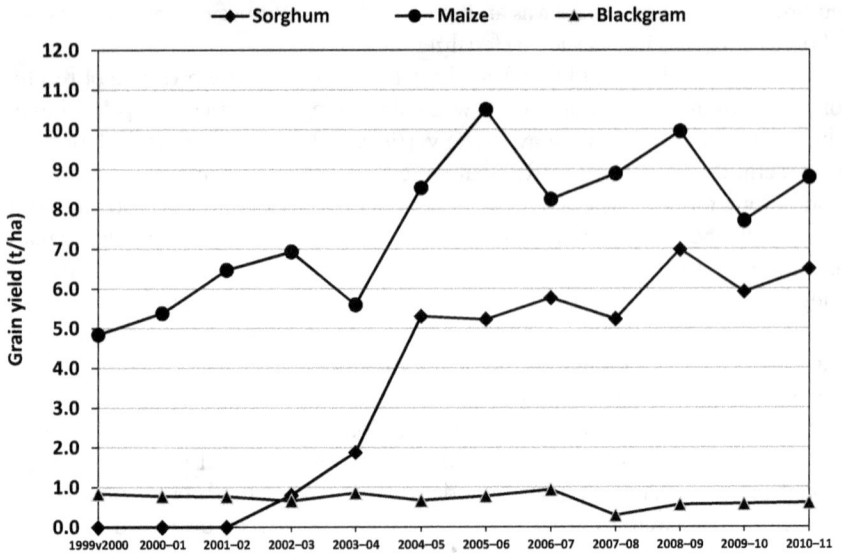

Figure 3.2 Success story of zero-till sorghum/maize revolution in rice fallows of coastal Andhra Pradesh.

Source: Chapke et al. (2017).

under ZT cultivation compared with <0.5 t/ha from blackgram (Figure 3.2). An area of about 150,000 ha is estimated to be under ZT maize in coastal Andhra Pradesh. This is often cited as one of the success stories of the adoption of ZT in coastal Andhra Pradesh and has immense potential for extension to other states including Odisha and West Bengal.

North-Eastern Hill (NEH) Region

Oilseed cultivation in the NEH region faces several constraints, such as water scarcity during the post-monsoon season, lack of irrigation facilities, short time lag after rice harvest for seed sowing, and a high incidence of pests and diseases in late-sown crops. As a result, only monocropping of rice is practiced and farmers leave their land fallow during the winter season.

Central Agricultural University, Imphal, in collaboration with the ICAR-Directorate of Rapeseed-Mustard Research, Bharatpur, implemented an extension project for augmenting rapeseed–mustard production for tribal farmers of the NEH region for sustainable livelihood security (ICAR 2012). The growth and yield parameters of rapeseed–mustard were better in ZT than CT due to residual soil moisture after rice harvest. Among the varieties, yellow sarson, Ragini, and NRCHB-101 gave average yields of 1.0 t/ha under ZT cultivation. The number of farmers, area covered, and the yields increased progressively over a period of

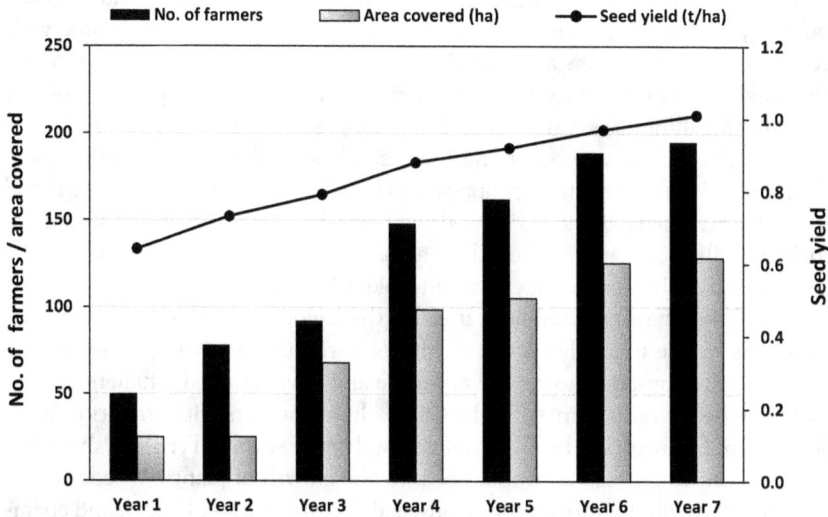

Figure 3.3 Progress of zero-till rapeseed–mustard in rice fallows in the NEH region.
Source: www.cau.ac.in/directorates.html.

seven years (Figure 3.3). Motivated by this success, many farmers in Manipur, Mizoram, and Arunachal Pradesh adopted this technology and the area coverage under ZT cultivation of rapeseed–mustard increased to >1000 ha over a period of two years.

This success story illustrated that rapeseed–mustard is a climate-resilient crop which can be grown without water in the residual soil moisture. By adopting ZT, the farmers could increase productivity, reduce the cost of cultivation, increase the cropping intensity, and earn an additional income with less effort. ZT also helped in timely sowing (October–November), conserved soil moisture and required less water, saved tillage cost and time, and the soil was protected from erosion due to the retention of surface residues. The improved version of ZT cultivation with bee pollination and no chemical method of plant protection has been recommended to the resource-poor farmers of the NEH region.

Konkan Region of Maharashtra

In the Konkan region of Maharashtra, including areas around Mumbai, zero-till broad-bed technology has been developed and promoted over the past decade by a local organization. Known as Shaguna Rice Technology (SRT), it is primarily meant for rice but can also be extended to other crops including groundnut, lablab bean, greengram, and vegetables grown in succession (https://srt-zerot ill.com/srt/). This technology involves preparing broad-beds (about 1 m wide) either manually with a spade or with a tractor-drawn bed maker, markings on the beds with a specially designed implement, placing the seeds and fertilizer manually, and using herbicides for weed control. This technology has found wide acceptance among farmers as it saved time and cost, and improved soil fertility, crop yields, and profitability compared with conventional transplanting of rice following puddling. Several farmers have adopted it, and more are following their example as they are becoming increasingly aware of it and learning more about. However, the technology appears to be labour-intensive as the seeds and fertilizers are placed manually. Power-driven raised-bed planters with a seed-cum-fertilizer drilling system are needed on a custom-hire basis to further accelerate the adoption of this technology by small-holder farmers.

Considering the erratic rainfall pattern of the region, it is advocated to advance the sowing of rice to the last week of May or early June so that seeds germinate with the early monsoon showers by mid-June and have attained sufficient growth before the heavy rains start from the end of June. Farmers with irrigation facilities can opt for irrigation before or immediately after sowing. Fertilizer should be basally placed to provide an initial boost to the growth of plants. Weeds can be controlled through the integration of physical, mechanical, ecological, and chemical methods. Light manual weeding should be undertaken at crop maturity or after harvest to avoid seed set from the leftover weed plants and to minimize the problem in the next season. Crabs are a serious problem in the early stages, and must be controlled using the appropriate insecticides. Similarly, wild boars, birds, rats, termites, and other pests should be controlled using the available technologies.

SRT appeared to be more suitable for small-holder farmers and those with family labour, as a team of 10–12 persons is required for sowing an area of one ha in a day. Large farmers owning >4–5 ha of land can use a tractor-drawn ZT seed-cum-fertilizer drill which will further reduce the cost/time and also ensure optimum crop stand, while small-holder farmers (1–2 ha) should have access to this machinery through custom-hiring centres. The benefits will multiply if a part of the crop residue is retained on the soil surface. Large increases in soil organic matter content over a short period of time and an increase in the earthworm population due to ZT cultivation and recycling of root biomass have been reported.

This technology has been adopted by over 5000 farmers who have reported very high rice yields of >10 t/ha. Based on the experiences of the farmers and also witnessing the excellent rice crops in the fields under SRT under aberrant weather conditions, this technology has the potential to replace conventional puddling/transplanting and revolutionize rice cultivation in the high rainfall areas of Konkan region of Maharashtra.

Practical Tips for Success of CA

CA requires greater skill and expertise in addition to equipment, training (especially in herbicide use), and markets for its successful adoption. The best management practices based on practical experiences on CA are described below.

Land Suitability

CA has been adopted across the globe on a variety of soils that vary from 90% sand to 80% clay. Soils with a very high clay content and that are sticky in nature have not been a hindrance to ZT adoption. Soils that are readily prone to crusting and surface sealing under tillage farming do not exhibit this problem under established CA systems. This is because minimum soil disturbance, mulch cover, and increased soil organic matter contribute to enhancement of soil quality, which prevents the formation of a surface crust. In some situations, CA has even allowed expansion to marginal soils in terms of rainfall or fertility. In highly degraded soils, it may not show encouraging results initially, but con-tinuous adoption essentially with residue recycling and inclusion of legumes in the system, shows soil improvement with CA. In India, CA has been suc-cessfully demonstrated in the light-textured soils of north-western India and the Indo-Gangetic plains, black cotton soils or vertisols in central India, and coastal alluviums of south India. Since the land is not ploughed, laser levelling of the land can provide a benefit in terms of water use efficiency under irrigated conditions. This will ensure placement of the seed and fertilizer at the correct depth in the soil. In fact, perfect land levelling is a pre-requisite for adoption of CA under irrigated conditions (Jat et al. 2006).

Suitability of Crops

Most research on CA in India has been carried out on wheat. Studies have also been made on rainy season crops like rice, maize, soybean, pigeonpea, and cotton, winter crops like chickpea, mustard, lentil, and pea, and summer crops like greengram. In the majority of these trials, as mentioned in the earlier sections, equal or higher yields under CA compared with CT have been obtained. This suggests that principles of CA are applicable for small-seeded crops like mus-tard as well as large-seeded crops like maize and soybean. Experience shows that winter crops are more suitable to be grown under CA, probably because of lower weed infestation in the growing season. After developing adequate expertise and confidence, especially with respect to weed management, the crops in the summer and rainy seasons can also be grown quite successfully under CA. It is better that where a system approach is followed, wherever possible, there should be inclusion of a legume or cover crop.

Residue Recycling

One of the essential requirements of CA is residue recycling, without which it may work for a few years initially, but long-term success is not assured. There have been questions raised on the availability of crop residues for recycling in crop production in India because these are an important fodder for animals. However, it is also a reality that crop residues in many areas are also available for recycling, where these are wasted or burnt, especially in combine-harvested crops. Since the land is not disturbed under CA, it is possible that it may become compact and hard if the residues are not retained on the soil surface. Residue retention promotes biological tillage through the activity of earthworms and soil microorganisms, and the soil in fact becomes more porous and friable over time than in the CT system (Hobbs 2007; Parihar et al. 2016). Residue retention as mulch modifies the micro-climate, and helps in soil moisture conservation, weed control, temperature moderation in the soil, as well as within the crop canopy, and enhances the fertility of the soil. It has been proved based on long-term experimentation that soil porosity and infiltration rates are much higher under CA than in tilled soils (Indoria et al. 2017). A rice residue load of 6–8 t/ha can be effectively managed while working with a happy seeder, provided it is spread uniformly over the entire field, and is not too wet. However, still better seeders are needed to tackle location-specific issues of residues such as groundnut and greengram in different regions.

Sowing and Crop Establishment

Ensuring an optimum plant population is the key to realizing potential product-ivity. Any loss in initial crop stand due to improper sowing is rarely compensated. Therefore, sowing under CA should be accomplished using an appropriate seed drill but also basally applied fertilizer uniformly. Most drills are designed so as to place the fertilizer sufficiently away from the seed to ensure germination. Zero-till seed-cum-fertilizer drills of normal type (knife-type types) for sowing in residue-free or in anchored residue conditions, and improved seed drills in anchored or loose residue conditions are available. Hand sowing with manually drawn seed drills often does not produce the desired results because of the power needed to pull through unploughed soil. The seeder needs to be well calibrated and proper spacing should be maintained for each crop. In some small-seeded crops like mus-tard, the seed and fertilizer (granulated SSP or DAP, not urea) can be mixed in definite proportion without any harmful effect on germination. If the sowing is done under optimum soil moisture and possible damage by termites, birds, and/or rodents is checked, a normal seed rate is adequate. However, it may be advisable to use a 20–25% higher seed rate than normal to offset the poor germination due to inadequate seed–soil contact likely under some field conditions.

Soil and Residue Conditions

It is essential to ensure relatively higher soil moisture at sowing under CA than for CT to obtain good germination. The happy seeder works efficiently only when the soil has optimum moisture and also the residue is dry – not too wet or green. In the absence of these conditions, the machine often creates problems due to choking of pores and collection of wet soil/residue within the tines. If the soil is too wet (moisture content more than field capacity), the soil gets collected with the tines and the pores also get clogged. Further, if the soil is too dry and hard, the seed and fertilizer are not placed at the correct depth and there will be too much energy required for the tractor, causing wear and tear of the machine. Dry residue of most crops is shredded well but the wet and fibrous residue of crops like rice, groundnut, greengram, *Sesbania* etc. may not be chopped by the rotating blades and may cause problems during sowing. Such problems can be overcome with experience and the farmer needs to understand the best soil moisture for planting in ZT in a particular crop and situation. It is advisable to have standing and anchored residue instead of cut and loose residue for efficient working of seeders.

Machine for Sowing

As emphasized earlier, manual sowing with hand-drawn seed drills on small plots for experimental purposes often does not work in untilled soils. Tractor-drawn seed drills having knife-type tines and separate boxes and tubes for dropping seeds and fertilizer are needed. The happy seeder is the most suitable machine for sowing in combine-harvested fields having standing/anchored as well as loose residue conditions. Recently developed seeders that work in fields with loose straw have been developed by local manufacturers in India. The happy seeder is a combination of two machines – one is the rotating device with blades which cuts the residue, and the other is the attached seed-cum-fertilizer drill. Therefore, a greater amount of energy is needed for sowing with the happy seeder for cutting the residue and simultaneously putting seed and fertilizer into the unploughed soil. A larger tractor of 50 HP or more is often required for carrying the load and for sowing with the happy seeder. New machines such as the super seeder which mixes the residue in the top soil layer before placing seed and fertilizer, and the roto-double disc drill which can work with less energy requirement and in heavy residue load like sugarcane trash are also being developed and are available for sowing.

Uniform spreading of the crop residue is needed for proper functioning of seeders for sowing in fields with loose residues. Previously available combine harvesters used to leave the residue in rows and heaps after threshing the grains. This residue had to be spread manually, but nowadays, combine harvesters have a straw management system (SMS) that spreads the residue uniformly during the harvesting operation. Sometimes, a mulcher or grass cutter is run for cutting and uniformly distributing the residue before sowing, which involves an extra cost. The government of India and most state governments provide a >50% subsidy on the procurement of these types of new-generation machinery to individual farmers as well as through custom hiring centres to ensure recycling of the residue *in situ* and to discourage burning.

Fertilizer Management

Efficient fertilizer management is an essential requirement for the success of CA. As far as possible, the basal fertilizer must be banded at the time of sowing (basal fertilizer). The happy seeder or even the normal seed-cum-fertilizer drill has separate boxes and tubes for seeds and fertilizer, which can effectively place the fertilizer at the correct depth and rate. Granulated fertilizers like urea, SSP, and DAP are now available which do not pose much of a problem through clogging the pipes or tines but ordinary prilled urea or MOP sometimes block the tubes, especially during the rainy season due to humidity. Mixing of these fertilizers with granulated ones often solves this problem. Proper calibration is required for the optimum fertilizer rate. Even for top dressing of urea fertilizer, some locally designed and crop-specific machines are available which can place the fertilizer below the soil surface close to the plant roots in the standing crop.

Crops under CA do not look very attractive and vigorous during the first month after sowing. This is partly because of the residue lying on the soil surface giving a shabby look and there may be some initial setback to the crop plants due to nutrient immobilization and inhibited growth. However, the plants pick up growth after irrigation and application of the first dose of top-dressed N, and soon overcome this initial setback. It is often advisable to use a 20–25% higher fertilizer dose, especially N, to promote the initial slow growth of plants and check any possible lock up of the nutrients due to immobilization. However, experience has shown that the fertilizer dose may have to be reduced after 3–4 years due to the build-up of soil fertility through the addition of crop residues which decompose over a period of time. Since the P is not lost from the system, and the fixed P becomes gradually available to the plants, the P requirement may decrease in successive cropping cycles. Furthermore, since a lot of K is recycled with the addition of crop residues, the requirement of K fertilizer may not be needed after some cropping cycles.

Weed Management

It has been conclusively proved based on long-term experiments under CA elsewhere that weed problems gradually decrease if all the principles of CA, along with the best management practices including crop, nutrient, weed, water, and other pest management methods are followed in a holistic manner (Baker and Saxton 2007; Chauhan et al. 2012; Fonteyne et al. 2020). Initially there may be some increase in herbicide use but it is possible to reduce dependence on herbicides after some years with decreased weed infestations. It is a fact that herbicides are an integral part of the integrated weed management strategy under CA, without which it cannot be practiced. However, it must be emphasized that herbicides should be used judiciously by following the principle of the 5 Rs: right source, right kind, right dose, right time, and right method of application. Weed management under CA requires greater skill and expertise, and this is not only science but an art as well. There is a need for training and the availability of suitable equipment for herbicide application as most farmers in India adopt the same method as for other pesticides.

Water Management

There is a saving in water required for irrigation by not tilling the land and planting earlier under CA than CT systems. In most cases the next crop can be sown under residual soil moisture. Even if the soil moisture at the time of crop harvesting and sowing of the next crop is not sufficient for germination, it is advisable to sow the crop in time, and apply light irrigation, preferably through a sprinkler after sowing. Also, when the land is not tilled, irrigation water moves across the field quicker and so allows the farmer to apply light irrigation, especially if the field is levelled. There is no greater problem of crust formation after irrigation to inhibit germination as the residue on the soil surface provides a congenial environment for the crop to emerge. Retention of crop residues on the soil surface prevents evaporation loss of soil moisture and keeps the soil in relatively wetter condition for a longer period of time. Experience has shown that the irrigation interval can be prolonged by 1–2 weeks under CA compared with the CT system (Gupta and Syre 2007; Jat et al. 2009b; Sharma and Singh 2018). This results in one or two irrigations less under CA than CT. Accordingly, there is a 20–30% saving of water under CA, and even more so under the sprinkler system practiced widely in central India.

It is a myth that soil becomes hard and irrigation or rain water does not percolate down the soil profile under CA. The evidence shows that water accumulates on the soil surface for a longer period under the conventionally ploughed systems, causing temporary waterlogging and also crop lodging under some situations. By not tilling, the root system and soil pores are not disturbed and so water infiltrates faster (Hobbs 2007; Parihar et al. 2016; Indoria et al. 2017). Also, the residue mulch prevents clogging of the soil pores in the untilled soil.

Pest and Disease Management

CA involves a total paradigm shift from conventional agriculture systems. Therefore, it is likely that pest and disease infestations will vary drastically when the land is not ploughed, and residues are retained on the soil surface.

Improper sowing in untilled soils, such as in dry soil or at shallow depth may result in some seeds lying at or near the soil surface, making them prone to attack by various pests like birds, termites, ants, and rodents. In fact, bird damage is considered to be a major limitation for adoption of CA in some situations where the seeds are not placed at the desired soil depth. Termites and other ants also take away seeds lying at or near the soil surface. Similarly, rodents hide beneath the residue mulch in the field or in the burrows on the bunds and cause damage to seeds. These problems are aggravated when the seeds are not placed at the proper depth in the soil or remain uncovered, and also where there is no effective seed–soil contact. In view of this, it is essential to treat the seeds with a suitable insecticide, sowing at the correct depth with a suitable machine and in proper soil moisture conditions. Also, suitable measures are needed for the control of birds and rodents during the first few days after sowing.

Some of pests and diseases are soil-borne. Hence, it is often recommended under CT systems to plough the fields during hot summer weather to expose the hidden egg masses and disease-causing pathogens to light and sun so that these are killed during hot weather cultivation. Furthermore, clean cultivation without any stubble is recommended as a preventive measure to avoid infestation of pests and diseases. However, under the CA system, all these practices are avoided, with the result that pest and disease infestation may undergo a complete change. Infestation of various insects in wheat, wilt, and root rot in chickpea and soybean, and nematodes in rice may be altered under CA systems. However, such problems may also be noticed even under CT systems, and CA alone cannot be blamed for that. In fact, the biological component of the soil is improved under ZT and residue mulch, and the microbial diversity may actually control pathogenic organisms and pests.

Experiences over the years from long-term CA studies have shown that pest and disease infestations are not insurmountable and are controllable with the available technologies. Suitable measures need to be undertaken as per the situation to keep a check on the incidence and proliferation of specific pests and diseases under CA systems. Application of need-based insecticides and fungicides may be required (Jaipal et al. 2005; Singh et al. 2005; Singh et al. 2014). The exudates from the decomposed residues also contain allelo-chemicals which keep a check on such pests and maintain a balance in the ecosystem.

Reasons for Low Yields Under CA

Different researchers have reported varied results on the performance of crops under CA. Observations from all over the country have shown that similar or higher yields can be obtained under CA as with CT systems. However, there are also good number of studies which have reported low yields under CA. This may be because of several deficiencies in the implementation of CA and due to not following the full protocol. Derpsch et al. (2014) enumerated several reasons for the low yields and advocated standardizing ZT research for a uniform and valid comparison of yields and other advantages of CA across locations, crops, soils, and weather conditions. Several factors may be responsible for low yields under CA, such as the following.

Lack of Period of Conversion

CT systems have been applied for decades or even centuries, therefore, the time to recover and realize the full benefits of CA may be longer. There is a requirement for a period between conversion from a conventional system and adoption of CA. The desired response to CA is normally not obtained in the initial years of conversion, and it is only after 3–4 years that the results of CA on yield performance of crops start to become noticeable.

Lack of Knowledge and Expertise

As mentioned earlier, CA is a knowledge-driven technology that requires expertise on how to manage the crops under this new system. Most researchers do not follow all the required protocols and adopt one of the three pillars of CA first and later add the other two. With knowledge and experience, confidence in CA improves and the results are far better in the later than in the earlier years as the farmer adapts CA to his own situation. There is a need for training and the availability of enabling factors such as equipment to farmers for promoting CA in a participatory manner rather than a top-down approach.

Lack of a Systems Approach

It is suggested to follow a systems approach under CA, and the findings based on single crop trials often lead to erroneous conclusions. The direct, residual, and cumulative effects of crops grown in sequence and their management practices must be considered. For example, in a rice-based cropping system, there may be some penalty on the yield of rice under CA but the loss is more than compensated for in the following crops in the winter and summer seasons; leading to higher overall system productivity. It is not sufficient to only stop tilling the land with all other management practices remaining the same as in the conventional systems. There is a need to bring about suitable changes and follow the best management practices with respect to sowing, residue mulching, fertilization, weed control, and pest, weed, and disease management.

Insufficient Soil Cover with Crop Residues

Surface residue cover is a key factor for the success of CA systems. Some researchers follow ZT with bare soil conditions or with insufficient residue cover. Studies have shown that removing residues can lead to reduced yields and lower economic returns with ZT alone. Therefore, it is advisable not to follow CA if the residues are not available for recycling as mulch cover on the soil surface. Farmers need to compromise and leave some part of the residues to feed the soil and some for their animals. Over time, when yields increase with CA, there will be more biomass residue available for recycling.

Lack of Experience of Tractor Drivers

CA requires a different approach and the experience of the tractor driver operating the machine for seeding is very important. Sowing cannot be done blindly or with less care as in the case of ploughed soils, with inadequate regulation of the seeding equipment, seed furrows staying open after seeding, too deep or too shallow seed placement, soil smearing because of excessive moisture, etc. The machine operator requires thorough knowledge and patience while undertaking sowing and should ensure proper drilling and placement of seeds at the required soil depth. This can be overcome by promoting service providers who are properly trained just like with combines.

Inappropriate Seeding Machinery

Poor crop stands and establishment are often due to inappropriate farm machinery used for sowing. Inadequate furrow closing and seed placement can lead to a poor crop stand; the seeds are placed too shallow or too deep or seed–soil contact is insufficient. Quite frequently researchers use CT seeding machinery in ZT experiments as well, which leads to poor results. New-generation seeding machines are available with separate boxes for seed and fertilizer metering mechanisms, tubes and pipes for dropping seeds, and their placement at proper depth. There may be a need for a mechanism to close the slot or press wheels to ensure good seed–soil contact.

Poor Weed Control

Poor weed control is the most often cited reason for poor yields under CA. Most researchers are not able to control weeds with the available technologies. There is either inadequate selection of herbicides or weed-suppressing technologies, or insufficient or excessive doses of herbicide causing weed escapes, crop injury, non-uniform coverage of the herbicide, etc. Frequently, researchers insist on using the same herbicide programme for all treatments (conventional, minimum, and zero-tillage), because tillage is the only variable they want to change. This may favour one system but be detrimental to the other. Therefore, training of farmers and applicators, and also the availability of suitable equipment assume greater importance for the promotion of CA.

Poor Disease and Insect Control

The CA system may favour or disfavour some diseases and pests differently in comparison with other tilled systems. There is likely to be a total paradigm shift in the infestation of diseases and insects under CA systems. Researchers often use calendar application of pesticides for all treatments instead of using system-specific pest management methods, which are needed under CA. An integrated approach for the management of pests and diseases is needed by ensuring training is provided to farmers and service providers.

Non-adjustment of Fertilizer Rate

CA involves minimum disturbance and retention of crop residues, and nutrient availability in the soil may be somewhat restricted in the early stages of plant growth. Therefore, a relatively higher fertilizer rate, especially of N, is needed in the early stages of transformation from CA to CT. It is likely that N fertilization may not have been adjusted during the first few years of ZT technology or a green manure legume crop may not have been seeded previously to provide the additional N needed initially to take care of immobilization of N in surface residues and soil organic matter. In a CT system, residues are incorporated and are more likely to tie-up N than in a CA system, where the residues are retained on the surface and the seed and fertilizer are placed at the proper soil depth.

Extremely Degraded Soils

The beneficial effects of CA have been reported across different soil types. However, low yields of crops may be obtained on extremely degraded and/or eroded soils with very low organic matter content, in which poor macro- and micro-biological activity and fertility may limit the initial success. In such degraded soils, an advantage may be confined to CT initially through continued mineralization of N with tillage disturbance until the organic matter is depleted. It is also possible that the yields are lower in newly reclaimed soils or recently levelled fields due to considerable shifting of the soil from one part of the field to the other. It may be essential to grow green manure or other legume crops for 1–2 seasons to build a uniform level of fertility over the entire area before regular cropping is taken up on these soils.

Inadequate Crop Rotation and Diversity

Dynamic crop rotation is the third major principle of CA, and it is essential to follow a judicious combination of cereals, pulses, oilseeds, and other cover crops in the system. Optimized crop rotation for CT may not be the same as for ZT. Additionally, a CA system may provide different opportunities for cover crop planting, whereas the conventional system may be limited due to time and moisture lost with tillage. The land should not remain fallow and must always be covered throughout the year under CA, and more so under irrigated conditions.

Constraints in the Adoption of CA

CA requires a different set of technologies and expertise for its success. It requires the application of all the principles of CA in a holistic manner coupled with other best management practices for maximum benefits. The major handicap in the adoption of CA is the conventional mindset of farmers who resist change. Despite proven benefits, the adoption of CA systems requires a total paradigm shift from conventional agriculture regarding crop management. The CA technologies are essentially machine- and knowledge-driven, and therefore require improved expertise, training, and resources for adoption. For wider adoption of CA, there is an urgent need for policy makers, researchers, and farmers to change their mindset and explore these opportunities in a site- and situation-specific manner for local adaptation. It also requires a more participatory extension approach with the farmer allowed to experiment and see for himself if it benefits him. This is a technology-driven agriculture and its very basic principles of sowing seeds in an untilled land and without removing crop residues are in sharp contrast to the traditional system.

Several factors are limiting the adoption of CA in India, such as the following:

- Typical mindset of continuing with business-as-usual, and a lack of will to innovate and adopt new technology
- Lack of suitable machines for sowing and fertilizer placement, and service providers
- Incomplete application of the CA principles and associated best management practices causing poor performance of crops
- Fragmented and small land holdings coupled with low purchasing power of farmers
- Farmer education and technical knowledge are insufficient to adopt CA technology
- Non-availability of crop residues and their competitive role for feed/fodder in some regions
- Lack of expertise in weed management and the use of suitable herbicides
- Possible infestations like nematodes, termites, birds, rodents, and other insects and diseases
- Lack of incentives and policy support to change from the traditional methods.

Roadmap for CA

CA-based technologies have been developed, and the benefits in terms of enhanced productivity, profitability, soil health, and climate resilience have been illustrated in most parts of the country. However, these technologies have been adopted on a limited scale due to some apprehensions, lack of will, inadequate incentives, and some operational constraints. While CA-based technologies can be adopted in most existing systems, there is immense scope for bringing the barren and fallow lands of central and eastern India under profitable cropping. There is required to be a coordinated effort involving multiple stakeholders to make farmers aware and demonstrate these technologies on a large scale. Further, necessary back-up in the form of suitable farm machinery is required to be provided to enable farmers to adopt these technologies.

There is a need for analysis of factors limiting the adoption and acceptance of CA among farmers. A lack of information on the effects and interactions of minimal soil disturbance, permanent residue cover, planned crop rotations, and integrated weed management, which are key CA components, can hinder CA adoption. Information has mostly been generated on the basis of research trials, but more on-farm research and development are needed. Farmers' involvement in participatory research and demonstration trials is crucial for accelerated adoption of CA. There is a need for a paradigm shift on how to extend CA technologies to farmers. More participatory approaches are needed, while allowing farmers to experiment and accept the approach when convinced. Service providers are needed for the promotion of CA, just as combine harvesters through custom-hiring centres have become popular all over the country at affordable prices to the small-holder as well as large farmers.

Following a stakeholders' meeting involving national and international organizations, a 10-point roadmap for the promotion of CA in India has been prepared (Jat et al. 2018).

i   CA contributes to at least eight of the UN's Sustainable Development Goals (SDGs) and should be valued by policy makers accordingly.
ii There is a need to better aggregate and map knowledge of CA across sites to define recommendation domains.
iii On-farm research-cum-demonstration with farmer participation involving Krishi Vigyan Kendras (KVKs) for validating CA performance on a broader scale.
iv Commercial availability of scale-appropriate machinery, and the establishment of CA mechanization hubs and service providers to benefit smallholder farmers.
v   Investigations on soil biology and pest dynamics (including insects, pathogens) under CA.
vi Creation of National Initiative on Conservation Agriculture (NICA) for evidence-based promotion of CA.
vii Development of scalable and sustainable business models for promoting adoption of CA and capacity building of all stakeholders.
viii CA should be part of the course curriculum, including a practical crop production programme for the agricultural universities.
ix Establishment of a learning platform/CA Community of Practitioners (CA-CoP) for regular interactions, knowledge sharing, and capacity development.
x   Establishment of 'India CA Center' – a Technical Working Group on Conservation Agriculture (TWGCA) involving key researchers from national and international organizations to promote CA in India.

Conclusions and Future Outlook

Research work on CA-based approaches involving ZT has been on-going in India since the early 1990s. Despite highly encouraging results in the research trials, there has not been greater adoption of these technologies, except in wheat in some areas of the Indo-Gangetic plains and central India, maize and sorghum in rice fallows of coastal Andhra Pradesh, rapeseed–mustard in rice fallows of NEH region, and also different crops in the Konkan region of Maharashtra. Analysis of research papers published in three leading national journals revealed that CA-based topics have been increasingly investigated by resource management scientists in recent times.

CA is a challenging and exciting subject for researchers, requiring a multi-disciplinary approach to address all the related issues in a comprehensive manner. Unlike CT systems, CA-based technologies are knowledge-driven, requiring skill for successful adoption. Resource management scientists should take the lead in working on CA and perfecting the technologies in their respective domains in collaboration with innovative farmers. Gone are the days of the conventional merry-go-round type of on-station research, for which there are no takers at present. The focus must be on reducing the cost of production while enhancing productivity, produce quality, and soil health. Scientists should generate the technologies by following a participatory approach with innovative farmers and demonstrate on large areas at research farms and in the farmers' fields. CA-based on-farm research should be an integral part of on-station research, without which the research findings cannot be transformed into technologies. CA has been the fastest adopted technology over the past decade, and shown wonders globally. All-out efforts are needed by all stakeholders to make it happen in India.

References

Baker, C.B. and Saxton, K.E. 2007. The what and why of no-tillage farming. In: *No-tillage Seeding in Conservation Agriculture*, 2nd Edn. (Eds.) Baker, C.J. and Saxton, K.E. FAO and CAB International, Wallingford, Oxon, U.K.

Bhullar, M.S., Pandey, M., Sunny Kumar, et al. 2016. Weed management in conservation agriculture in India. *Indian Journal of Weed Science* 48(1): 1–12.

Chapke, R.R. and Babu, S. 2016. Promising practices for highest sorghum productivity in rice fallow under zero tillage. *Indian Farming* 65(12): 46–49.

Chapke, R.R., Babu, S., Subbarayudu, B. et al. 2017. Growing Popularity of Sorghum in Rice Fallows: An IIMR Case Study, 40 p. Bulletin, ICAR-Indian Institute of Millets Research, Hyderabad, India.

Chapke, R.R., Mishra, J.S., Subbarayudu, B., et al. 2011. Sorghum Cultivation in Rice Fallows: A Paradign Shift, 31 p. Bulletin. ICAR-Directorate of Sorghum Research, Hyderabad, India.

Chauhan, B.S., Singh, R.G. and Mahajan, G. 2012. Ecology and management of weeds under conservation agriculture: A review. *Crop Protection* 38: 57–65.

Derpsch, R., Franzluebbers, A.J., Duiker, S.W., et al. 2014. Why do we need to standardize no-tillage research? *Soil and Tillage Research* 137: 16–22.

Fonteyne, S., Singh, R.G., Govaerts, B., et al. 2020. Rotation, mulch and zero tillage reduce weeds in a long-term conservation agriculture trial. *Agronomy* 10(7): 962.

Gupta, R. and Syre, K. 2007. Conservation agriculture in South Asia. *Journal of Agricultural Science*, Cambridge 145: 207–214.

Hobbs, Peter R. 2007. Conservation agriculture: What is it and why is it important for future sustainable food production? *Journal of Agricultural Science*, Cambridge 145(2): 127–137.

ICAR. 2012. Zero tillage cultivation of rapeseed mustard in Imphal East district by tribal farmers, p. 13. *ICAR Reporter*, April–June, 2012.

Indoria, A.K., Srinivasa Rao, Ch., Sharma, K.L., et al. 2017. Conservation agriculture – a panacea to improve soil physical health. *Current Science* 112(1): 52–61.

Jaipal, S., Malik, R.K., Yadav, A. 2005. IPM Issues in Zero-tillage System in Rice-Wheat Cropping Sequence. Technical Bulletin (8), 32 p. CCS Haryana Agricultural University, Hisar, India.

Jat, M.L., Biswas, A.K., Pathak, H., et al. 2018. *Conservation Agriculture Roadmap for India*. Policy Brief, 4 p. ICAR-CIMMYT, New Delhi.

Jat, M.L., Chakraborty, D., Ladha, J.K., et al. 2020. Conservation agriculture for sustainable intensification in South Asia. *Nature Sustainability* 3: 336–343.

Jat, M.L., Chandna, P., Gupta, Raj, et al. 2006. Laser Land Leveling: A Precursor Technology for Resource Conservation, p. 48. Rice-Wheat Consortium Technical Bulletin Series 7. Rice-Wheat Consortium for the Indo-Gangetic Plains, New Delhi, India.

Jat, M.L., Dass, S., Sreelatha, D., et al. 2009a. Corn Revolution in Andhra Pradesh: The Role of Single Cross Hybrids and Zero-tillage technology, 16 p. Technical Bulletin 2009/5. ICAR-Directorate of Maize Research, Pusa, New Delhi.

Jat, M.L., Singh, R.G., Saharawat, Y.S. et al. 2009b. Innovations through conservation agriculture: progress and prospects of participatory approach in the Indo-Gangetic Plains, pp. 60–64. 4th World Congress on Conservation Agriculture held on February 4–7, 2009, New Delhi, India.

Keil, A., Krishnapriya, P.P., Mitra, A., et al. 2021. Changing agricultural stubble burning practices in the Indo-Gangetic plains: Is the happy seeder a profitable alternative? *International Journal of Agricultural Sustainability* 19(2): 128–151.

Malik, R.K., Kumar, V., Yadav, A., et al. 2014. Conservation agriculture and weed management in south Asia: Perspective and development. *Indian Journal of Weed Science* 46(1): 31–36.

Mohanty, M., Painuli, D.K., Mishra, A.K., et al. 2007. Soil quality effects of tillage and residue under rice–wheat cropping on a Vertisol in India. *Soil and Tillage Research* 92(1–2): 243–250.

Om, H., Kumar, S. and Dhiman, S.D. 2004. Biology and management of *Phalaris minor* in rice–wheat system. *Crop Protection* 23(12): 1157–1168.

Parihar, C.M., Yadav, M.R., Jat, S.L., et al. 2016. Long term effect of conservation agriculture in maize rotations on total organic carbon, physical and biological properties of a sandy loam soil in north-western Indo-Gangetic Plains. *Soil and Tillage Research* 161: 116–128

Patil, M.D., Wani, S.P. and Garg, K.K. 2016. Conservation agriculture for improving water productivity in Vertisols of semi-arid tropics. *Current Science* 110(9): 1730–1739.

Saharawat, Y.S., Gill, M., Gathala, M., et al. 2021. Conservation agriculture in South Asia: Present status and future prospects. In: Advances in Conservation Agriculture, Volume 3 – Adoption and Spread. Kassam et al. (Eds.). Burleigh Dodds, Cambridge, UK.

Sharma, A.R., Jat, M.L., Saharawat, Y.S., et al. 2012. Conservation agriculture for improving productivity and resource use efficiency: Prospects and research needs in Indian context. *Indian Journal of Agronomy* 57(3rd IAC Special Issue): 131–140.

Sharma, A.R., Mishra, J.S. and Singh, P.K. 2018. Conservation agriculture for improving crop productivity and profitability in the non-Indo-Gangetic regions of India. *Current Advances in Agricultural Science* 9(2): 178–185.

Sharma, A.R. and Singh, V.P. 2014. Integrated weed management in conservation agriculture systems. *Indian Journal of Weed Science* 46(1): 23–30.

Sharma, A.R. and Singh, P.K. 2018. Adoption of conservation agriculture-based technologies in the vertisols in Madhya Pradesh. *Indian Farming* 68(5): 12–15.

Singh, R. Dabur, K.R. and Malik, R.K. 2005. Long-Term Response of Plant Pathogens and Nematodes to Zero-Tillage Technology in Rice-Wheat Cropping System. Technical Bulletin 7. Directorate of Extension Education, CCS Haryana Agricultural University, Hisar, India.

Singh, B., Kular, J.S., Hari Ram, et al. 2014. Relative abundance and damage of some insect pests of wheat under different tillage practices in rice–wheat cropping in India. *Crop Protection* 61: 16–22.

Sreelatha, D., Reddy, M.L., Reddy, V.N., et al. 2015. *Production Technology for Zero-Till Maize*. Bulletin, 10 p. Maize Research Centre, PJTSAU, Rajendranagar, Hyderabad.

www.cau.ac.in/directorates.html. Zero tillage cultivation - a viable option for large scale production of rapeseed-mustard in rice fallow.

www.srt-zerotill.com/srt

4 Conservation Agriculture in Rainfed Areas

G. Pratibha, K.V. Rao, I. Srinivas, K.L. Sharma,
M. Srinivasa Rao, H. Arunakumari,
K. Sammi Reddy, and V.K. Singh

Introduction

Current agricultural practices are characterized by high fossil fuel and energy consumption, excessive nutrient use, soil degradation, water pollution, massive exploitation of natural resources, and increased greenhouse gases emissions (GHGs). Under the current scenario of natural resource degradation and poor response to nutrients, attaining food security for the growing population is a major concern in Indian agriculture. Thus, it is imperative to produce more, with a less negative environment impact. Indian agriculture has reached a point where it must seek new directions which call for new production strategies, policies, and actions that are different from those adopted in the 'green revolution' era to improve crop productivity. Therefore, a paradigm shift in farming practices is required through eliminating unsustainable parts of conventional agriculture which is crucial for future productivity gains while sustaining natural resources.

India must achieve a growth rate of 3–4% per annum in the agricultural sector and food grain production of 400 M tonnes by 2030. The past strategies which ushered in the green revolution of the 1970s and 1980s are no longer working due to reduced responsiveness of the agricultural inputs and degradation of natural resources (Indoria et al. 2018). In view of the stagnating productivity from irrigated agriculture, the contribution from rainfed agriculture should increase to meet the requirements of the growing population. India has the largest irrigated area in the world. However, 55% of its cropped area is rainfed, which contributes 87% of the production of nutri-cereals and pulses, 77% of oilseeds, and 67% of fibres. The state of rainfed agriculture is precarious and the problems associated with it are multifarious. Variable rainfall (quantum and distribution), highly variable soil resource base, lack of sufficient infrastructure, and low adoption of improved management practices are responsible for low crop productivity. A higher degree of poverty, food insecurity, hunger, and malnutrition are the significant characteristics of poor rainfed smallholder agriculture-based communities in India. Hence,

DOI: 10.4324/9781003292487-5

sustainable agricultural intensification is a must to lift resource-poor farmers out of hunger and poverty.

In India, the extent of rainfed area to the total cultivated area is the highest in Jharkhand (91%), followed by Maharashtra (81%) and Chhattisgarh (74%). On the other hand, agriculturally important states in the Deccan plateau, include Andhra Pradesh and Karnataka which have 53% and 71% area under rainfed agriculture, respectively. Rainfed agriculture is practiced under a wide variety of soils, and agro-climatic and rainfall conditions ranging from 400–1600 mm per annum. In terms of the distribution of rainfall, 15 M ha of the land receives an annual rainfall of <500 mm, 15 M ha receives 500–750 mm, while 42 M ha and 25 M ha of land receive 750–1150 mm and >1150 mm, respectively. The characteristics and dryland farming practices for these regions are well known (Indoria et al. 2018). The annual rainfall and potential evapo-transpiration ranges from >500 mm in arid to 1000 mm in dry sub-humid regions. In the rainfed region, soils are poor in organic carbon (OC) and are starved of nutrients. Carbon storage not only improves fertility and productivity of soils, but also abets global warming.

Soil Types

Major soil types in rainfed regions are: vertisols (35%), alfisols (30%), and entisols (10%). Alfisols are light-textured, shallow, and have low water-retention capacity due to the higher percolation rate. Soil crusting is a major problem in these soils. Vertisols or black soils are deeper, heavier with a high clay content, and hold more water than alfisols. However, they are highly erodible and runoff can be as high as 40% or more depending on the volume and intensity of rainfall and the slope. It has been estimated that 68.5 t/ha per year soil is lost from the vertisols. Due to their high clay content, black soils swell, become sticky, and shrink rapidly when dry, leaving large clods and deep fissures. This creates considerable problems in field preparation and timely planting operations and stress to crops. Entisols are generally loamy-sand or sandy-loam, and depth is not a constraint (Indoria et al. 2017). These soils have a low clay content and nutrient status, and hold water up to 200 mm/m³ of soil profile. In fact, the rainfed regions, especially arid and semi-arid areas, are particularly represented by alfisols and aridisols, which experience a diversity of soil-related constraints.

The intensity and type of tillage in dryland regions depend on cropping intensity, soil type, and rainfall patterns. In semi-arid India, animal-drawn wooden or iron ploughs consisting of an iron crow bar with a hardened point (e.g. non-inverting desi plough) is traditionally used for tillage. Other special tillage tools include the blade harrow used for smoothening the soil surface, weed control, and breaking the soil crust, especially at the early stage of plant growth in order to improve infiltrability of the soil. The traditional system of tillage on vertisols in most parts of India involves fallowing the land in the rainy season and growing a post-rainy season crop on stored moisture in the profile. In this system, the land is harrowed occasionally using animal traction during the rainy season to control weeds. In high-rainfall zones with black soils, rainy season fallowing is practiced because of risky cropping due to flooding and waterlogging. Furthermore, difficulty encountered in tilling the hard clay soils prior to the commencement of rains or sticky wet soil after the onset of rains are other reasons for the use of rainy season fallow in these areas. Generally, tillage operations are done without recycling crop residues. Usually, the crop residues are either used for cattle feed or fuel. Burning of certain crop residues like cotton and maize has also increased.

In recent years, draft animals have been steadily replaced by tractors. The total energy input in Indian agriculture has increased several-fold during the past six decades. Among the various implements used, the mould-board plough has been found to be beneficial in alfisols with a hard murrum layer at a sub-surface depth. This was beneficial in the inversion of clay argillic horizons to the top and its mixing. Apart from bringing the fertile layer to the top, it also temporarily alters the relative proportion of textural constituents in the surface layer. Contrary to this, mould-board ploughs and disc ploughs have been ranked as major detractors of soil quality since the mixing up of surface residues results in fast decomposition which emits more CO_2 to the atmosphere from the agricultural lands. This, in combination with imbalanced fertilization and poor recycling of crop residues, results in deterioration of soil quality leading to low crop productivity in rainfed regions. The quantum of GHG emission, carbon efficiency (CE), and other associated parameters as influenced by different tillage implements are presented in Table 4.1. Apart from loss of soil organic carbon (SOC), intensive tillage has contributed to increasing the risk of wind and water erosion of soil due to the formation of a hard pan in the sub-soil layer. Tillage is mostly done for seedbed preparation and as intercultural operations to avoid weed competition. Summer tillage with a blade harrow is mostly practiced for weed control and for capturing early-season rainfall to increase soil moisture retention. This off-season tillage helps in better establishment of seedlings, weed control, and *in situ* conservation of rainwater (Pratibha et al. 2019).

Crops and Cropping Systems

The major rainfed crops grown in India are rice, sorghum, pearlmillet, maize, sunflower, soybean, rapeseed, mustard, groundnut, castor, pigeonpea, and cotton in the rainy season, and linseed and chickpea in the winter season. These crops are grown either as sole crops or in intercropping combinations.

The rainfed cropping systems are mostly single-cropped in red soil areas, while in black soil regions, a second crop is taken on the residual soil moisture. In the post-rainy season cropped black soils, farmers keep lands fallow during the rainy season and grow winter crops on conserved moisture. Recently, the area under maize, soybean, and cotton has increased over a short period at the cost of the total area under coarse cereals like sorghum and pearlmillet. Such changes in cropping patterns affect fodder availability for livestock and resource use. In addition, continuous monocropping degrades soil fertility, depletes groundwater, and increases the build-up of pests and diseases. These issues have to be addressed

Table 4.1 Global warming potential and carbon efficiency of tillage implements used for seed bed preparation

Treatment	Soil-based GHG emissions (CO₂ equivalents) (kg CO₂ eq/ha)			Tillage and sowing diesel + implement energy		Total GHG emissions (soil-based + tillage + sowing + production)	Total CI (kg C eq/ha)	CO (kg C eq/ha)	CE
	CO_2	N_2O	CH_4	Tillage	Sowing				
CV	1217[e]	9.4[b]	-0.13[b]	25[b]	14.72[a]	1599[d]	799[d]	1780[e]	2.23[b]
CVH	996[c]	11.2[d]	-0.17[c]	54[c]	14.72[a]	1409[bd]	704[bc]	1736[e]	2.46[c]
DP	1106[d]	8.4[b]	-0.23[d]	54[c]	14.72[a]	1516[cd]	758[cd]	1480[c]	1.95[a]
DPH	1052[c]	11.8[e]	-0.11[b]	84[d]	14.72[a]	1495[cd]	748[cd]	1680[d]	2.25[b]
MP	1120[d]	8.4[b]	0.08[a]	59[c]	14.72[a]	1535[d]	767[d]	1472[c]	1.92[a]
MPH	933[bc]	8.1[b]	-0.02[d]	85[d]	14.72[a]	1374[c]	687[c]	1650[d]	2.4[c]
RO	1161[d]	10[c]	-0.1[a]	45[c]	14.72[a]	1564[d]	782[d]	1960[f]	2.51[d]
DH	912[b]	7.3[a]	-0.13[b]	25[b]	14.72[a]	1295[b]	648[b]	1700[de]	2.62[e]
BP	501[a]	7.1[a]	-0.14[b]	15[a]	3.8[b]	860[a]	430[a]	1080[b]	2.51[d]
BH	548[a]	8.9[b]	-0.18	4.5[a]	3.8[b]	899[a]	449[a]	1120[b]	2.49[cd]
NT	543[a]	8.4[b]	-0.11[b]	0[a]	14.72[a]	900[a]	450[a]	1000[a]	2.22[b]

Source: Pratibha et al. (2019).

Notes: Means followed by same superscript letter are not significantly different at p=0.05. CV, cultivator; CVH, cultivator fb disc harrow; DP, disc plough; DPH, disc plough fb harrow; MP, mould board plough; MPH, mould-board plough fb disc harrow; RO, rotavator; DH, disc harrow; BDP, bullock-drawn plough; BPH, bullock-drawn plough fb harrow; NT, no-tillage; CI, carbon inputs; CO, carbon outputs; CE, carbon efficiency.

through both technological and policy interventions. Simultaneously, need-based policy incentives are required to encourage farmers to adopt agro-ecology-compatible cropping patterns so that the farmer's income is enhanced and the resource base is also restored and sustained.

Tillage in Rainfed Areas

Tillage is both a contributor to and cause of higher and lower yields of rainfed crops. Tillage helps in germination and establishment of crops, due to increased infiltration through the larger clods which results in enhanced moisture retention in the soil profile. Under semi-arid conditions, summer tillage helps with higher soil moisture retention and reduces weed infestation. Improper tillage, lack of desired land configuration, and residue management lead to lower productivity and deterioration of soil because of higher runoff loss. Conventional tillage (CT) exposes the soil to rain, which results in high runoff and erosion. Further, it is a high energy-demanding, time-consuming, labour-intensive, and high-cost cultural operation. Hence, issues related to resource conservation in the semi-arid tropics (SAT) have assumed importance in view of widespread natural resource degradation, and the need to reduce production costs and make agriculture profitable. In this context, reduced or zero tillage is important in rainfed areas for efficient management of soil and water, and to reduce the cost of cultivation.

Long-term studies with different tillage intensities have been conducted in different climatic regions and soil types involving cereals and oilseeds. The results indicated that CT recorded higher rice yields at Varanasi (dry sub-humid inceptisols) and Phulbhani (moist sub-humid oxisols), wheat at Ballowal Saunkhri (dry sub-humid inceptisols), fingermillet at Bengaluru (semi-arid alfisols), maize at Rakh Dhiansar (dry-sub-humid inceptisols), pearlmillet at Agra (arid inceptisols), and soybean at Indore (semi-arid vertisols) (Maruthi Shankar et al. 2006). The higher yields in CT were due to better weed control and high soil moisture retention. The differential response of different crops to tillage was due to variation in the amount of rainfall during cropping period and soil type.

A relationship between rainfall and different tillage practices was observed on the productivity of different crops (Table 4.2). The advantage of deep tillage (DT) on crop yields was found to be dependent on rainfall pattern and crop. In a normal rainfall season, the yield advantage with DT was observed for both short-duration (sorghum and pearlmillet) and long-duration (castor) crops. However, in an above-normal rainfall season, the advantage of DT was restricted to long-duration crops like castor and pigeonpea. On the other hand, in sub-normal rainfall years, the yield advantage was not observed in either short- or long-duration crops. This differential response of crops to tillage was due to interaction between profile recharge and rooting depth. In the semi-arid alfisols of Hyderabad, by considering the historical rainfall data, it was observed that the benefit of DT was about 63% and 91% for short- and long-duration crops, respectively. However, DT requires high energy and monetary input, hence it is advisable to adopt this practice only for castor and pigeonpea crop.

Table 4.2 Correlation between rainfall and crop yield attained with tillage practices

Crop	Location	Climate/soil type	Correlation between yield and rainfall under		
			Conventional tillage	Low tillage+ interculture	Low tillage+ herbicide
Rice	Varanasi	Dry sub-humid inceptisols	-0.852	-0.817	-0.592
Rice	Phulbani	Moist sub-humid oxisols	0.586	0.767	0.664
Rice	Ranchi	Moist sub-humid oxisols	-0.771	-0.441	-0.047
Wheat	B. Saunkhri	Dry sub-humid inceptisols	-0.275	-0.214	-0.241
Maize	B. Saunkhri	Dry sub-humid inceptisols	-0.940*	-0.872	-0.605
Maize	R. Dhiansar	Dry sub-humid inceptisols	-0.976*	0.646	-0.026
Pearlmillet	Agra	Arid inceptisols	-0.898*	-0.711	-0.717
Pearlmillet	Solapur	Semi-arid vertisols	-0.521	-0.488	-0.102
Pearlmillet	Hisar	Arid aridsols	0.847	0.931*	0.851
Fingermillet	Bangalore	Semi-arid alfisols	0.979**	0.928*	0.379
Groundnut	Anantapur	Arid alfisols	0.829	0.956*	0.909*
Soybean	Indore	Semi-arid vertisols	0.735	0.525	0.824
Soybean	Rewa	Dry sub-humid vertisols	0.006	0.546	-0.012
Clusterbean	Dantiwada	Semi-arid aridsols	-0.776	-0.539	-0.175

Source: Maruthi Shankar et al. (2006).

In arid regions (<500 mm rainfall), MT was found to be on a par with CT, while in semi-arid (500–1000 mm) regions, CT was superior (Maruthi Shankar et al. 2006). Reduced tillage + interculture was superior in semi-arid vertisols, and MT + herbicide was superior in aridisols. In sub-humid (>1000 mm) regions, the weed problem was severe, depending on the rainfall distribution. In this zone, there is a possibility of reducing weed growth with low tillage intensity by using herbicides. These studies revealed that ZT or MT alone without residue retention reduced the yields. Therefore, recycling of residues along with ZT is appropriate rather than tillage alone for success of CA in rainfed regions.

While highlighting the importance of tillage in dryland crops such as sorghum and pearlmillet, it was reported that DT up to 24 cm helped in improving grain yield by better moisture recharge of soil profile and enhancing the rooting depth in alfisols (Venkateswarlu et al. 2010). Hence, the rainfall pattern and crop requirements should be considered to select an appropriate tillage practice under rainfed farming. In the short term, the yield gains are higher with DT or with inversion tillage with implements like the mould-board plough but this may be deleterious to soil quality. Intensive tillage causes a gradual decline in soil organic matter content through accelerated oxidation, resulting in reduced capacity of the soil to regulate water and nutrient supplies to plants.

In a long-term experiment (eight years) in the sorghum–mungbean system carried out under semi-arid conditions in Hyderabad, CT was superior to MT, which could be attributed to more weed growth and less infiltration of water due to compaction of the surface soil (Sharma et al. 2009). Conventional tillage and MT (plough planting) were compared for their effect on yields and sustainable yield index (SYI) of sorghum and castor grown in rotation (Table 4.3). In this case, the surface residue applied at 2 t/ha was inadequate to create the desirable soil ameliorative effect. However, adverse effects such as hard setting tendencies of soil and poor infiltration can be mitigated by building-up more and more

Table 4.3 Long-term effects of tillage and residue on crop yields and sustainable yield index (SYI)

Residue	Sorghum (t/ha)		Castor (t/ha)		SYI	
	Conventional tillage	Minimum tillage	Conventional tillage	Minimum tillage	Conventional tillage	Minimum tillage
Sorghum	1.13	0.81	0.82	0.48	0.49	0.35
Gliricidia loppings	1.20	0.90	0.93	0.51	0.50	0.37
No residue	1.10	0.84	0.84	0.45	0.48	0.31
Mean	1.14	0.85	0.86	0.48	0.49	0.34
Tillage (T)	**	**	**			
Residue(R)	**	**	**			
T*R	NS	NS	NS			

Source: Sharma et al. (2009).

organic carbon by employing various C sequestration approaches. It is anticipated that if residue levels are enhanced, the beneficial effect of MT could be seen over a few years. Rainfall and soil type have a strong influence on the performance of reduced tillage. Therefore, Indian agriculture can accrue benefits of conservation farming, especially reduced tillage if followed appropriately depending upon the type of soils.

Improper tillage, lack of desired land configuration, and residue management are important factors which lead to lower productivity and deterioration of soil because of higher runoff losses in rainfed region. Therefore, it is imperative that some vegetation or stubble is maintained on the soil to prevent degradation. The presence of residue on the soil surface prevents aggregate breakdown by direct raindrop impacts as well as by rapid wetting and drying of soil. Results from rainfed and irrigated long-term trials have shown that the combination of ZT with retention of sufficient soil-surface crop residue resulted in improved physical, chemical, and biological soil quality (Indoria et al. 2017a). It was also reported that ZT without residue retention resulted in deteriorated soil beyond the CT practices. Straw mulch consistently reduced runoff compared with bare plots. Adding organic residues to a ZT system could significantly lower the runoff and increase the amount of water available for the crop. Runoff from tilled plots declined sharply after tillage operation and reverted back to that from an untilled plot after a few storms totalling about 150 mm of rainfall, suggesting breakdown of soil aggregates in tilled plots. Runoff from a tilled system may be reduced from 35% to 10% of rainfall using straw. On average, straw mulch and tillage increased infiltration by 127 mm and 26 mm, respectively.

Crop diversification and crop rotation reduce the risk of crop failure, and are recognized as a cost-effective solution to build resilience into an agricultural production system. A crop rotation is a sequence of crop types grown in succession on a specific field. Crop rotation also brings stability to soil fertility through cultivating legumes with cereals in a rotation or intercropping system. In addition to the build-up of soil fertility, crop rotations are important to manage pest, disease, and weed problems. Rotation with deep-rooted cover crops, such as hairy vetch, helps in biological loosening of compact soils and allows the cultivation of more than one crop so that farmers can spread the risk of fluctuating prices. Experiments at Hyderabad have shown that a maize + pigeonpea intercropping system is more sustainable and associated with less risk compared to a maize–chickpea sequential cropping system or maize alone.

The results of most research station studies as well as prediction by models such as DSSAT have established that the ZT approach can make the desirable differences over a period of time if the crop residue is allowed to remain on the surface and appropriate crop rotations are followed (Patil et al. 2016). ZT systems without crop residues left on the soil surface have no particular advantage because much of the rainfall is lost as runoff, probably due to rapid sealing of the soil surface. Mulching with crop residues along with ZT practice helps in controlling the soil moisture by reducing the impact of rainfall on runoff. Minimum tillage + crop residue has been found to be beneficial for conserving water and improving crop productivity (Jat et al. 2012). Conservation tillage in a maize–wheat cropping system involving MT (in wheat) with *Lantana camara* mulch (in standing maize or at its harvest) conserved more moisture, and resulted in a higher grain yield of wheat in a hill ecosystem as compared to DT (Sharma and Acharya 2000). In addition, soil compaction was observed in ZT where the soil porosity decreased because of ZT practice. Therefore, recycling of crop residues along with tillage is beneficial in restoring the soil structure. Crop rotation with deep root crops also helps in improving the soil structure because of the substantial rooting system, especially in the 0–40 cm soil layer. The residues and crop rotation increase the macro-fauna population and thus the soil biological activity, which is beneficial for the growth of the plant (Sharma et al. 2016).

Benefits of CA in Rainfed Areas

The benefits of CA can be seen at the farm, regional, and national levels. The benefits can be classified into three broad categories: (i) agronomic benefits that improve soil productivity, (ii) economic benefits that improve the production efficiency and profitability, and (iii) environmental and social benefits that protect the soil and make agriculture more sustainable.

Rainfed agriculture in the SAT is typically characterized by low crop yields and a high risk of crop failure. This is due to frequent dry spells, extreme rain events which often cause a water stress situation, and land degradation. To counter all these adverse effects, CA with all three principles has emerged as an alternative strategy to sustain agricultural production, particularly in rainfed alfisols. However, in the past many studies in rainfed agriculture reported lower yields in the CA system as compared to CT. Further, it was observed in rainfed systems that the yield gap narrowed between CA and CT systems over a period of 8–10 years. Our research experiences have indicated that CA practices aimed at conserving the top soil improve the organic matter content and soil health.

In a 13-year experiment conducted in a sorghum–mungbean system under rainfed conditions in alfisols, CT recorded 11.0% higher yield (1.53 t/ha) over reduced tillage (1.38 t/ha) (Sharma et al. 2015). In sorghum, reduced tillage could not establish its superiority, whereas in mungbean, the yield was close to that of CT with a slight fluctuation depending upon the rainfall distribution during the cropping season. In sorghum–blackgram rotation, an increase in sorghum residue with manipulation of the harvesting height at 60 cm and 35 cm recorded 32.8% and 14.8% higher blackgram grain yields over no residue. Similarly, in a ten-year study on a fingermillet+pigeonpea cropping system, a yield reduction was observed with a reduction in tillage intensity (Prasad et al. 2016).

Experiments conducted on alfisols at Hyderabad in a maize–horsegram system revealed that there was no significant difference between CA and CT on maize seed and stover yields (Kundu et al. 2013). However, the effect of CA was more pronounced in succeeding horsegram. The yield of horsegram in CA was 32% higher than CT due to an increase in water retention with residue application. However, the yields of maize were higher under a CA system with balanced fertilization in low rainfall years than using CT due to an increase in water retention and soil carbon. Long-term application of crop residue can increase water retention by up to 2–4% in semi-arid alfisols, which helps in mitigating intermittent dry spells or terminal water stress impact on the yield of rainfed crops.

Simulation results based on rainfed maize in medium black soils revealed that CA practices increased the maize yield by 46% in low rainfall years over CT (Patil et al. 2016). The yield increase during high rainfall years was between 2% and 15%. However, no significant difference in crop yields between the CT and CA systems was observed during normal and wet years. CA has the potential to build system resilience for alleviating water scarcity in rainfed areas and reducing the risk of crop failure.

Field studies in terrace uplands in eight cropping systems, viz. upland rice–mustard, upland rice–pea, ricebean–mustard, ricebean–pea, maize–mustard, maize–pea, soybean–mustard, and soybean–pea revealed that CT with residues increased the crop yields in all cropping systems (Ghosh et al. 2010). Similarly, in lowland and valley upland situations, the yields increased in ZT with residue retention. The added advantage with CA practice was a second crop of wheat, linseed, and mustard in lowland, and field pea and mustard in valley upland and terraced upland, respectively. ZT with residue retention not only moderated the soil rhizosphere and produced a higher grain yield over the long term, but also made water available for crops during dry periods by permitting the downward movement of water across the root boundary.

Soil Quality

Soil quality plays an important role in improving crop productivity. CT helps in good seed-bed preparation and weed control, which results in higher yields over the short term. However, in the long term, DT hastens the decomposition of soil organic matter and nutrient mineralization. Furthermore, these practices increase soil erosion, ultimately causing soil quality deterioration.

Crop residues on the soil surface in rainfed regions have beneficial effects, such as the following: (i) reducing the striking impact of raindrops on soil particles, (ii) acting as an insulator for soil and thus minimizing evaporation loss, (iii) decreasing the impact of wind erosion, (iv) enhancing water productivity, (v) minimizing soil loss and water runoff, and (vi) regulating the hydrothermal regime by freeze–thaw and wet–dry cycles (Indoria et al. 2017a). Thus, land cover by crop residue in CA improves long-term productivity.

Besides tillage and land cover components of CA, the third important component is crop rotation. The nature and type of crops in crop rotation determine the extent and magnitude to which soil physical health can be modified. In CA, crop rotations (inclusion of shallow- and deep-rooted crops alternately, and leguminous crops) help in improving soil structure, organic matter content, water infiltration and its retention in soil, and other associated soil properties (Indoria et al. 2017a).

An improvement in soil quality helps in achieving sustainable system productivity through better ecosystem services. The soil quality index was the highest under minimum tillage combined with maize–cowpea intercropping, followed by mustard residue retention, and the lowest was under CT (Pradhan et al. 2018).

Soil Chemical Properties

CA practices improve soil health and increase carbon sequestration by reducing the tillage intensity. Permanent raised-beds (PRB) with full residue retention increased soil organic matter content 1.4-fold in the 0–5 cm layer compared to conventionally tilled raised-beds with straw incorporated (Srinivasarao et al. 2012). The response was significant, with increased amounts of residue retained on the soil surface for permanent-raised beds. Prasad et al. (2016) reported a higher rate of carbon sequestration from 62–186 kg C/ha/year, assuming that carbon sequestration was linear from 2000–10 with the recommended dose of fertilizers in CA. The rate of carbon sequestration in MT was 169 kg C/ha/year, as compared to 62 kg C/ha/year in CT.

The combination of ZT with crop residue retention increases chemical properties by improving the SOC storage and macro- and micronutrient dynamics. It is clear that when CA systems are well-designed and adapted to local conditions, they can improve the SOC content of many soils compared to CT systems, and this can lead to significant improvements in soil physical, chemical, and biological properties and productive capacity.

In vertisols of central India, the SOC content increased under CA over the initial status considerably after nine years of the study (Somasundaram et al. 2018). The SOC was higher in CA as compared to CT, but decreased with soil depth, which was due to lower crop residue addition to the deeper layers of the soil profile. The higher SOC in CA practices in surface layers was due to minimum soil disturbance coupled with the retention of residues. CA practices showed a positive impact on the aggregate-associated C and different carbon pools.

The continuous application of manure and fertilizer in tropical soils of India has shown that the SOC and MBC increased with balanced fertilization (Goyal et al. 1999; Sharma et al. 2018). The larger amount of C immobilized in microbial biomass suggests that soil organic matter under CA systems provides higher levels of more labile C than CT systems. CA improved the SOC, MBC, available water, and available nutrient status, particularly K, in the soil. Improved water retention by improvement of SOC can mitigate intermittent drought or terminal water stress, which is a major gain for dryland farmers in view of climate change. Higher nutrient use efficiency of N, P, K, S, and Zn was recorded through adoption of CA practices in maize compared to the conventional system (Kundu et al. 2013).

Soil Physical Properties

CA influences soil physical health and efficient utilization of nutrients. The impact of CA technologies on soil physical health is site-specific and varies with location, inherent soil properties, site limitations, period of time under CA system, extent of soil disturbance, nature of crop, intensity of crop rotation, proportion of total surface area covered by crop, and the prevailing climatic factors.

Sharma et al. (2011) found that MT in combination with mulches had a pronounced effect on soil physical properties, productivity, energy requirement, and monetary returns of the maize–wheat system in sub-humid inceptisols. In CT soil, dry bulk density increased with an increase in the number of traffic passes during tillage operations, but lower bulk density was recorded in CA compared to CT, suggesting that the effect was not immediate but that it took some years to decrease the bulk density compared to conventional tillage. At Vasad, ZT tillage increased surface SOC, WSA, and bulk density, but this effect was negated quickly with CA (Kurothe et al. 2014).

CA plays an important role in improving soil–water characteristics, and stabilizing the soil temperature, reduction in water runoff and soil loss, increase in soil water infiltration, and decrease in evaporation loss. Zero tillage increased the soil hydraulic conductivity compared to CT (Indoria et al. 2017a). This could be due to improved pore characteristics of soil such as pore continuity, pore diameter, and an increase in the number of macropores. The role of higher fungal activity and accumulated organic matter by the addition of crop residues has also been highlighted as increasing the hydraulic conductivity.

CA is an alternative strategy for reducing runoff and increasing infiltration. Runoff and soil loss are invariably tightly linked quantitatively. Therefore, any intervention which effectively restricts runoff also reduces soil loss. Soil loss under ZT is less than CT due to less sediment concentration in similar volumes of runoff, because disturbed soil is more easily eroded. Experiments in pigeonpea–castor system have shown that ZT recorded 20 and 17% lower soil and nutrient losses (N, P, K, C) as compared to CT and RT (Kurothe et al. 2014). Varied crop performance under different tillage systems and their environmental effect support the hypothesis that higher crop productivity can be achieved on a sustainable basis through tillage manipulations under rainfed semi-arid conditions. Intercropping of long-duration crops like castor or pigeonpea with a short-duration crop like cowpea or pearlmillet appears to have greater potential to utilize conserved moisture and stabilize crop production (Yadav et al. 2015). Thus, CA improved soil physical quality by favouring soil aggregation, soil hydraulic conductivity, and BD compared to CT.

Soil Biological Activity

Enzymatic activity generally increases under CA as compared to CT. Soil organic matter (SOM), an important biological indicator of soil quality, is a direct product of the biological activity of plants, micro-flora, and fauna, and numerous biological factors which affect soil functions like aeration and fertility. Enzyme activities were found to be significantly correlated with labile carbon and organic carbon (Sharma et al. 2016). The activities of dehydrogenase and arylsulfatase were found to be significantly higher under MT compared to CT. Urease activity in MT was 25% higher than CT. This may be due to a relatively higher amount of crop residue on the soil surface in MT. In tropical soils, rapid decomposition of the organic residues owing to the prevailing high temperatures and erratic rainfall hinders the beneficial effects of low tillage.

Energy Use

Zero tillage can reduce the negative effects on the environment by reducing fossil fuel consumption, which in turn reduces energy input, CO_2 emissions, wind and water erosion of soil, along with a reduction in the cost of cultivation. ZT saved 68% fossil fuel as compared to CT in rainfed alfisols in a pigeonpea–castor system (Pratibha et al. 2015). Energy-use efficiency (EUE) was influenced by different tillage treatments and anchored residue height in both pigeonpea and castor. Higher EUE was observed in CA as compared to conventional practices in pigeonpea as well as in castor crop. Averaged over treatments, castor recorded higher EUE (9.43 EUEt) as compared to pigeonpea (8.11 EUEt) (Table 4.4). Research on energy use for some of the rainfed crops for a seed-bed preparation has been reviewed by Maruthi Sankar et al. (2006). A long-term study in

Table 4.4 Energy-use efficiency under different tillage and residue treatments in pigeonpea–castor cropping system

Crop/tillage	Residue	EUEg	EUEt
Pigeonpea			
Conventional tillage	0	3.29[Aa]	6.69[Aa]
	10	3.94[Ab]	7.39[Aab]
	30	3.89[Ab]	7.50[Ab]
Reduced tillage	0	3.82[Aa]	8.10[Ba]
	10	4.18[Ab]	7.93[Aab]
	30	4.52[Ab]	8.91[Bb]
Zero tillage	0	3.86[Aa]	8.39[Ba]
	10	4.67[Bb]	9.12[Bab]
	30	4.49[Ab]	8.98[Bb]
Castor			
Conventional tillage	0	5.3[Aa]	8.14[Aa]
	10	5.77[Ab]	8.78[Ab]
	30	5.97[Aab]	8.9[Ac]
Reduced tillage	0	5.80[Ba]	8.83[Aa]
	10	6.44[Ab]	9.77[Bb]
	30	6.42[Aab]	9.91[Bc]
Zero tillage	0	6.61[Ba]	9.85[Ba]
	10	7.00[Bb]	11.11[Cb]
	30	6.16[Aab]	9.57[Bc]

Source: Pratibha et al. (2015).

sorghum–mungbean in alfisols of semi-arid regions revealed that reduced tillage recorded higher EUE compared to conventional tillage (Sharma et al. 2015).

Mitigation and Adaptation Potential of CA

CA acts as a strong adaptation strategy to manage extreme climatic events such as wind and rainfall by reducing wind and water erosion. The crop residue surface cover and increase in aggregate stability due to the addition of residues helps to reduce wind and water erosion. In ZT wheat, residue retention lowers the canopy temperature by 1.0–1.5°C during the grain filling stage because of cooling due to higher water retention in the soil, and reduced water losses due to weeds and evaporation. In the absence of residue retention, farmers need to apply irrigation at the grain filling stage to avoid terminal heat stress. Thus, CA can help a great deal in climate change mitigation and adaptation.

In the rainfed alfisols of Hyderabad, CT recorded 26% and 11% lower indirect GHG emissions over ZT and RT, respectively, in a castor–pigeonpea sequence (Pratibha et al. 2016). Castor grown on pigeonpea residue recorded 20% higher GHG emissions over pigeonpea grown on castor residue. The GHG emissions depend on the type of implement used for field preparations. Hence, selection of an environment-friendly farm implement for land preparation is also very important. The cultivator, mould-board plough, and rotavator recorded higher soil-based GHG emissions and indirect GHG emissions from fuel consumption. Furthermore, ZT and animal-drawn implements recorded lower soil-based GHG emissions, and fuel consumption-based CO_2 emissions for preparatory cultivation and sowing. These were 92%, 81%, 60%, 60%, and 40% lower in bullock-drawn plough, bullock-drawn harrow, tractor-drawn cultivator, disc harrow, and rotavator, respectively, as compared to mould-board plough followed by disc harrow (Pratibha et al. 2019).

Prospects of CA in Rainfed Areas

In India, efforts to adapt and promote resource conservation technologies have been underway for more than a decade, but it is only in the past few years that the technologies have been finding acceptance by farmers using a rice–wheat system in the states of Haryana, Punjab, and Western Uttar Pradesh. Efforts to develop and promote CA technologies have been made through the National Agricultural Research and Education System (NARES) and the Consultative Group on International Agricultural Research System (CGIAR). Unlike the rest of the world, the success in India is largely confined to the irrigated belt of the IGPs where the rice–wheat cropping system dominates. However, CA systems have not been promoted in rainfed semi-arid tropics, arid regions, or the mountain agro-ecosystem, which have great potential for a further increase in crop productivity. Conservation agriculture in rainfed regions is a win–win situation through promoting more efficient crop production, reducing soil degradation, and making more efficient use of natural resources. The beneficial effect of CA is greater in rainfed regions than irrigated areas. Zero tillage, systematic crop rotations, and *in situ* crop residue management are the pillars of CA to obtain the maximum benefit. However, these are also the most pronounced barriers to its widespread practice especially on smallholder farms in the rainfed regions.

Residue Availability

Rainfed areas are single-cropped and the length of the growing period is only 4–6 months. About 6–8 months is the off-season, when the land is barren and exposed to weathering and soil loss. Therefore, it is imperative that some vegetation or stubble is retained on the soil to prevent weathering. The minimum surface coverage required for the effective control of soil erosion and degradation in CA systems is 30%. Hence, the success of CA in rainfed areas depends on two critical elements, namely residue retention on the surface and weed control. The crop residues should be protected from burning, grazing, displacement from the field due to wind, runoff, and decomposition by termites. Hence, the goal of ZT should be to achieve organically rich surface cover, which can be achieved by either recycling of crop residues or use of animal-based manures. The CA should not ignore the farming system mode that provides food, fibre, fodder, fish, and fuel, which will lead to economic prosperity, environment protection, and conservation of energy.

In rainfed agriculture, moisture stress is common. Moreover, the rainfall is unimodal and erratic, with high variability both within and between seasons. Droughts are common due to which the crop yields and residue production are low. Therefore, limited availability of crop residues is a major constraint as the residues have competing uses (e.g., fodder, fuel, or construction material). Crop residues, particularly of cereals and legumes, provide high-value fodder for livestock in smallholder farming systems in rainfed agriculture. In developing countries like India, demand for animal products is increasing. In fact, this will continue to increase in future with a high demand for livestock products along with fruits, vegetables, and fish. Feed is in short supply due to typical small farm sizes, and limited common land for grazing. Apart from the short supply of fodder, other problems like termite infestation and community grazing are also encountered with crop residue retention on the soil.

One possibility for increasing crop residues and fodder is manipulation of the harvest height of crops. In the case of cereals like sorghum and maize, the crops can be harvested at 60 cm since the crop residue above 60 cm is nutritious and palatable to livestock. In crops like pigeonpea and castor, a 30 cm harvest height is recommended (Pratibha et al. 2015). Cultivation of cover crops as pre- or post-monsoon crops with grain legumes is a viable strategy since growing legumes provide residues and also help in conserving moisture and suppression of weeds. Post-monsoon crops like horsegram are successful if the rainfall is around 70 mm during October–December after the harvest of short-duration crops like maize and sorghum. These crops can also be grown as cover crops during pre-monsoon. Intercropping with legumes is another strategy to increase crop residues and surface cover in rainfed situations. Alley cropping in rainfed areas offers productivity, economic, and environmental benefits. Arable crops can be also sown between widely spaced tree species.

Live Fencing

Live fencing provides nutritive fodder and combines well with CA. This can be adopted in areas with a slope of 0.4–0.8%. These fences include rows of perennial grasses, hedges, wind breaks, and shelter belts on contours. The ideal characteristics of the species are that they should be bushy, non-grazable, and have some economic value, such as fuel and fodder. The major precaution is that the roots of the barriers should not compete with the main crop. An additional advantage of species like Gliricidia is that the leaves contain 2.4% N, 0.1% P, 1.8% K, besides Ca, Mg, and micronutrients. Hence, the leaves can be used as a green leaf manure and also as a mulching material. One row of Gliricidia of 700 m length and 0.5 m width can provide three cuttings per year, supplying 30–45 kg N/ha/year.

Community Grazing

Individual farmers cannot restrict grazing, even on their own land. In semi-arid areas, residual biomass is the main source of forage during the dry season. In some regions, this leads to tension between crop and livestock producers. Under these circumstances, where the society places no monetary value on crop residues, it is difficult for individual farmers to retain the crop residues on their own fields. Community grazing can be stopped by adopting social fencing by creating self-help groups.

Incidence of Termites

Besides the production and competing demands for crop residues, termite infestation on crop residues is a major problem in rainfed alfisols. Due to termite attack, residue cover is not seen on the surface for a long time. Termites enhance the decomposition of surface-applied organic materials, stimulating nutrient release, which can be used by the growing plants. This suggests that termites are not only pests but can also be highly beneficial biological agents. In rainfed alfisol, the intensity of termite infestation differs with the type of crop residue. A higher level of termite infestation is observed in maize crop residue as compared to pigeonpea and castor. Termite control can be achieved by using cow dung application and chloropyriphos spray.

Weed Management

One of the major challenges for low adoption of CA in rainfed regions is crop–weed competition. The benefits of CA systems in irrigated regions in general and rainfed regions in particular are offset by heavy weed infestation and shifts in weed communities. Weed infestations during and between two crop seasons are high. Although some proponents of CA argue that good ground cover due to mulching or cover crops results in less weed pressure under CA, higher weed intensity is a reality in many situations. Tillage and residue management are crucial factors determining the weed species diversity. A shift in the weed flora of crop lands from easy-to-control annual weed species to perennial weeds which are difficult-to-control, and broad-leaved weeds to grasses, since the annual broad-leaved species adapt better to frequently disturbed habitats or changes in the rank order of weeds occurrence, are observed. Moreover, greater weed diversity is observed with tillage, which prevents the domination of a few awkward weeds. It was observed that there were smaller and more diverse weed communities in conventionally tilled than minimum-tilled plots. The shift in weed population may affect competitive interactions between crops and weeds, and also crop growth and productivity. The weed problem can be reduced through the use of herbicides. With decreasing labour availability and low productivity, herbicides are attracting the attention of farmers to control weeds.

A major criticism of CA is its enhanced reliance on herbicides as compared to tilled systems. In particular, glyphosate is heavily used, especially to control perennial weeds. Lack of availability of certain herbicides and knowledge about their proper use is the major constrain. Moreover, the efficacy of herbicides in rainfed regions depends on the weather conditions, specifically the quantity and timing of rainfall. The lower efficacy of herbicides in rainfed regions is due to low soil moisture and uncertain rainfall. For pre-emergence herbicides like pendimethalin, optimum soil moisture is required after application to dissolve the herbicide in the soil-water solution so that it can be taken up by the emerging weeds during germination. Inadequate or delayed precipitation can reduce herbicide effectiveness and therefore decrease weed control. Depending on the soil type, higher precipitation (>25 mm) within 24 h after application, can cause herbicides to leach through the soil profile and consequently reduce efficacy. However, there is an opportunity for post-emergence herbicide application. Most post-emergence herbicides require no precipitation for several hours after application to ensure that movement across the leaf membrane can occur. Hence, farmers need effective and economical alternative weed management strategies. The other limitation in rainfed regions is the use of a single herbicide without rotation, which may create herbicide resistance in weeds and a shift in the weed population (Pratibha et al. 2021).

Considering the challenge of weed control and herbicide resistance of weeds in CA, integrated weed management is often proposed as the fourth crucial component for successful adoption of CA. Practices such as the use of herbicides in combination with mechanical weeding to reduce the weed density over time are recommended. Economically feasible integrated weed management strategies such as the use of herbicide combinations, herbicide rotation, mechanical methods, and development of new techniques like allelopathy and crop nutrition may be effective tools in CA, which are required to reduce herbicide resistance and weed shift, and to sustain crop productivity. These practices aim at depleting the weed seed bank and also reducing the species richness. The integration of leguminous cover crops in CA suppresses weed growth, reduces herbicide dose, and improves soil quality in the long term.

Machinery

Conservation agriculture involves a different set of operations and management. Therefore, it requires specialized implements and management skills than those used in conventional agriculture. Farm implements are needed for seed and fertilizer placement simultaneously to ensure optimum plant stand and early seedling vigour of rainfed crops. The major requirement of the CA system is the development and availability of equipment that places the seeds and fertilizers at the desired depth by cutting through the surface-retained residue in ZT fields and promotes good germination of crops. The implements required for rainfed and irrigated agriculture differ as rainfed regions have undulated topography. The seed does not require any covering mechanism in irrigated regions because of the high quantity of residues, but in rainfed regions the residue load is lower, hence covering of seed is required. In addition, the depth of seed placement is also different from irrigated agriculture. The implements are available for ZT sowing of crops in an irrigated system. Often, farmers lack the necessary skills to operate the implements, and local artisans and machinery manufacturers for repairing and maintenance are not available. Appropriate policies that support quality machinery and open access to quality machinery are limited since subsidies are available only to limited vendors. A precision planter (zero-till planter with a herbicide and fertilizer applicator) has been designed and developed at CRIDA (Srinivas et al. 2014). This planter, like in conventional planters, has seed, fertilizer boxes, seed metering device, seed and fertilizer delivery tubes, and seed depth control wheels, in addition to a herbicide tank. The wide furrow openers are replaced with inverted T-type openers to place seeds and fertilizers in narrow slits with minimal soil disturbance. The advantage of this implement is that the seed has better seed–soil contact as compared to traditional disc openers. Germination is better when used for sowing the crop at the onset of monsoons when there is heavy demand on the available draft power.

Compaction

There are reports of increasing bulk densities and soil compaction under CA, particularly in rainfed regions during the initial years where the soil porosity decreases because of ZT. However, this can be minimized if farm operations are carried out using animal or power-driven implements and controlling traffic. Mulching with crop residues along with ZT practice helps in preventing soil compaction, improving the soil moisture from the deeper root zone, and controlling soil moisture by reducing rainfall impact and runoff loss. Infiltration of water in CA systems is greater than in CT plots as long as ZT is combined with residue cover due to greater earthworm activity facilitating biological tillage. Sharma et al. (2018) established that MT in combination with surface residue application on a long-term basis can play an important role in improving the physical, chemical, and biological soil quality parameters.

Fourth Principle of CA

Like other revolutionary concepts, CA is prone to an overlay of dogma. However, due to the variability of agro-ecological environments, cropping systems, and farmer capacities and preferences, the principles of CA are not applicable for all situations. Water conservation is considered a key element of CA, especially in dryland areas exposed to erratic and unreliable rainfall. To demonstrate the effects of mulching on infiltration of rainfall, water balance studies are needed to analyse rainfall capture, soil storage, and crop water use, including simple measurements of rainfall productivity of CA compared to conventional farming methods. The effect of cover cropping or crop residue mulch on the soil–water balance components has not been explored in dryland production systems. Researchers and practitioners should keep an open mind for the local evolution of CA and consider the inclusion of the fourth principle such as *in situ* moisture conservation, nutrient management, or weed management for successful implementation of CA.

In situ moisture conservation has the potential to reduce substantially the degree of soil erosion. The permanent-bed and furrow method, and paired row and permanent conservation furrow can be used as *in situ* moisture conservation methods under rainfed conditions. This method has several advantages, such as timely sowing of the crop, low fuel and labour costs, improved soil health and quality, reduced erosion, and conserved soil moisture with higher water-use efficiency and yield enhancement. In permanent-bed and furrow planting, the furrow acts as a drainage channel in high rainfall areas and provides a better microclimate for plant growth and root development, also acting as a conservation furrow in low-rainfall areas. A bed planter was fabricated at CRIDA to prepare beds and furrows in rainfed regions. With this bed planter, 100 cm and 35 cm wide beds and furrows, respectively, and 15–20 cm deep furrows, are prepared. After preparing the fresh beds during the first year, these beds can be kept as permanent beds for the subsequent year with retention of the crop residue, and reshaping of the beds, if required, during the sowing of the next crop. Depending on the spacing, 2–3 rows of crops can be sown on the beds. This planter can be used both in irrigated and rainfed conditions. The implement can carry out four operations simultaneously, i.e. reshaping of the bed, sowing, and herbicide and fertilizer application.

Another method of *in situ* moisture conservation is paired row planting and permanent conservation furrow. In this method, the recommended row spacing of the crop is reduced and the optimum plant population is maintained. For example, in pigeonpea the recommended spacing of 90 cm between rows is reduced to 60 cm, with both rows together known as paired rows. The next pair of rows is spaced at 120 cm. The inter-space between two pairs can be utilized for sowing intercrops, and the conservation furrow is made between two crop rows (60 cm). For efficient planting and formation of conservation furrows, an implement was designed and developed at CRIDA with which sowing of crop, formation of conservation furrows, and fertilizer application can be done simultaneously. A furrow of 60 cm width and 20–30 cm depth can be made with this implement. A furrow of this dimension can store 250 m^3 water and reduce the runoff by one-sixth. The implement can be used to sow all crops, since it can be adjusted according to the required spacing of the crop. Besides conserving water and increasing the yields, it increases profitability as saves labour and input cost also.

Appropriate use of fertilizer is required to be defined in CA to enhance both crop productivity and produce. The threshold limits for adequacy of crop residues to ensure soil cover under smallholder conditions under rainfed conditions need to be worked out. Integrated weed management may also be used as the fourth component that is crucial for successful implementation of CA. This is because weeds have been highlighted as one of the most difficult management issues within this system.

Conclusion and Future Outlook

The rainfed regions of India are characterized by erratic and poor distribution of rainfall, degraded soils, and the low socio-economic condition of farmers. Crop productivity in these regions is low, with no or very low surplus produce. CA has been suggested as a sustainable and eco-friendly crop production technique in these fragile ecosystems. CA practices reduce wind and water erosion, increase water-use efficiency through improved water infiltration and retention, increase nutrient-use efficiency through enhanced nutrient cycling and fertilizer placements adjacent to seed, reduce oscillation of surface soil temperatures, increase soil organic matter and diverse soil biology, reduce fuel, labour, and overall crop establishment costs, and ensure timely operations. These benefits make CA a viable alternative in dry areas, where it can help to address the challenges of scarce and degraded natural resources. However, adoption of this technology in rainfed regions is low. Local production constraints often limit implementation of CA by resource-poor farmers. CA components like crop residue management, cultivar selection, and crop choice for rotation, nutrient, weed, disease, and pest management, and soil water management practices for crop and agro-ecological resource management are complex and location-specific. Hence, for better adoption of CA, these components must be managed and fine-tuned based on the cropping systems and agro-ecoregions.

Significant efforts for the development and dissemination of CA technologies have been made but adoption in rainfed regions is low in India. The knowledge of the effects and interactions of minimal soil disturbance, permanent residue cover, crop rotations, integrated weed management, *in situ* rain water conservation, fertilizers, and socio-economic factors, which are key CA components, is low, which seriously limits its adoption. The existing studies from different regions of the world where CA-related practices have been widely adopted have shown that extension and diffusion processes are needed for extensive use of CA. The adaptation of CA practices to various farming systems requires a shift in research paradigms and in favour of cost-effective technologies and agricultural implements specifically suited for small-scale farm enterprises and resource-conserving practices. Moreover, successful implementation of site-specific CA practices demands strong support for farmers and awareness-raising activities among policy makers. There is a need for an active farmer participatory approach in the development of location-specific CA technologies so that they can be adapted, especially under smallholder farming systems.

References

Baker, J.M., Ochsner, T.E., Venterea, R.T., et al. 2007. Tillage and soil carbon sequestration – what do we really know? *Agriculture, Ecosystems and Environment* 118(1–4): 1–5.

Ghosh, P.K., Das, A., Saha, R., et al. 2010. Conservation agriculture towards achieving food security in North East India. *Current Science* 99(7): 915–921.

Goyal, S., Chander, K., Mundra, M.C., et al. 1999. Influence of inorganic fertilizers and organic amendments on soil organic matter and soil microbial properties under tropical conditions. *Biology and Fertility of Soils* 29: 196–200.

Indoria, A.K., Sharma, K.L., Sammi Reddy, K., et al. 2017. Role of soil physical properties in soil health management and crop productivity in rainfed systems. I: Soil physical constraints and scope. *Current Science* 112: 2405–2414.

Indoria, A.K., Sharma, K.L., Sammi Reddy, K., et al. 2018. Alternative sources of soil organic amendments for sustaining soil health and crop productivity in India – impacts, potential availability, constraints and future strategies. *Current Science* 115: 2052–2062.

Indoria, A.K., Srinivasarao Ch., Sharma K.L., et al. 2017a. Conservation agriculture – apanacea to improve soil physical health. *Current Science* 112: 52–61.

Jat, M.L., Malik, R.K., Saharawat, Y.S., et al. 2012. Proceedings of Regional Dialogue on Conservation Agriculture in South Asia, 32 p. New Delhi, India, APAARI, CIMMYT, ICAR.

Jat, R.D., Nanwal, R.K., Jat, H.S., et al. 2017. Effect of conservation agriculture and precision nutrient management on soil properties and carbon sustainability index under maize–wheat cropping sequence. *International Journal of Chemical Studies* 5(5): 1746–1756.

Kundu, S., Srinivasarao, Ch., Mallick, R.B., et al. 2013. Conservation agriculture in maize (*Zea mays* L.)–horsegram (*Macrotyloma uniflorum* L.) system in rainfed Alfisols for carbon sequestration and climate change mitigation. *Journal of Agrometeorology* 15: 144–149.

Kurothe, R.S., Kumar, G., Singh, R., et al. 2014. Effect of tillage and cropping systems on runoff, soil loss and crop yields under semiarid rainfed agriculture in India. *Soil and Tillage Research* 140: 126–134.

Maruthi Sankar, G.R., Vittal, K.P.R., Ravindra Chary, G., et al. 2006. Sustainability of tillage practices for rainfed crops under different soil and climatic situations in India. *Indian Journal of Dryland Agriculture Research and Development* 21(1): 60–74.

Patil, M.D., Wani, S.P. and Garg, K.K. 2016. Conservation agriculture for improving water productivity in Vertisols of semi-arid tropics. *Current Science* 110 (9): 1730–1739.

Pooniya, V., Zhiipao, R.R., Biswakarma N., et al. 2021. Long-term conservation agriculture and best nutrient management improves productivity and profitability coupled with soil properties of a maize–chickpea rotation. *Scientific Reports* 11: 10386.

Pradhan, A., Chan, C., Roul, P.K., et al. 2018. Potential of conservation agriculture (CA) for climate change adaptation and food security under rainfed uplands of India: A transdisciplinary approach. *Agricultural Systems* 163: 27–35.

Prasad, J.V.N.S., Srinivasa Rao, Ch., Srinivas, K., et al. 2016. Effect of ten years of reduced tillage and recycling of organic matter on crop yields, soil organic carbon and its fractions in Alfisols of semi-arid tropics of southern India. *Soil and Tillage Research* 156: 131–139.

Pratibha, G., Srinivas, I., Rao, K.V., et al. 2015. Impact of conservation agriculture practices on energy use efficiency and global warming potential in rainfed pigeonpea–castor systems. *European Journal of Agronomy* 66: 30–40.

Pratibha, G., Srinivas, I., Rao, K.V., et al. 2016. Net global warming potential and greenhouse gas intensity of conventional and conservation agriculture system in rainfed semi-arid tropics of India. *Atmospheric Environment* 145: 239–250.

Pratibha, G., Srinivas, I., Rao, K.V., et al. 2019. Identification of environment friendly tillage implement as a strategy for energy efficiency and mitigation of climate change in semi-arid rainfed agro ecosystems. *Journal of Cleaner Production* 214: 524–535.

Pratibha, G., Rao, K.V., Srinivas, I., et al. 2021. Weed shift and community diversity in conservation and conventional agriculture systems in pigeonpea-castor systems under rainfed semi-arid tropics. *Soil and Tillage Research* 212: 105075.

Sharma, P., Abrol, V. and Sharma, R.K. 2011. Impact of tillage and mulch management on economics, energy requirement and crop performance in maize–wheat rotation in rainfed sub-humid inceptisols, India. *European Journal of Agronomy* 34(1): 46–51.

Sharma, P.K and Acharya, C.L. 2000. Carry-over of residual soil moisture with mulching and conservation tillage practices for sowing of rainfed wheat (*Triticum aestivum* L.) in north-west India. *Soil and Tillage Research* 57: 43–52.

Sharma, K.L., Grace, J.K., Srinivas, K., et al. 2009. Influence of tillage and nutrient sources on yield sustainability and soil quality under sorghum–mungbean system in rainfed semi-arid tropics. *Communications in Soil Science and Plant Analysis* 40: 2579–2602.

Sharma, K.L., Sammireddy, G., Ravindra Chary, G., et al. 2018. Effect of surface residue management under minimum tillage on crop yield and soil quality indices after 6 years in sorghum (*Sorghum bicolor* (L.) Moench)–cowpea (*Vigna ungiculata*) system in rainfed alfisols. *Indian Journal of Dryland Agricultural Research & Development* 33(1): 64–74.

Sharma, A.R., Singh, R and Dhyani, S.K. 2005. Conservation tillage and mulching foroptimizing productivity in maize–wheat cropping system in the outer western Himalaya region – areview. *Indian Journal of Soil Conservation* 33(1): 35–41.

Sharma, K.L., Srinivasarao, Ch., Suma Chandrika D., et al. 2016. Assessment of GMean biological soil quality indices under conservation agriculture practices in rainfed Alfisol soils. *Current Science* 111: 1383–1387.

Sharma, K.L., Suma Chandrika, D., Munna Lal, et al. 2015., Long term evaluation of reduced tillage and low cost conjunctive nutrient management practices on productivity, sustainability, profitability and energy use efficiency in sorghum (*Sorghum bicolor* (L.) Moench)–mungbean (*Vigna radiata* (L.) Wilczek) system in rainfed semi-arid Alfisol. *Indian Journal of Dryland Agricultural Research and Development* 30(2): 50–57.

Somasundaram, J., Chaudhary, R.S., Awanish Kumar, A., et al. 2018. Effect of contrasting tillage and cropping systems on soil aggregation, carbon pools and aggregate-associated carbon in rainfed Vertisols. *European Journal of Soil Science* 69: 879–891.

Srinivas, I., Sanjeeva Reddy, B., Adake, R.V., et al. 2014. Farm Mechanization in Rainfed Regions: Farm Implements Developed and Commercialized, p. 31. ICAR–Central Research Institute for Dryland Agriculture, Hyderabad, Telangana.

Srinivasarao, Ch., Venkateswarlu, B., Lal, R., et al. 2012. Soil carbon sequestration and agronomic productivity of an Alfisol for a groundnut–based system in a semi-arid environment in southern India. *European Journal of Agronomy* 43: 40–48.

Venkateswarlu, B., Sharma, K.L. and Prasad, J.V.N.S. 2010. Conservation agriculture – constraints, issues and opportunities in rainfed areas, pp. 80–84. 4th World Congress on Conservation Agriculture – Innovations for Improving Efficiency, Equity and Environment, February 4–7, 2009, New Delhi.

Yadav, B.L., Patel, B.S., Ali, S., et al. 2015. Intercropping of legumes and oilseed crop in summer pearlmillet [*Pennisetum glaucum* (L.) R. Br. Emend. Stuntz]. *Legume Research* 38: 503–508.

Part II

Management Options for Higher Resource Use Efficiency

5 Nutrient Management in Conservation Agriculture

C.M. Parihar, H.S. Nayak, D.M. Mahala, S.L. Jat, Yadvinder-Singh, M.L. Jat, K. Patra, K. Majumdar, T. Satyanarayana, M.D. Parihar, and Y.S. Saharawat

Introduction

Agriculture continues to be the backbone of the Indian economy, contributing about 18% of the country's GDP and employing the vast majority of the country's population. Although Indian agriculture has made impressive progress in achieving food self-sufficiency, challenges have been raised for a sustainable increase in food production to meet future food security needs. Nearly 94% (143 M ha) of agriculturally suitable land is already under cultivation and there is very limited scope for further horizontal expansion. Hence, in the future, the pressure on land will increase to produce more from the same cultivated area by increasing the input use-efficiencies and by use of good agronomic practices (GAP).

Conservation agriculture (CA)-based crop management technologies quickly address two critical concerns of today, i.e. low farm income and natural resources degradation (Parihar et al. 2016a). Soil organic carbon (SOC) content in most cultivated soils is less than 5 g/kg (0.5%) compared with 15–20 g/kg (1.5–2.0%) in the uncultivated virgin soils of India (Bhattacharyya et al. 2010). This is attributed to intensive tillage, removal/burning of crop residues, and intensive monotonous cropping. Global research-based evidence suggests that CA-based crop management practices enhance soil health with lower environmental footprints, leading to positive repercussions on soil, water, and nutrient holding capacity, which ultimately improve crop production. Although CA is gaining momentum in South Asia, nutrient management is still largely focused on blanket recommendations similar to those offered for the conventional tillage (CT) system. This may lead to, in many cases, sub-optimal crop yields, low nutrient-use efficiency, lower economic profitability, and greater environmental footprints.

Nutrient management is the science and practice of applying nutrients to crops that link soil, crop, and weather factors to achieve optimal nutrient use-efficiency, crop yields, and economic returns, while reducing nutrient losses and negative impacts on the environment. It involves harmonizing the right source, rate, time, and place (commonly known as the 4Rs nutrient stewardship) of nutrient application with site-specific soil, climate, and crop management conditions. The current fertilizer recommendations are based upon crop response

DOI: 10.4324/9781003292487-7

data aggregated across large geographic areas, without considering the indigenous nutrient-supplying capacity of soils (Majumdar et al. 2013). Such blanket fertilizer application for multi-nutrient-deficient soils results in under-fertilization in some cases and over-fertilization in others, ultimately compromising the soil health and crop productivity. The availability and cost of fertilizers necessitate that every unit of fertilizer should be used effectively. Because of location-specific nutrient needs, fertilizer recommendations based on target yield, fertilizer use-efficiency, and inherent soil nutrient-supplying capacity provide options to remove the soil fertility constraints.

Conventional fertilizer recommendations, which have been calibrated mainly based on traditional tillage-based systems, are not necessarily appropriate for CA systems and 4Rs nutrient stewardship must be formulated taking into account the specific nutrient dynamics of CA systems. Site-specific nutrient management (SSNM) is a set of nutrient management principles, which aims to supply a crop's nutrient requirements tailored to a specific field or growing environment (Majumdar et al. 2012; Sapkota et al. 2014; Jat et al. 2016). Under varying soil conditions (within and between fields), SSNM-based tools can give appropriate fertilizer prescriptions on a cost- and time-effective basis. However, these tools may need calibration and validation for diverse management scenarios and varying crops and genotypes. Nutrient management is an important aspect of CA for crop productivity and for the adoption of CA by farmers. CA improves NUE as it reduces soil erosion and prevents nutrient loss from the field. Developing effective nutrient management strategies in CA requires: (i) better understanding of the nutrient dynamics in reduced tillage or zero-till systems; (ii) proper assessment of the nutrient contribution from different levels of crop residues to supplement external nutrient inputs; (iii) developing scalable precision nutrient management strategies and supporting tools; and (iv) quantifying and conveying the economic and environmental benefits of these new tools and techniques of nutrient management to appropriate stakeholders.

This chapter discusses the effects of CA on soil nutrient availability and methods of managing fertilizer nutrients to improve the use efficiency in different cropping systems. The focus is on irrigated rice–wheat (RW), rice–maize (RM), and maize–wheat (MW) systems; however, many of these nutrient management practices can also be applied to other crops and cropping systems under CA.

Nutrient Dynamics in CA-Based Cropping Systems

Being a biologically based management practice with an agro-ecological perspective, CA does not focus on a single commodity or species. Instead, it addresses the complex interactions of crops with local conditions capitalizing on the physical, chemical, and biological interactions involved when managing soil systems productively and sustainably. In general, four important chemical and biochemical processes, often working simultaneously, are involved in influencing the nutrient dynamics of the soil system. These are: mineralization–immobilization, sorption–desorption, dissolution–precipitation, and oxidation–reduction, and most of the dynamic behaviour of soil nutrients can be explained by one or a combination of these processes. Among these, mineralization–immobilization and sorption–desorption seem to play a more dominant role in governing the source–sink interactions characterizing the nutrient dynamics (Sanyal and Majumdar 2009). Conservation agriculture, through its three key principles, is expected to influence the above-mentioned chemical and biochemical processes considerably. The changes in physical and biological properties of the soil associated with CA practices are expected to modify the direction and kinetics of the chemical and biochemical processes, leading to altered nutrient dynamics in the soil.

Crop residue addition leads to a continuous increase in soil surface biomass and soil biological processes under CA. In a fully established CA system, fertilizer nutrient management aims to balance or maintain soil nutrient levels, replacing the losses resulting from the nutrients exported by the crops. As CA systems have a diverse crop mix, including legumes and nutrients that are recycled through residue biomass, nutrients and their cycles must be managed more at the cropping system or crop mix level. Thus, fertilization would not be strictly crop-specific, except for N top dressing (if required at all), but should be given to the soil system at the most convenient time during the crop rotation. Top-dressing of N can be supplemented or replaced by the N captured by the legumes and released during the subsequent cropping cycle at the required time (more legume content – earlier release, more grass content in cover crop – later release) by the inclusion of legumes, either as a previous crop in the rotation or as a component in a cover crop before the next cash crop. Additionally, undisturbed soils are habitats for free-living N_2-fixing bacteria and there is rhizospheric fixation of N_2. Conservation agriculture and conventional systems differ in nutrient dynamics and have been well illustrated by Majumdar et al. (2012) through a nutrient omission study in maize. They recorded lower yield in N omission plots under ZT spring maize as compared to conventionally grown maize, but not in winter maize (Figure 5.1). This is probably due to either greater immobilization of available N, losses of N through leaching and denitrification, lower mineralization of soil organic N, or some combination of these factors that reduced the availability of N to ZT spring maize, particularly in the initial growing phase of the crop. However, over the longer duration of winter maize, the yield in ZT practices increased over conventionally grown maize, which may be due to mineralization of the immobilized N under the ZT system which supplied more N in the later stages.

The nutrient transformation and distribution in soils under CA-based systems are related to the SOC content and its pools (Yadvinder-Singh et al. 2005). Under CA-based systems, ZT with crop residue retention acts as a biological mulch and immobilizes a portion of applied inorganic N during the initial years (0–5 years), but supply additional N and other plant nutrients through mineralization in the subsequent years. In the transition phase (5–10 years), amounts of crop residues, as well as carbon and nutrient contents in soil, start to increase (Jat et al. 2018). It is only in the maintenance phase (>20 years) that the ideal situation with the maximum benefits from CA is achieved and the need for external nutrient supply is reduced. These changes in soil with time need to be factored

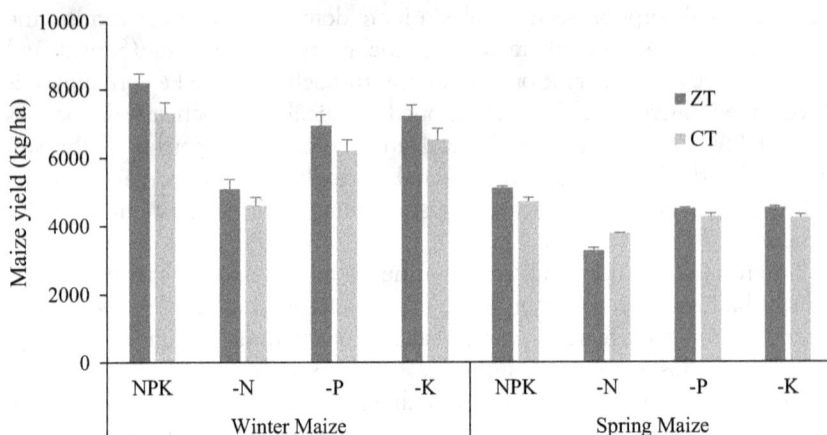

Figure 5.1 Average yields of winter and spring maize in omission plot trials under ZT and CT systems. The bars represent the standard error.

Source: Majumdar et al. (2012).

Table 5.1 Soil N content in the 0–30 cm layer after harvesting of wheat as influenced by tillage treatments

Tillage	Depth (cm)	Soil total N content		Fertilizer N remaining in soil	
		(%)	(kg/ha)	(% of soil N)	(kg/ha)
Conventional till	0–15	0.0444	889	2.080	17.84
	15–30	0.0311	622	0.427	2.65
	0–30	–	1511	–	20.49
Zero-till	0–15	0.0489	978	2.494	24.39
	15–30	0.0322	644	0.441	2.89
	0–30	–	1622	–	27.23

Source: Pasricha (2017).

for while designing nutrient management in CA. The distribution of nutrients in soil under CA differs from that in CT as CA practices enhance the stratification of nutrients and their availability near the soil surface (Table 5.1). The altered nutrient availability, cycling, and flows under CA are probably due to the surface placement of crop residues as opposed to the incorporation of crop residues with CT. The slow decomposition of crop residues left on the soil surface can prevent faster leaching of nutrients through the soil profile, which is more likely when the residues are incorporated into the soil (Kushwaha et al. 2000).

Under CA, continuous pores between the surface and sub-surface lead to more rapid passage of soluble nutrients deeper into the soil profile than in tilled soil.

Thus, in CA-based practices, the root density is usually closer to the soil surface compared to CT as more nutrients are taken up from the soil surface. Furthermore, CA-based practices have implications on the soil moisture regime and nutrient dynamics, whose interaction effect influences nutrient response, nutrient availability, and use efficiency, all of which differ greatly between conventional and CA-based practices (Jat et al. 2011; Majumdar et al. 2012). The soil sampling procedure also needs modification for CA systems as it results in a highly concentrated layer of soil test extractable nutrients (e.g. P and K) in the surface 0–7.5 cm with much lower concentrations below 7.5 cm as compared to CT. It is therefore suggested that conventional soil analysis data might not necessarily be a valid basis of fertilizer recommendations for CA, since the available soil volume and the mobility of nutrients through soil biological activities tend to be higher than for tillage-based systems against which the existing recommendations have been calibrated. '*Last year's residue – this year's nutrients*' is the general proverb used as crop residues are good sources of plant nutrients. About 30–40% of the N, 25–35% of the P, 70–85% of the K, and 35–45% of the S absorbed by cereals remain in the vegetative parts at maturity. Typical amounts of nutrients in rice straw at harvest are: 5–8 kg N, 0.7–1.2 kg P, 15–25 kg K, 0.5–1 kg S, 3–4 kg Ca, and 1–3 kg Mg per tonne of straw on a dry weight basis. Besides NPK, one tonne of rice and wheat residues also contains about 100 g Zn, 777 g Fe, and 745 g Mn (Yadvinder-Singh and Sidhu 2014).

Nutrient Management in CA

Nitrogen Management

Mineralization is the biochemical transformation of nutrients from an organic to an inorganic state, while immobilization is the reverse process. These two processes significantly influence the dynamics of several nutrients, namely N, P, S, and micronutrients. Both mineralization and immobilization have fundamental functions in the universal N cycle. These two opposing processes either result in net mineralization or net immobilization depending on the difference in the rate of the normal and dominating reaction. Such continuing transfer processes of mineralized N into organic products of synthesis and of immobilized N back into the inorganic decay product are defined as mineralization–immobilization turnover (MIT).

Whether N is mineralized or immobilized depends on the C:N ratio of the organic matter being decomposed by soil microorganisms. There is a rapid increase in the number of heterotrophic organisms during the initial stages of fresh organic matter decomposition as indicated by elevated CO_2 evolution. If the C:N ratio of the residue is >20:1, net immobilization will occur. Insufficient N in the substrate will induce the organisms to draw on the mineral nitrogen from the soil, leading to immobilization of N. The residue C:N ratio will however decrease as the decay proceeds because of decreasing C (respiration as CO_2) and increasing N (N immobilized from soil solution). A combination of high C:N ratio plant residues and low soil N is expected to reduce N availability to plants, at least at the initial phases of crop growth, which may be compared to the initial phase of CA adoption. The retention of cereal straws under CA, with the C:N ratio ranging between 60:1 to 100:1 and low available N in soils, prolongs the stage of N immobilization and reduces N availability to plants at least during the initial phases of crop growth. The crops planted immediately after cereal residue incorporation in such soils may face N deficiency and will require sufficient external N application to satisfy the needs of microorganisms and the growing crops. The presence of crop residues with a high C:N ratio transforms fertilizer or soil N into slowly available forms, which may act as slow-release fertilizer (Yadvinder-Singh et al. 2005). It has been suggested that the net immobilization phase during the initial years of ZT adoption is transitory, and the immobilization of N under ZT systems in the long term reduces the possibility of leaching and denitrification losses of soil mineral N. Crop residues that were incorporated into the soil decomposed 1.5 times faster than the surface-placed residues (Kushwaha et al. 2000). Due to the lack of soil mixing of crop residues in CA, N mineralization from crop residues is rather slow. Therefore, under CA, N management needs to be done carefully to avoid N deficiency due to slow mineralization in the initial period and to avoid excess N fertilization during later periods. A significant increase in total N is observed with increasing additions of crop residue and the amount of straw retained under permanent raised-beds.

Jat et al. (2018) conducted a nutrient omission plot experiment to study the effects of CA-based practices in rice–wheat (RW) and maize–wheat (MW) systems on soil quality parameters and the response of wheat to varying levels of N and K fertilization after four years of continuous cultivation. The N response was observed to be up to 100% of the recommended N (160 kg/ha) in conventional RW system (Sc1), while in other scenarios, it was observed to be up to 85% of the applied N (136 kg/ha) (Table 5.2). The grain yield at the 100% N rate in Sc1 was similar to 70% of applied N in other scenarios. Meanwhile partial and full CA-based scenarios saved 30% N (48 kg N/ha) compared to farmers' practice after four years of continuous cultivation with the same management practices. The saving of fertilizer N was attributed to the improvement of soil properties and N availability due to higher solubility of nutrients through moderation in soil moisture and temperature by crop residue retention under ZT conditions.

The different soil microclimates in ZT and CT are also responsible for the altered loss pattern of N from the soil. The soils in ZT systems tend to contain more moisture than under CT due to low evaporation from the soil. Thus, ZT soils generally tend to be wetter and cooler. Cooler soil temperatures slow down nutrient release from soil organic matter, reduce diffusion of nutrients to the plant roots, and can affect root growth. In the absence of frequent tillage, mineralization is slowed and the release of plant nutrients starts to decline, making fertilization more important for producing a higher yield. It has been observed that 15–25% extra available soil moisture during the growing season under ZT favours N losses from the system through leaching and gaseous emissions than CT. This happens

Table 5.2 Grain yield of wheat (t/ha) in N and K omission plots under different scenarios

Treatment	Sc1	Sc2	Sc3	Sc4
N (% of 160 kg N/ha)				
100	5.33 ± 0.80[a]	5.23 ± 0.5[b]	4.99 ± 0.50[bc]	5.30 ± 0.16[ab]
85	5.12 ± 0.07[a]	5.63 ± 0.17[a]	5.48 ± 0.11[a]	5.42 ± 0.04[a]
70	4.63 ± 0.06[b]	5.37 ± 0.03[ab]	5.32 ± 0.18[ab]	5.16 ± 0.12[b]
55	3.56 ± 0.06[c]	4.92 ± 0.09[c]	4.62 ± 0.06[c]	4.98 ± 0.14[c]
0	2.41 ± 0.02[d]	3.89 ± 0.06[d]	3.83 ± 0.10[d]	3.68 ± 0.10[d]
K (% of 60 kg K/ha)				
100	5.00 ± 0.05[a]	5.5 ± 0.10[a]	5.01 ± 0.05[a]	5.35 ± 0.07[a]
50	4.52 ± 0.04[b]	5.10 ± 0.05[b]	5.06 ± 0.02[a]	5.40 ± 0.05[a]
0	4.36 ± 0.04[c]	4.46 ± 0.02[c]	4.50 ± 0.05[b]	5.05 ± 0.10[b]

Source: Jat et al. (2018).

Note: a Sc1, conventional rice–wheat system; Sc2, partial CA-based rice–wheat–mungbean system; Sc3, CA-based rice–wheat–mungbean system; Sc4, CA-based maize–wheat–mungbean system. For all variables ±standard deviation. Values within the same column differ significantly at P=0.05 when not followed by the same small letter(s) according to the Duncan Multiple Range Test for separation of mean.

due to the higher number of aerobic and anaerobic microorganisms in ZT than in CT soils. In addition, a higher amount of organic substrate, nitrate content, and soil moisture, together with the existence of large aggregates under ZT than in the CT system, contribute to higher N loss through denitrification. Nitrate reduction takes place only under low O_2 supply conditions. The soils which appear well-aerated may yet reduce nitrate, particularly if the organic substrate level is high enough to create microsites where the O_2 demand by the microbial population exceeds supply from the soils. The size of the zones increases with an increase in temperature because of increased respiration causing larger gradients in oxygen concentration and thus rendering more of the soil volume devoid of oxygen. The development of larger aggregates with a diameter of >9 mm is likely to have such sites within them, even in soils where aeration around the aggregates appears satisfactory.

Requirements for active denitrification, i.e. easily available organic substrate, nitrate, suitable organisms, the existence of large aggregates, and high moisture content, are prevalent in ZT soils and denitrification can contribute to lower N availability in such soils. Lower mineralization in ZT soils, deep placement of N at crop establishment when the water content is high in soils, and splitting of N to match crop demand can considerably decrease the denitrification potential in ZT systems. Imperfectly drained ZT soils have much greater potential for denitrification than well-drained ZT soils. There is immense scope for coated fertilizers in terms of reducing losses in CA-based systems. Under residue retention, the 100% basal application of coated fertilizer like neem and sulphur-coated urea was effective for enhancing NUE compared to the conventional split application of prilled urea in the maize system (Jat et al. 2014).

In the CT system, basal fertilizer is applied by broadcasting the fertilizer followed by some form of tillage to incorporate the fertilizer into the soil. In the case of rice, the basal application is done by broadcasting the fertilizer over the puddled soil. The remaining dose of N is almost always applied by broadcasting urea at later stages. In the CA system, on the other hand, the basal dose of fertilizer is drilled just below the seed row using a seed-cum-fertilizer drill. By and large, under the CA system, the remaining dose of N is also applied by broadcasting urea. Surface application of urea fertilizer can lead to substantial losses of N through ammonia (NH_3) volatilization and gaseous loss to the atmosphere. This loss is indirectly related to surface-retained crop residues under CA. The performance of urea is very poor in terms of more losses as compared to any other sources of N such as ammonium nitrate in a residue-retained ZT system. The rate of urea hydrolysis, mediated through the soil enzyme urease, has a direct bearing on losses of N via NH_3 volatilization. The overall rate of urea hydrolysis is decided by the combined effect of lower soil temperature, increased urease activity, and moisture regime under surface residue-retained condition in CA. When urea is broadcast-applied in ZT systems, the potential for volatilization is greatest with warm, moist soils and drying weather conditions. Drying of the soil surface promotes the movement of ammonia away from the surface. Therefore, urea application must be carried out with due care in a CA-based system, especially because the enzyme (urease) that breaks down urea is primarily found in crop residues.

Volatilization loss of applied N can also be minimized by activities that move N into the soil so that ammonia formed from urea hydrolysis can be attached to soil CEC. These activities include: (i) applying N just before a rain, (ii) irrigating after fertilizer application, and (iii) deep placement of the urea into the soil. Ammonium nitrate fertilizer is not subject to loss by the volatilization process. Soil drainage and crop rotation are the most important factors to consider while deciding a recommended rate of N for CA-based systems. Because crops grown under ZT on well-drained light-textured soils may have increased yields because of the additional soil moisture retained by the crop residue the N requirement of such soils may be increased by 10–15%. However, imperfectly drained heavy-textured ZT soils may have the potential for denitrification and under such situations, the high residue levels can result in a great deal of N loss. Drilling of urea before irrigation is better under black cotton soils and after irrigation in loamy soils. The use of information about soil physical and chemical properties and experience on the behaviours of each crop species under fertilization conditions with variations in soil chemical properties allow for adjustment of the nutrient rates for different production situations.

There have been reports which indicate that more fertilizer N is required for optimum ZT maize production than for CT maize. The additional N is expected to compensate for the high risk of leaching losses of NO_3-N and for the lower rate of mineralization of residual soil N. In general, the broadcasting of fertilizer nutrients enables the plant roots to become a surface feeder, whereas drilling facilitates roots to grow deeper. Deeper root systems efficiently forage nutrients available in the deeper layer, which reduces the leaching losses of the nutrients and the deeper root system reduced crop lodging. It is important to consider 'the suitable soil nitrate N level', at which the lower limit does not restrict grain yield and the upper limit does not lead to unacceptable N losses to the environment.

Any measure to reduce fertilizer–residue contact, such as injection of N fertilizer, band application, sub-surface placement, and drilling of fertilizer is desirable for improving the NUE. The N loss can be reduced considerably by sub-surface application of urea beneath the soil surface and crop residues. Surface application of urea and urea-containing fertilizers results in severe loss of N under the ZT system, particularly at the early phases of crop establishment when there is ample moisture and a substantial amount of non-decomposed organic substrate at the surface of the soil. Incorporating coated urea fertilizer deep into the soil can reduce N losses through NH_3 volatilization. Therefore, such urea-containing fertilizers can be used in the ZT system through injection or banded placement (Nayak et al., 2022). Delaying the application of N fertilizer when a significant portion of residues have undergone decomposition will help in improving the NUE. With a high residue load, the use of nitrate fertilizers should be given preference over ammonium fertilizers because nitrate dissolves easier and is more mobile in soil, with lower surface immobilization during sowing. There are several options available to enhance N use efficiency: (i) controlled-release coated urea products, (ii) slow-release urea–aldehyde polymer products, (iii) urea super-granules for deep placement, (iv) nitrification inhibitors to reduce nitrate leaching and denitrification, (v) urease inhibitors to reduce ammonia volatilization from urea, and (vi) ammonium sulphate to enhance N efficiency over urea, etc.

Yadvinder-Singh et al. (2015) showed that a better fertilizer N management strategy in terms of achieving higher grain yield and N use efficiency for ZT wheat sown into rice residue with drilling of 25 kg N/ha as diammonium phosphate into the soil at seeding followed by two top dressings of 48 kg N/ha each just before the first and second irrigations compared to the presently recommended N fertilizer recommendation for CT wheat; applying 60 kg N/ha at sowing and the remaining 60 kg N/ha with first post-sowing irrigation (Table 5.3).

Table 5.3 Effect of method and time of N application on yield and nitrogen-use efficiency of applied N in ZT wheat sown into rice residues

N applied (kg/ha) at			Grain yield (t/ha)	Recovery efficiency of N (%)
Sowing	Before 1st irrigation	Before 2nd irrigation		
25D+35B	60	0	4.42	45.0
25D+35B	30	30	4.29	44.1
25D+65B	0	30	4.27	41.9
25D+95B	0	0	4.02	39.1
25D	48	48	4.79	56.7

Source: Yadvinder-Singh et al. (2015).

Note: D, drill, B; broadcast at sowing.

The efficient use of N fertilizer is important for crop yield and the environment and depends on the level of available N in the rooting zone. Applied N fertilizer rates should consider the available N in soils and other factors that affect the crop response to N fertilization. Despite the importance of soil tests for N application, the adjustment of fertilizer rates as a result of soil tests is rare. Apart from inorganic N, organic soil N mineralized during crop growth can provide N for the crop. In addition to soil N status measurements, several other diagnostic tools have been developed to determine N deficiency, which is used to improve N management and decrease the risk of N loss to ground and surface waters.

Phosphorus Management

In CA systems, broadcast and seed-placed P and K fertilizer applications lead to the accumulation of these nutrients in the surface and depletion in the deeper soil profile. Therefore, sub-surface banding of P and K with the seed, or ideally about 6–10 cm below the seed, is highly recommended to promote deeper root growth and avoid stranding these nutrients near the soil surface under the CA system. Further, P is strongly held by the soil. Therefore, band application of P fertilizer confines its interaction with a smaller soil volume.

In general, net immobilization of P occurs when residue P is less than 0.2–0.3% P, while net mineralization will occur with >0.3% P contents. The application of rice and wheat straw (with C:P ratio >300) caused immobilization of P during the first 15 days and then progressively increased the available P content in soil from the 30th day onward when the ratio declined below 200 (Yadvinder-Singh et al. 1988). During the early stages of residue decomposition, net immobilization of P can conserve a substantial amount of P in slowly available organic forms in CA. In CT systems, P is remixed into the soil profile, whereas in CA-based systems, P accumulates at the soil surface. Plants take up P from sub-surface soil layers by mining and deposit it on the surface. After 20 years of ZT, extractable P was 42% higher at 0–5 cm, but 8–18% lower at 5–30 cm depth compared with conventional tillage treatments in a silt loam soil (Ismail et al. 1994). Generally, concentrations of P are higher in the surface layers of all tillage systems compared to deeper layers but are most striking in ZT.

Reduced mixing of fertilizer P with the soil leads to lower P-fixation. This is an important benefit when P is limited, but may be a threat when there is excess P due to the possibility of soluble P losses in runoff water. However, when P fertilizer is applied on the soil surface, a part of P will be directly fixed by soil particles, making it unavailable for the crop plants. However, when P is banded as a starter application under the soil surface, there is P stratification which is taken up by the crop plants subsequently. This suggests that there is less need for P starter fertilizer in long-term ZT because of the high levels of P available in the topsoil seed placement zone. Deeper placement of P in ZT soil may be beneficial if the surface soil dries out frequently during the cropping season. However, if mulch is present on the soil surface in no-tilled condition, the surface soil is likely to be moister than conventionally tilled soils and the need for deep P placement is unlikely, especially in humid areas. The surface placement of crop residues leads to the accumulation of SOM and higher microbial biomass near the soil surface. Decomposition of crop residues releases humic molecules and low-molecular-weight aliphatic acids which block Al-oxide adsorption sites and reduce overall adsorption of P so that extractable P is better redistributed in ZT as compared to CT. This effect is dependent on the quality of the residue; legumes are generally more effective due to increased decomposition rates. Higher extractable P levels below the tillage zone occur probably due to the accumulation of P in senescent roots. The role of crop residues in enhancing the availability of P in the soil and improving P nutrition of the rice–wheat system has been well established (Gupta et al. 2007). Crop residue retention every year increased the availability of P and helped to reduce the fertilizer P requirement of the rice–wheat system.

Potassium Management

As 80–85% of absorbed K remains in the cereal straw, residue recycling can markedly increase K availability in CA-based systems. Recycling of crop residues can improve crop yields at low rates of K application and decrease the crop response to K applications. Potassium is not bound in any organic compound in the plant material, and thus its release does not involve microorganisms. In CA-based systems, K stays at the soil surface because it is not remixed by tillage. Decreased tillage intensity has larger extractable K levels in surface soil. The increased K concentration was more pronounced for wheat than for maize because wheat takes up large amounts of K, and most of this remains in harvest residues. Jat et al. (2018) reported that the application of 50% of the recommended K (60 kg K/ha) produced similar yields as 100% K in CA-based scenarios (Table 5.2). In Sc1 and Sc2, the application of 50% K recorded a lower yield compared to 100% K. At 0% K, Sc4 recorded similar yields to 100% K in Sc1 and 50% K in Sc3 and Sc2. CA-based scenarios produced wheat grain yields similar to that with 100% K/ha, while saving 30 kg K/ha. It was also concluded that an appreciable amount of N and K fertilizers, to the tune of 30% and 50%, can be saved under a CA-based management system after four years of continuous cultivation. By following CA-based practices, we can save precious nutrient resources through building soil quality along with the well-established advantage of higher productivity (yield, water, and energy) and profitability.

Other Nutrients

Information about the management of nutrients other than N, P, and K under CA is not yet available from India. Jat et al. (2018) reported that CA-based cropping systems improved soil properties and the availability of macro- (N, P, K) and micro-nutrients (Zn, Fe, and Mn) in the surface soil layer compared to conventional farmer practice.

Nutrient-Use Efficiency

The nutrient-use efficiency (NUE) can be defined as the yield increase of a fertilized plot over a control plot (N0) per unit of nutrient applied, absorbed, or utilized by the plant. It is an important index that can be used in CA to quantify the different nutrient management practices and to determine which one works best. It has two components: (i) N uptake efficiency (crop N uptake per unit of nutrient available from the soil and fertilizer), and (ii) N utilization efficiency (which is the grain yield per unit crop N uptake at harvest). The term NUE has been used extensively to describe nutrient use by a crop and is defined as grain yield divided by the supply of available nutrient from the soil and added fertilizer. In general, cereal crops are quite inefficient in using N and other nutrients as only 33% of the applied N as fertilizer can be recovered in the grain. More than 50% of the N not assimilated by plants becomes a potential source for environmental pollution – groundwater contamination, eutrophication, acid rain, ammonia re-deposition, global warming, and stratospheric ozone depletion.

While N losses cannot be avoided completely, there is certainly scope to min-imize losses with new and innovative precision N management technologies. CA is known to improve NUE as it reduces losses from the field by recycling leached nutrients from the topsoil with the use of deep-rooting cover crops. This leads to the greater availability of both native and applied nutrients to crop plants which can have a significant effect on fertilizer efficiency. Further significant increases in NUE can be achieved through fine-tuning of nutrient management practices. There are several methods and technologies to increase the use efficiency of various sources of nutrients. Climate-smart management techniques include appropriate timing of nutrient application, fertigation, integration of nitrogen-fixing plants as cover crops and crop rotation. Employing advanced fertilization techniques such as controlled-release fertilizers and nitrification inhibitors sus-tainably enhances crop productivity. All these management techniques need to be tested through different quantitative measures of nutrient-use efficiency of crops under CA under variable soil and climatic conditions.

Measures to Improve NUE Under CA

NUE is affected by leaching, denitrification, volatilization, and immobilization in the soil; fertilizer rates; source, placement, and timing of fertilizer application; climatic conditions; plant type which can influence absorption, translocation, assimilation, and re-translocation of the nutrient; and plant characteristics such as tissue nutrient concentration, size, and the number of reproductive sinks.

There are several ways to increase NUE:

- Crop rotations with the inclusion of legumes.
- Use of crop cultivars/genotypes produced by genetic selection under low nutrient inputs to increase NUE. These cultivars/genotypes usually have a higher harvest index and low nutrient loss.

- Surface retention of crop residue reduces erosion and sub-surface placement of fertilizers such as N has the potential to significantly improve N availability and NUE.
- The form of fertilizer can affect NUE in both CT and CA systems, e.g. N as NH_4-N is more efficient as plants require energy to assimilate NO_3 compared with NH_4 form.
- Precision and site-specific nutrient management (SSNM)-based fertilizer application practices can improve NUE.
- Controlled and slow-release fertilizers reduce N losses and increase NUE.

Precision Nutrient Management

Precision nutrient management is the science of using advanced, innovative, cutting-edge, site-specific technologies to manage spatial and temporal variability in inherent nutrient supply from soil to enhance productivity, efficiency, and profitability of agricultural production systems. It requires an understanding of the spatial soil fertility variability. Precision nitrogen management practices can efficiently reduce or increase fertilizer N use in comparison to conventional nitrogen management when larger variability exists. Conventionally, the spatial and temporal variability of nutrients in soils is assessed based on a rigorous field sampling followed by soil testing, both of which involve time and energy. Some non-destructive optical methods have been developed to estimate the chlorophyll content of plant leaves based on leaf greenness, absorbance, and/or reflectance of light by the intact leaf. These include chlorophyll meters, leaf colour charts (LCCs), optical sensors, and ground-based remote sensors. In India, chlorophyll meters, LCCs, and GreenSeeker optical sensors have been extensively tried to improve NUE in rice, wheat, and maize (Varinderpal-Singh et al. 2016). Site-specific nutrient management (SSNM) provides an approach to 'feeding' crops with all the required nutrients based on the crop's needs and thus improves the crop yield and nutrient use efficiency. Precision nutrient management can be accomplished by different methods, tools, and techniques for increasing nutrient-use efficiency.

Site-Specific Nutrient Management

SSNM is a set of nutrient management principles that aims to supply a crop's nutrient requirements tailored to a specific field or growing environment. It is an approach to supplying plants with nutrients to optimally match their inherent spatial and temporal needs. It accounts for indigenous nutrient sources, including crop residues and manures; and ensures optimal rates of fertilizer application at critical growth stages to meet the deficit between the nutrient needs of the crop and the indigenous nutrient supply. The SSNM approach does not necessarily aim to either reduce or increase fertilizer use. Instead, it aims to recommend nutrients at optimal rates and times to achieve high profit for farmers, with high efficiency of nutrient use by crops across spatial and temporal scales, thereby preventing loss of excess nutrients to the environment. The basic steps to be followed in the SSNM approach are: (i) set an attainable yield target (generally 75–80% of the yield potential), (ii) estimate the indigenous supply of nutrients through nutrient omission trials, (iii) estimate response to nutrients (difference between target yield and yield in nutrient omission plot), and (iv) estimate the nutrient rate based on response and agronomic use efficiency of nutrients. The results of Parihar et al. (2017a,b) revealed that SSNM-based nutrient application coupled with CA-based tillage practices in a maize–wheat–mungbean (MWMb) system have complementarity to attain higher system productivity. The SSNM uses a nutrient balance technique, in which the amount of N to be applied at the time of crop establishment is determined by using within-season nutrient estimation.

Soil Plant Analysis Development (SPAD)

Leaf chlorophyll content can be linked with leaf N content because the major fraction of leaf N is contained in chlorophyll molecules. Therefore, measurement of leaf greenness by a chlorophyll meter such as SPAD throughout the growing season can signal potential N deficiency from an unplucked leaf tissue early enough to correct it without reducing yields. Most of the research directly evaluating the usefulness of chlorophyll meters in improving NUE has been conducted with rice and wheat (Bijay-Singh et al. 2002). Chlorophyll meter-based N management led to significant increases in NUE when compared with the farmers' fertilizer practices. To use a chlorophyll meter, the critical value that is unique to genotype and environment, below which N-use efficiency and/or crop yields are likely to be adversely affected, was determined (Bijay-Singh et al. 2006). Two approaches have been used to guide fertilizer N applications to rice: (a) when the SPAD value is less than a set critical reading (Bijay-Singh et al. 2002; Maiti et al. 2004), and (b) when the sufficiency index (defined as the SPAD value of the plot in question divided by that of a well-fertilized reference plot or strip) falls below a predetermined value; such as 0.90 in the case of rice (Hussain et al. 2000). Despite greater reliability of the sufficiency index or dynamic threshold value approach, the fixed threshold value approach is more practical as it does not require a well-fertilized or N-rich plot. Shukla et al. (2004) observed a linear correlation between SPAD values and rice leaf N concentration for all the growth stages and lines tested. It has also been suggested that different threshold SPAD values may have to be used for different varietal groups (Balasubramanian et al. 2000). For rice cultivars grown in the Indo-Gangetic plains in India, a threshold SPAD value of 37 or 37.5 is appropriate for optimum rice yields, whereas, for rice cultivars grown in south India, the threshold SPAD value was found to be 35 (Bijay-Singh et al. 2002; Maiti et al. 2004).

GreenSeeker (GS)

GreenSeeker is a variable rate application and mapping equipment designed for use throughout a growing season. Here, crop vigour measured as the normalized difference vegetative index (NDVI) is used as the basis for N prescription rates. Dividing NDVI (estimate of total biomass) by the number of days from planting to sensing (or emergence to sensing), gives an estimate of biomass produced per day (to count a day, growing degree day must be >0). This index (NDVI/days from planting to sensing or emergence to sensing) is called INSEY (In Season Estimated Yield), a predictor of yield (grain or forage depending on the system) with no added inputs YP_0. A critical component of the algorithm is to precisely predict whether or not there will be an in-season response to applied fertilizer N and the magnitude of that response. The Response Index (RI) to added fertilizer N expected is calculated by dividing the average NDVI in the Nitrogen Rich Strip (NRS) by the average NDVI in the test plot. The ability of the environment to supply N (via mineralization of soil organic matter and/or deposited in rainfall) is quite variable and we need to take this amount of N supplied by the environment into consideration when deciding on mid-season fertilizer N recommendations.

Using the GreenSeeker optical sensor, Bijay-Singh et al. (2013) observed robust relationships between in-season GreenSeeker optical sensor-based estimates of yield at Feekes 5–6 and 7–8 growth stages and actual wheat yields. The study showed that the optical sensor-guided fertilizer N applications resulted in high yield levels and high N-use efficiency. The greenness of wheat leaves at the maximum tillering stage was found to be a function of N applied at the planting and crown root initiation stages. For rice, 49 days and 56 days after transplanting were the stages earmarked when optical sensors can guide mid-season fertilizer N applications (Bijay-Singh et al. 2013). Bijay-Singh et al. (2015) reported significant improvements in rice yield, agronomic efficiency, and recovery efficiency of N through GreenSeeker optical sensor-based N application in rice.

Leaf Colour Chart (LCC)

The leaf colour chart, an alternative to SPAD, is also used to measure the relative greenness of the crop leaf. It is a cost-effective tool for real-time or crop-need-based N management in rice, maize, and wheat. It is used to rapidly monitor leaf N status at the tillering to panicle initiation stage, and thereby guide the application of fertilizer N accordingly. There are two major approaches to the use of the LCC. In the real-time approach, a prescribed amount of fertilizer N is applied whenever the colour of rice leaves falls below a critical LCC value. The fixed splitting pattern approach provides a recommendation for the total N fertilizer requirement and a plan for splitting and timing of applications per crop growth stage, cropping season, the variety used, and crop establishment method. The LCC is used at critical growth stages to decide whether the recommended standard N rate will need to be adjusted up or down based on leaf colour (Bijay-Singh et al. 2012). Following LCC-based N management, the rice and wheat yields were similar to those with farmer's practice but with less N fertilizer application. In another scenario, an increase in grain yield with a reduction in N fertilizer use was observed by following the LCC method. The precision N management strategy using LCC was evaluated at 23 on-farm locations in Punjab (Varinderpal-Singh et al. 2011). After applying a basal dose of 30 kg N/ha at planting, 30 kg N/ha was applied as top-dressing every time the leaf greenness was less than LCC 5 starting from 21 days after sowing to initiation of the silking stage. On average, the LCC-based fertilizer N management produced a grain yield equivalent to or more than the blanket N recommendation with 20 kg N/ha less fertilizer N. Although most of the above research is related to conventional agriculture the same principles apply to CA.

Nutrient Management Models

Nutrient Expert® (NE), a decision support system (DSS), was developed by the International Plant Nutrition Institute (IPNI) and its partners for the smallholder production system of South Asia (http://software.ipni.net). It is an easy-to-use, interactive, computer-based decision tool that can rapidly provide nutrient recommendations for individual farmers' fields in the absence of soil testing data. The Nutrient Expert DSS modules for wheat and hybrid maize for South Asia were developed by IPNI and CIMMYT, validated in close partnership with NARES, and released in 2013 for public use (http://blog.cimmyt.org/tag/nutri ent-expert). This tool estimates the attainable yield for a farmer's field based on the growing conditions, determines the nutrient balance in the cropping system based on yield and fertilizer/manure applied and residue retained in the previous crop, and combines such information with expected N, P, and K responses in the concerned field to generate a location-specific nutrient recommendation for crops (wheat, maize). Satyanarayana et al. (2012) evaluated NE maize nutrient recommendations against farmers' practice (FP) and blanket state recommenda- tion (SR) during both (rainy and winter) the growing seasons. Across seasons, NE recorded 11% higher grain yield in CA (9.3 t/ha) in comparison to CT (8.4 t/ha) in South India (Figure 5.2). The NE maize tool was able to capture the inherent differences between conventional and CA practices of crop manage- ment, and NE-based fertilizer recommendations generated on the principles of SSNM performed better than FP and SR for maize.

The higher maize productivity with SSNM compared with the RDF (recommended fertilizer dose) or FFP (farmers' fertilizer practices) was reported by different researchers (Singh et al. 2016; Parihar et al. 2016b, 2017a) and also in wheat (Bhende and Kumar 2014; Parihar et al. 2016b, 2017a). The International Rice Research Institute (IRRI), together with its partners, has developed Crop Manager for Rice (released in the Philippines and Bangladesh and under evaluation in different states of India), Maize and Rice-Wheat System (http://cropmanager.irri. org/home), which can also be used by extension workers, crop advisors, and service providers provide farmers with advice that is specific to their growing conditions.

Fertilizer Placement

Simultaneous banding of seed and fertilizer is more important in ZT than in tilled soils, and follows somewhat different principles. It is especially important

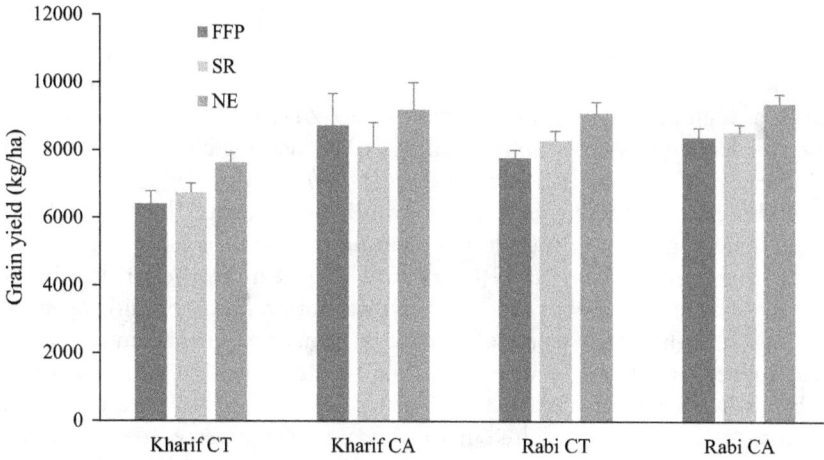

Figure 5.2 Nutrient Expert® improved grain yield in ZT and CT maize in southern India. The bars represent the standard error.

Source: Satyanarayana et al. (2012).

in ZT to sow fertilizer at the same time as seed, but only if the fertilizer can be placed in a separate band from the seed. Growth and yield advantages from fertilizer banded near the seed at the time of sowing have been well documented (Baker et al. 2007). Broadcasting of fertilizers during ZT often results in poor crop responses. However, there are some considerations when applying fertilizer in banded placement: (i) possible toxicity of fertilizer to the seeds and seedlings, often referred to as 'seed burn', and (ii) yield responses of the growing plants to the placed fertilizer. Mixing of fertilizer with seed is a risky undertaking at any time because of potential toxic chemical damage to the seeds and seedlings. In tilled soils, a measure of dilution of the fertilizer with loose soil will often reduce the risk of 'seed burn'. But in an untilled soil, soil dilution by mixing becomes minimal. Therefore, seed and fertilizer must be placed in different positions in the soil and remain in these positions after the openers have passed and the slot has been closed.

Mechanization for Nutrient Application in CA

Proper placement of fertilizer is crucial to ensure that plant roots can absorb the required nutrients during the growing period and thereby increase NUE in the CA system. Placement near the seed row may increase access of crops to the nutrient early in the growing season and provide a 'starter' effect that improves early growth (Nayak et al. 2022). Applying nutrients at the right place (both in the horizontal and vertical dimensions) in the soil ensures that plant roots can absorb enough of each nutrient at all times during the growing season. This is particularly important for P and K as they are less mobile. With the development of the 'happy seeder', farmers are now able to drill fertilizer P (as DAP) and K directly in soil under the residue-retained condition at the appropriate depth (Sidhu et al. 2015). By and large, farmers normally apply N through broadcasting urea in moist soil which is vulnerable to volatilization loss. NH_3 volatilization from urea can be as high as 42% and could be more under ZT(Patra et al. 2004). In CA-based permanent raised bed and zero tilled flatbed planting NUE increased with point placement of first and second N splits (Nayak et al. 2022).

Machinery needs to be developed for drilling of fertilizer N in ZT and surface residue conditions either during planting or post-emergence application in standing crops. Drilling of N in standing crops can precisely place the N near the root zone, thereby reducing the volatilization losses. Seeding openers' adjustment in the happy seeder allows the placement of fertilizer in a separate band between two rows of wheat to avoid seed germination and emergence problems. Kumar et al. (2010) found that placement of N between rows of wheat *vis-à-vis* broadcast resulted in a significant yield increase. In general, the broadcasting of fertilizer nutrients enabled the plant roots to become surface feeders, whereas drilling facilitates roots to grow at a deeper soil layer. Deeper root systems efficiently forage nutrients available in the deeper layer, which reduces leaching losses of the nutrients and the deeper root system reduces crop lodging.

Fertigation Under CA

The rapid adoption of micro-irrigation in agriculture has been largely due to the efficiencies from more precise delivery of water, and the multiple benefits of fertigation (simultaneous delivery of water and nutrients) are also widely recognized. A sub-surface drip system (SDI) limits evaporation from the soil and allows delivery of water and nutrients directly to the root zone. Since SDI can restrict the size of the root system to the wetted volume of soil, it is essential to maintain a continuous supply of moisture and nutrients during the entire growth cycle. Simultaneous delivery of water and nutrients directly to roots has proven to be advantageous for a variety of crops while minimizing nitrate-leaching losses.

A field experiment conducted in Punjab showed that sub-surface drip irrigation as well as fertigation in a CA-based maize–wheat system saved 60% irrigation water in maize and 50% in wheat compared to conventional flood irrigation. Application of 75% of recommended N fertilizer in fertigation produced grain yields similar to those obtained with the application of 100% of recommended N in CT, thereby resulting in a saving of 25% of fertilizer N in both maize and wheat. Similarly, in the rice–wheat system, NUE described as partial factor productivity of applied N (PFP_N) for both rice and wheat in a sub-surface drip irrigation system was significantly higher by 24.2 and 4.2% in rice and 33.1 and 51.3% in wheat compared to flood-irrigated ZT and CT rice–wheat systems, respectively (Table 5.4).

Table 5.4 Effect of tillage, residue mulch, and irrigation system on partial factor productivity of nitrogen (PFP$_N$) in a CA-based rice–wheat system

Treatment	Fertilizer N applied (kg/ha)			PFPN (kg grain/kg N applied)		
	Rice	Wheat	RWS	Rice	Wheat	RWS
ZTDSR-ZTW+R (SSD 67.5cm)	60	96	156	79.1ab	57.5a	65.8a
ZTDSR-ZTW+R (FI)	75	120	195	63.7c	43.2b	51.1b
CTRR-W	60	120	180	75.9b	38.0c	50.7b

Source: Yadvinder-Singh (unpublished data).

Note: ZT, zero tillage; DSR, dry seeded rice; SSD, sub-surface drip irrigation; FI, flood irrigation; +R, with residue; W, wheat; CTRRW, puddled rice conventional till wheat system; PFP$_N$, partial factor productivity.

Conclusion and Future Outlook

CA is increasingly being advocated as a management strategy for saving production resources, and improving farm profitability and soil health on a sustainable basis. With the development of planting equipment that can handle loose straw left in the field after combine harvesting of rice that can drill seed and fertilizer directly through the residues at appropriate depth (e.g. the turbo happy seeder), farmers are also retaining previous crop residues over the soil surface and moving towards full CA-based rice–wheat and other cropping systems. Effective plant nutrient management plays an important role in increasing and sustaining agricultural productivity under CA. The CA-based practices substantially influence nutrient dynamics in the soil, thereby requiring a paradigm shift in nutrient management practices and strategies. The management of nutrient input–output relationships in CA systems must balance the nutrient accounts, which means that the levels of outputs of biological products that are aimed at will dictate the levels of inputs, and ongoing nutrient balances must remain positive.

The major difference with CA systems, as opposed to CT, is that the management of the multiple sources of nutrients and the processes by which they are acquired, stored, and made available to crops are more biologically mediated. For example, increased soil organic matter content and soil porosity, increased biological nitrogen fixation by legumes in rotation, or exploitation of the deeper soil layers through crops with deep and dense root systems, have a significant bearing on nutrient dynamics across the soil profile. Some researchers claim a greater likelihood of more immobilization, denitrification, or leaching of applied N in CA systems requiring higher initial N fertilizer application. However, the immobilization phase in CA is transitory and the build-up of a larger pool of readily mineralizable organic N eventually requires less N in the CA system than in conventional systems on a long-term basis. Sub-surface banding of P and K in the seed row but below the seed promotes deeper root growth and avoids stranding these nutrients near the soil surface. Similarly, precise placement of N-fertilizer through side-banding in the CA system reduces immobilization (as it separates fertilizer and residue) and volatilization loss.

Efficient nutrient management under CA systems involves maintaining the nutrient levels in soil considering the plant extraction and losses/addition resulting from various soil processes induced by CA systems. Because the CA system involves diverse crop mixtures including legumes, nutrients and their cycles must be managed at the cropping system level rather than at a single crop level. An opportunity exists to further enhance the yield, profitability, and nutrient-use efficiency of these systems through SSNM. Research findings show that, in CA systems, nutrient requirements are generally lower, nutrient efficiencies are higher, and risks of environmental pollution are lower. The current fertilizer application practices under CA need revision, with a thrust in nutrient management research to improve NUE, soil health, and crop productivity while reducing the environmental footprint. All the fertilizer best management practices need to be tested through different quantitative measures of NUE of crops under different soil conditions in CA. Tailoring of higher NUE cultivars of different crops should be accelerated. Improved mechanizations for fertilizer application at the sub-surface and residue-retained condition both for basal application and at later crop growth stages for split application in a tall crop like maize are the need of the hour. Further, the scope and economics of fertigation and the use of real-time N management tools need to be explored under CA-based systems.

References

Baker, C.J. 2007. Fertilizer placement, pp. 118–133. In: *No-tillage Seeding and Conservation Agriculture*, 2nd Edn. (Eds.) Baker, C.J. and Saxton, K.E. FAO and CAB International, Wallingford, Oxon, U.K.

Balasubramanian, V., Morales, A.C., Cruz, R.T., et al. 2000. Adaptation of the chlorophyll meter (SPAD) technology for real-time N management in rice: a review. *International Rice Research Notes* 25(1): 4–8.

Bhattacharyya, R., Ved Prakash, Kundu, S. et al. 2010. Long-term effects of fertilization on carbon and nitrogen sequestration and aggregate associated carbon and nitrogen in the Indian sub-Himalayas. *Nutrient Cycling in Agroecosystems* 86: 1–16.

Bhende, S.N. and Kumar, A. 2014. Nutrient Expert®-based fertilizer recommendation improved wheat yield and farm profitability in the Mewat. *Better Crops-South Asia* 8: 21–23.

Bijay-Singh, Gupta, R.K., Singh, Y., et al. 2006. Need-based nitrogen management using leaf color chart in wet direct-seeded rice in northwestern India. *Journal of New Seeds* 8(1): 35–47.

Bijay-Singh, Varinderpal-Singh, Purba, J., et al. 2015. Site-specific nitrogen management in irrigated transplanted rice (*Oryza sativa*) using an optical sensor. *Precision Agriculture* 16: 455–475.

Bijay-Singh, Varinderpal-Singh, Yadvinder-Singh, et al. 2012. Fixed-time adjustable dose site-specific fertilizer nitrogen management in transplanted irrigated rice (*Oryza sativa* L.) in South Asia. *Field Crops Research* 126: 63–69.

Bijay-Singh, Varinderpal-Singh, Yadvinder-Singh, et al. 2013. Supplementing fertilizer nitrogen application to irrigated wheat at maximum tillering stage using chlorophyll meter and optical sensor. *Agricultural Research* 2: 81–89.

Bijay-Singh, Yadvinder-Singh, Ladha J.K., et al. 2002. Chlorophyll meter- and leaf color chart-based nitrogen management for rice and wheat in North western India. *Agronomy Journal* 94: 821–829.

Gupta, R.K., Yadvinder-Singh, Ladha, J.K., et al. 2007.Yield and phosphorus transformations in a rice-wheat system with crop residue and phosphorus management. *Soil Science Society of America Journal* 71: 1500–1507.

Hussain, F., Bronson, K.F., Yadvinder-Singh, et al. 2000. Use of chlorophyll meter sufficiency indices for nitrogen management of irrigated rice in Asia. *Agronomy Journal* 92: 875–879.

Ismail, I., Blevins, R.L. and Frye, W.W. 1994. Long-term no-tillage effects on soil properties and continuous corn yields. *Soil Science Society of America Journal* 58: 193–198.

Jat, M.L., Bijay-Singh and Gerard, Bruno. 2014. Nutrient management and use efficiency in wheat systems of South Asia. *Advances in Agronomy* 125: 171–259.

Jat, R.D., Jat, H.S., Nanwal, R.K. et al. 2018. Conservation agriculture and precision nutrient management practices in maize-wheat system: Effects on crop and water productivity and economic profitability. *Field Crops Research* 222: 111–120.

Jat, H.S., Jat, R.K., Yadvinder-Singh, et al. 2016. Nitrogen management under conservation agriculture in cereal-based systems. *Indian Journal of Fertilizers* 12(4): 76–91.

Jat, M.L., Saharawat, Y.S. and Gupta, R. 2011. Conservation agriculture in cereal systems of south Asia: Nutrient management perspectives. *Karnataka Journal of Agricultural Science* 24(1): 100–105.

Kumar, M., Sheoran, P., and Yadav, A. 2010. Productivity potential of wheat (*Triticum aestivum*) in relation to different planting methods and nitrogen management strategies. *Indian Journal of Agricultural Sciences* 80: 427–429.

Kushwaha, C.P., Tripathi, S.K. and Singh, K.P. 2000. Variations in soil microbial biomass and N availability due to residue and tillage management in a dryland rice agroecosystem. *Soil and Tillage Research* 56: 153–166.

Maiti, D, Das, D.K., Karak, T., et al. 2004. Management of nitrogen through the use of leaf color chart (LCC) and soil plant analysis development (SPAD) or chlorophyll meter in rice under irrigated ecosystem. *Scientific World Journal* 4: 838–846.

Majumdar, K., Jat, M.L. and Shahi, V.B. 2012. Effect of spatial and temporal variability in cropping seasons and tillage practices on maize yield responses in eastern India. *Better Crops-South Asia* 6(1):8–10.

Majumdar, K., Johnston, A.M., Dutta, S., et al. 2013. Fertilizer best management practices: Concept, global perspectives and application. *Indian Journal of Fertilizers* 9: 14–31.

Nayak, H.S., Parihar, C.M., Mandal, B.N., et al. 2022. Point placement of late vegetative stage nitrogen splits increase the productivity, N-use efficiency and profitability of tropical maize under decade long conservation agriculture. *European Journal of Agronomy* 133:126417.

Parihar, C.M., Jat, S.L., Singh, A.K., et al. 2016a. Conservation agriculture in irrigated intensive maize-based systems of north-western India: Effects on crop yields, water productivity and economic profitability. *Field Crops Research* 193: 104–116.

Parihar, C.M., Jat, S.L., Singh, A.K., et al. 2016b. Long term effect of conservation agriculture in maize rotations on total organic carbon, physical and biological properties of a sandy loam soil in north-western Indo-Gangetic Plains. *Soil and Tillage Research* 16: 116–128.

Parihar, C.M., Jat, S.L., Singh, A.K., et al. 2017a. Effects of precision conservation agriculture in a maize-wheat-mungbean rotation on crop yield, water-use and radiation conversion under a semiarid agro-ecosystem. *Agricultural Water Management* 192: 306–319.

Parihar, C.M., Jat, S.L., Singh, A.K., et al. 2017b. Bio-energy, water-use efficiency and economics of maize-wheat-mungbean system under precision-conservation agriculture in semi-arid agro-ecosystem. *Energy* 119: 245–256.

Pasricha, N.S. 2017. Conservation agriculture effects on dynamics of soil C and N under climate change scenario. *Advances in Agronomy* 145: 269–312.

Patra, A.K., Chhonkar, P.K. and Khan, M.A. 2004. Nitrogen loss and wheat (*Triticum aestivum* L.) yields in response to zero tillage and sowing time in a semi-arid tropical environment. *Journal of Agronomy and Crop Science* 190: 324–331.

Sanyal, S.K. and Majumdar, K. 2009. Nutrient dynamics in soils. *Journal of Indian Society of Soil Science* 57(4): 477–493.

Sapkota, T.B., Majumdar, K. Jat, M.L., et al. 2014. Precision nutrient management in conservation agriculture based wheat production of Northwest India: Profitability, nutrient use efficiency and environmental footprint. *Field Crops Research* 155: 233–244.

Satyanarayana, T., Majumdar, K., Pampolino, M., et al. 2012. Nutrient Expert®: A tool to optimise nutrient use and improve productivity of maize. *Better Crop-South Asia* 6: 18–21.

Shukla, A.K., Ladha, J.K., Singh, V.K., et al. 2004. Calibrating the leaf color chart for nitrogen management in different genotypes of rice and wheat in a systems perspective. *Agronomy Journal* 96: 1606–1621.

Sidhu, H.S., Manpreet-Singh, Yadvinder-Singh, et al. 2015. Development and evaluation of the Turbo Happy Seeder for sowing wheat into heavy rice residues in NW India. *Field Crops Research* 184: 201–212.

Singh, V.K., Yadvinder-Singh, Dwivedi, B.S., et al. 2016. Soil physical properties, yield trends and economics after five years of conservation agriculture based rice-maize system in north-western India. *Soil and Tillage Research* 155: 133–148.

Varinderpal-Singh, Kaur R., Bijay-Singh, et al. 2016. Precision nutrient management: A review. *Indian Journal of Fertilizers* 12(11): 1–15.

Varinderpal-Singh, Yadvinder-Singh, Bijay-Singh, et al. 2011. Calibrating the leaf colour chart for need based fertilizer nitrogen management in different maize (*Zea mays* L.) genotypes. *Field Crops Research* 120: 276–282.

Yadvinder-Singh, Bijay-Singh, Maskina, M.S., et al. 1988. Effect of organic manures, crop residues and green manure (*Sesbania aculeata*) on nitrogen and phosphorus transformations in a sandy loam at field capacity and under waterlogged conditions. *Biology and Fertility of Soils* 6: 83–187.

Yadvinder-Singh, Bijay-Singh and Timsina J. 2005. Crop residue management for nutrient cycling and improving soil productivity in rice-based cropping systems in the tropics. *Advances in Agronomy* 85: 269–407.

Yadvinder-Singh, Manpreet Singh, Sidhu, H.S., et al. 2015. Nitrogen management for zero-till wheat with surface retention of rice residues in north-west India. *Field Crops Research* 184: 183–191.

Yadvinder-Singh and Sidhu, H.S. 2014. Management of cereal crop residues for sustainable rice-wheat production system in the Indo-Gangetic plains of India. *Proceedings of Indian National Science Academy* 80(1): 95–114

6 Water Management in Conservation Agriculture

M.L. Jat, Naveen Gupta, H.S. Jat, C.M. Parihar,
H.S. Sidhu, and Yadvinder-Singh

Introduction

Water is the most critical resource for the sustainable development of any nation. Food security depends on the ability to increase production with decreasing irrigation water availability to grow crops. It is widely known that the application of irrigation maintains adequate soil moisture supply throughout the growing period and results in higher crop yields. It enables a significant production response from the associated use of high-yielding varieties, fertilizer, and crop establishment methods. Agriculture is the largest user of water, accounting for about 70% of the water withdrawals in India. There are pressures for diverting water from agriculture to other sectors. However, re-allocation of water out of agriculture could have a dramatic impact on global food markets. It is projected that the availability of water for agricultural use in India is likely to decrease significantly, resulting in a drop in the yields of irrigated crops, especially rice, thus resulting in price rises and a reduction of food available for the plates of the poor. India has about 140 M ha of cultivable land and 54% of the net sown area is dependent on rain. Irrigated agriculture accounts for about 60% of the food production in India. The population of India is likely to be 1.6 billion by 2050, resulting in increased demand for water, food, energy, and shelter. This calls for infrastructure expansion and improved water resource utilization.

Per capita water availability in India is lower than the global average, is extremely stressful in the Indus Basin, and has more than halved over the last two decades. Rainfall, a major source of freshwater, is also expected to be severely affected by projected changes in global climate, caused by an increase in the concentration of greenhouse gases. In addition, projected climate change impacts are expected to reduce water supplies and increase water demand. For example, in north-western (NW) India, the rainfall trends over the last 40 years have shown not only a considerable reduction in rainfall amount, but also in its irregular distribution over time and space (Prabhjyot-Kaur et al. 2013).

Enhancing agricultural water productivity (WP) is now a priority due to the growing physical shortage of water on the one hand, and the scarcity of economically accessible water on the other. Currently, the rapid population growth along with the extension of irrigation agriculture, industrial development, and climate

DOI: 10.4324/9781003292487-8

change, are stressing the quantity and quality aspects of the natural systems. Rice–wheat (RW), followed by cotton–wheat (CW), maize–wheat (MW), and millet–wheat are important cropping systems grown on an area of about 14.6 M ha in India. The high water input and low water productivity of conventional irrigated RW systems has led to the depletion of surface water and ground water in the Indo-Gangetic plains (IGP) of India, which has forced work on the development of new crop management technologies to counter this alarming situation. To meet the increasing food demand, the productivity of the wheat-based cropping system needs to be increased significantly in the near future. However, a major obstacle to boosting crop yields is the availability and efficient use of water, since there is increasing stress on water resources used for agriculture in India. Among the different cereals, irrigated rice production requires large amounts of water – to produce 1 kg of rice requires 2500 L of water and it loses a larger amount of water through evapotranspiration (ET) and soil percolation due to the cultivation practices entailed (Kukal and Aggarwal 2002).

Improvements in crop WP and the amount of food produced per unit of water consumed have the potential to improve both food security and water sustainability in many parts of the world. Earlier, with plentiful water supplies, the emphasis was to maximize yield per unit land. Water management has played a pivotal role in enabling the green revolution in the NW Indo-Gangetic plains. Now, with shrinking freshwater supplies, the aim has shifted to maximizing WP with as little an effect on the environment as possible. In NW parts of India, rice–wheat (RW) is the dominant cropping system, and is the principal determinant of a country's food security. However, due to serious groundwater depletion, its future is under threat (Hira 2009). Hence, there is an urgent need to manage water efficiently by using cost-effective and eco-friendly techniques. The WP can be maximized either by producing more with a given amount of water or by producing the same amount with lower water use. Water management by proper irrigation scheduling in combination with better crop management techniques (viz. efficient agronomic techniques) is a potential option to save water and to increase WP. In rainfed agriculture, technologies to increase WP include increasing groundwater storage and recharging, and reducing technological constraints to achieve the potential in rainwater harvesting. A range of strategies needs to be implemented to reduce the impacts of water scarcity in agriculture. Recognizing the several reviews that have already appeared in improving WP in rice or RW system (Ali and Talukder 2008; Yadvinder-Singh et al. 2014), the main focus of this chapter is to synthesize new information and present the research findings for efficient water management under CA-based management systems.

Water Management

Sustainable water management in agriculture aims to match water availability and needs in quantity and quality, in space and time, at reasonable cost with acceptable environmental impact. The economic health and sustainability of irrigated agriculture depend on the ability of producers to adapt to growing constraints on water, particularly through improved WP. Efficient water use not only reduces production costs, but also increases the profitability by reducing the nutrient losses.

Management of irrigation water at the field scale can be improved by quantifying the water balance and using advanced techniques for irrigation scheduling for more effective and economic use of limited water supplies. The irrigation scheduling is not necessarily based on full crop water requirement, but designed to ensure the optimal use of allocated water. In irrigated areas, the WP can be improved with better systems for water conveyance, allocation, and distribution, and water losses can be drastically reduced by using advanced irrigation methods/practices including drip irrigation systems that allow water to be delivered precisely when and where it is needed (Yadvinder-Singh et al. 2014). Uniformity distribution of water in the field is also very important for enhancing WP and is favoured by laser land levelling and precise delivery systems. Irrigation water saving does not necessarily help in reducing the ET that represents the depletion of aquifer. In a sub-surface drip irrigation system, drippers are located in the subsoil, which helps in reducing ET considerably.

Approaches for Higher Water Productivity

Smart Seeding Method in Rice

Rice, as a submerged crop, is a prime target for water conservation because it consumes great deal of water. A fundamental approach to reduce water inputs in rice is to grow the crop similarly to an irrigated wheat or maize, known as direct-seeded rice (DSR). Instead of trying to reduce the water input in lowland paddy fields, the concept of having the field flooded or saturated is abandoned altogether. The main driving force behind DSR is economic water use. Studies showed that yields varied between 4.5–6.5 t/ha, which were about 20–30% lower than those of lowland varieties grown under flooded conditions (Kumar and Ladha 2011). However, water use was ~60% less, total water productivity was 1.6–1.9 times higher, and net returns to water use were two-fold higher than that of lowland rice. In other studies, a water saving of 25–30% in DSR compared to flooded transplanted rice was reported in NW India under silty loam soils (Gathala et al. 2014). In DSR, water use decreased relatively more than yield, and water productivity under aerobic cultivation increased by 20–40% over that under flooded conditions.

Zero-Tillage in Wheat

Elements of CA began to be introduced in the RW systems of the IGP in the late 1990s, with the introduction of ZT wheat sown after rice (Erenstein and Laxmi 2008). In the IGP across India, Pakistan, Nepal, and Bangladesh, there has been considerable adoption of ZT wheat with some 5 M ha, but only marginal adoption of permanent ZT systems. ZT has been evaluated in wheat because of wet soil conditions after rice and long time (10–15 days) needed for seed-bed preparation resulting in poor wheat yields. Studies (Humphreys et al. 2005) showed 15–30% savings in irrigation water (IW) under ZT wheat compared to conventionally tilled in the RW belt of the IGP of India. Erenstein and Laxmi (2008) reported a 5–7% yield increase in ZT wheat after rice, mostly due to timely planting of wheat compared to CT. Using ZT for wheat in the RW system of the IGP can save 1.0 million L of IW and 98 L of diesel per ha of land (Pathak et al. 2009).

Surface Mulching

Soil evaporation generally accounts for about one-third of crop ET with a relatively small contribution to crop yield. In fact, apart from adjusting the growing period of crops, mulching is the only practice that reduces the ET by decreasing evaporation (Balwinder-Singh et al. 2011). Gupta et al. (2021) reported that mulching of ZT wheat with rice straw decreased total system soil evaporation by around 45 mm.

Raised-Bed Planting

The use of raised-beds for the production of irrigated wheat in the RW system of the IGP was inspired by the success of beds for wheat–maize (MW) systems in Mexico (Sayre and Hobbs 2004). There are also several reports of reduced irrigation amounts or time by up to 30–40%, with similar or higher yields for wheat on raised-beds compared with conventional flat-sown wheat (Ram et al. 2011). Jat et al. (2015) recorded 24.5% higher WP in an MW system because of less irrigation water applied in maize and wheat under permanent raised-beds (PRBs) than ZT flat. This is evident from the relatively lower application of irrigation water but proportionally higher yields of individual crops and cropping system in PRBs compared with ZT flat. Similar results of lower water use and higher WP of maize on PRBs were also observed by Jat et al. (2013). On a system basis, PRBs saved 29.2% irrigation water compared with ZT flat.

Approaches for Irrigation Scheduling

Irrigation scheduling is the decision-making process for determining when and how to irrigate the crops and how much water to apply. It forms the sole means for optimizing agricultural production and for conserving water. It is the key to improving the performance and sustainability of the irrigation systems. Irrigation scheduling is conventionally based on 'soil water balance' where the soil moisture status is estimated by the water balance approach, in which the change in soil moisture storage over a period of time is given by the difference between the inputs (irrigation + precipitation) and the losses (runoff + drainage + ET). Irrigation scheduling involves the timing of irrigation and the amount of water applied.

Climate-Based Approaches

Climate-based approaches to irrigation scheduling involve the use of a measure of cumulative potential evaporation. Potential evaporation is determined in a range of ways including pan evaporation and reference ET calculated from meteorological data in a variety of ways but the modified Penman–Monteith method is generally preferred.

Evaporativity-Based Approach

This approach is based on the concept of applying IW when the profile-stored water gets depleted to such an extent that it may start affecting crop growth. Long ago, Prihar et al. (1974) suggested a simple concept for irrigation scheduling to wheat based on the ratio of fixed depth of IW to CPE (cumulative pan evaporation) since previous irrigation [open pan evaporation (Pan E) minus rain amount]. The irrigation water amount is estimated on the basis of allowable water depletion in the soil profile. This is, in fact, the deficit irrigation (DI) technique that promotes the use of profile-stored water by encouraging deeper rooting in crops. This practice was reported to save two out of six irrigations applied to wheat at fixed growth stages without an adverse effect on crop yield. In rice, higher yields could be maintained by intermittent irrigation at 2-day intervals after the disappearance of ponded water either due to evaporation or infiltration. This could save a considerable amount of IW and thereby improve irrigation WP compared to that when continuously flooded throughout or part of the crop season.

Soil-Based Approach

Soil-based irrigation scheduling is based on the determination of soil water status (volumetric soil water content or matric potential) within the root zone, and knowledge of the critical threshold for irrigation. When based on volumetric soil water content, the threshold for irrigation is generally expressed as percentage depletion from full of the total plant available soil water-holding capacity (PAWC, the amount of water held in the soil between field capacity and permanent wilting point) of the root zone. For example, a common recommendation is to irrigate when the soil water content of the root zone decreases to 50% of PAWC, then apply enough water to replenish the deficit. This method can be used to calculate both when and how much to irrigate. In practice, when the rate of soil water extraction (soil drying) starts to decline is a good indication of the time to irrigate. A range of techniques can be used to determine volumetric soil water content, including neutron attenuation, time-domain reflectometry (TDR), and capacitance. In the second approach, irrigation is scheduled according to the soil matric potential (SMP), usually at a particular soil depth. The soil matric potential is directly related to the energy required by the crop to extract water from the profile.

Rice is a very sensitive crop to water deficit stress as the soil starts to dry below saturation (as soil water tension starts to increase above 0 kPa), with significant and increasing yield penalty as the root zone dries beyond a soil water tension of about 10 kPa at about 15 cm depth (Bouman and Tuong 2001). Under different tillage and mulch treatments, Gupta et al. (2016) reported SMP for scheduling irrigation to be −35 kPa at 32.5 cm and −15 kPa at 17.5 cm soil depth in wheat and dry-seeded rice, respectively.

Plant-Based Approach

Plant-based irrigation scheduling is based on the physiological and phenological conditions of the crop. The physiological condition (water stress level) can be judged from canopy temperature depression relative to air temperature (measured by infrared thermometry), and then calculation of cumulative stress degree days (SDD) and crop water stress index used for irrigation scheduling. Phenological stages can also be used to determine when to irrigate. In wheat, critical growth stages for irrigation are crown root initiation (CRI), tillering, jointing, flowering, and grain filling. Water stress at any of these stages may result in a loss of yield depending upon the severity of the stress. However, this technique, being costly, may not be economically feasible for smallholder farmers.

Deficit Irrigation Approach

Deficit irrigation (DI) is a water-saving irrigation strategy used in many parts of the world in which IW is applied at lower amounts than full crop water requirements (i.e. ET). It has been argued that the level of irrigation supply under DI should be between 60% and 100% of ET. The WP increases under DI, relative to its value under full irrigation, since small irrigation amounts increase crop ET more or less linearly up to a point when yield reaches its maximum value, and additional amounts of irrigation do not increase it any further. In particular, many crops have different sensitivities to water stress at various stages of growth and development, and the DI programme must be designed to manage the stress so that the yield decline is minimized. Information regarding the crop response to DI is essential to achieve such objectives when water is limited.

In DI, irrigation water is applied during drought-sensitive growth stages of a crop to maximize the productivity of water by allowing the crops to sustain some degree of water deficit and yield reduction. For DI, there is a need to identify the critical growth stage(s) of various crops in respect to their water demand, and irrigation needs to be applied at the critical growth stages to realize the maximum water use efficiency (WUE). Crown root initiation (CRI) and booting to heading are the two most sensitive growth stages of wheat, and drought stress should be avoided during these stages. To quantify the level of DI, it is necessary to define the full crop ET requirements. When irrigation is applied at rates below the ET, the crop extracts water from the soil reservoir to compensate for the deficit. Two situations may then develop. In one case, if sufficient water is stored in the soil and transpiration is not limited by soil water, even though the volume of IW is reduced, the ET and yield are unaffected. In the other case, if the soil water supply is insufficient to meet the crop demand, growth and transpiration are reduced, and DI induces an ET reduction below its maximum potential (Yadvinder-Singh et al. 2014). If the stored soil water that was extracted is replenished by seasonal rainfall, the DI practice is sustainable and has the advantage of reducing IW use. Under some situations, both water use and ET are reduced by DI but yields may be negatively affected. In an RW system in Punjab, four irrigations are recommended for wheat crops at different growth stages, which generally come out to be lower than the actual ET of the crop. It is perhaps the water stored in the profile during the preceding rice crop season which is used to compensate for the water to be used for meeting the ET requirements of wheat.

A study at Dhenkanal, Odisha, showed that with two supplemental irrigations, WP of maize, groundnut, sunflower, wheat, and potato was 0.55, 0.22, 0.23, 0.41, and 2.27 kg/m^3, and resulted in WP enhancement by 40%, 14%, 22%, 38%, and 7%, respectively, when three irrigations were applied (Kar et al. 2004). It has been suggested that yields and WP could increase even more if DI was used in combination with water conservation practices (i.e. mulching) or rainwater-harvesting techniques (Ali and Talukder 2008). Another form of DI, partial root-zone drying (PRD), increases WP while margining the crop yield under limited water resources. From an exhaustive review, Sepaskhah and Ahmadi (2012) concluded that in comparison to the DI strategy, PRD is a successful alternative irrigation technique compared to full irrigation that can save IW up to approximately 50% without significant yield losses, and thereby improve WP. Therefore, it is necessary to alternatively irrigate the two sides of the root system to keep roots in dry soil alive and fully functional and to sustain the supply of root signals (Kang et al. 2004).

Alternate, or every-other furrow, irrigation is also considered as PRD irrigation. Alternate furrow irrigation was successfully used as a water-saving irrigation. PRD has been adopted for different crops by using alternate furrow irrigation, resulting in higher WP.

Drip Irrigation Systems

In regions of water scarcity, drip irrigation has great potential to save large quantities of IW, which may help in bringing more area under irrigation and so resulting in increased crop productivity. A majority of farmers in the irrigated areas use a traditional surface (flood) irrigation method. Most of this flood IW goes as the deep drainage component, which is considered to be an energy-consuming process when the source is groundwater. In canal-irrigated areas, it could result in waterlogging and salinization. To overcome these situations, the drip irrigation system is the best option. Drip irrigation systems can be surface or sub-surface, and are more water-efficient than conventional flood irrigation systems. These drip irrigation systems have a conveyance efficiency of 100% and an application efficiency of 70–90%, while the corresponding figures for basin irrigation are 40–70% and 60–70%, respectively (Narayanamoorthy 2006). These irrigation systems have the potential to increase irrigation WP by providing water to match crop water requirements, reducing deep drainage losses, and generally keeping the top soil surface drier, thereby reducing soil evaporation. Agronomic manipulation, such as paired row sowing in cotton and the other wide-row crops like maize and sugarcane, can further improve productivity and reduce the cost of a drip irrigation system.

Many researchers have reported substantial improvements in yield and WP in cotton, sugarcane, soybean, maize, and wheat under a drip system over conventional irrigation methods (Aujla et al. 2007). Drip irrigation increased the wheat yield by 16% and 28%, and saved IW by 43% and 24% when irrigated at 0.6 and 0.8 IW/CPE ratios, respectively. Sharda et al. (2017) observed similar water savings in rice under surface drip irrigation compared to flood irrigation due to lower evaporation and drainage losses. They further reported that drip irrigation in DSR resulted in higher irrigation WP (0.85–0.90 kg/m^3) than FI (0.42–0.52 kg/m^3) on a sandy loam soil.

In an effort to lower the operational cost of drip systems, the planting pattern (normal, paired row, and pit method) has been changed to reduce the initial investment on laterals and drippers. The cotton seed yield was 19% higher under a drip system over a check basin method with the application of a similar amount of IW in both systems. A further increase of 6% in seed cotton yield was achieved along with a reduction of 25% IW as well as the cost of laterals. Furthermore, a 50% saving in IW and cost of laterals can be realized under paired sowing by sacrificing 12% seed cotton yield as compared to normal sowing. Seed cotton yield was increased by 32% and WP by 26% using drip irrigation compared to flood irrigation with the same quantity of water used (Aujla et al. 2005). Furthermore, under paired-row sowing there was a yield enhancement of 20% with 50% water saving and similar saving on cost of laterals.

With increasing adoption of CA, surface drip irrigation (SDI) and sub-surface drip irrigation (SSDI) provide an exceptional opportunity for complementing water-saving benefits. Sandhu et al. (2019) reported that SDI with residue retention saved 88 mm and 168 mm of water, and increased water productivity by 66% and 259% in wheat and maize on permanent beds compared to the conventional furrow irrigation system with residue removed, respectively (Table 6.1). On an average, after the harvesting of maize, soil moisture conservation up to 90 cm soil depth for wheat crop was 108% higher under CA than conventional agriculture plots. They also compared different combinations of lateral spacing and depth of SDI and SSDI with conventional till flood and ZT flood in a rice–wheat system. Irrigation water savings of 48–53% in rice and 42–53% in wheat were obtained under combination of SSDI and CA compared to a flood irrigation system (Figure 6.1a,b).

Table 6.1 Effect of irrigation method and residue on water productivity (WP) of a wheat–maize system

Treatment	Wheat		Maize	
	Irrigation water use (mm)	WP (kg/m³)	Irrigation water use (mm)	WP (kg/m³)
Furrow irrigation – residue removed (FI–R)	278[a]	1.32[c]	286[a]	1.63[c]
Drip irrigation – residue removed (DI–R)	202[b]	1.96[b]	122[b]	4.25[b]
Drip irrigation – residue retained (maize 50% upper + wheat 25% lower) (DI + R)	190[c]	2.19[a]	98[c]	5.85[a]

Source: Sandhu et al. (2019).

Note: In a given column, the values with same superscript are at par and with different superscript are significantly different at $P \leq 0.05$.

CA and Soil Water Balance

The productivity benefits from CA are mainly associated with its positive environmental and soil effects compared to conventional systems, including reduced erosion, runoff, and surface crusting, increased aggregate distribution and stability, and increased infiltration, soil water content, and WUE. Water conservation is a key element of CA, especially in rainfed/dryland areas exposed to erratic and unreliable rainfall. A general tendency of improved rainfall productivity was reported under CA in dry locations, which could be explained by a water-harvesting effect, leading to a strategy of *in situ* moisture conservation. The major advantages reported for CA are reduced wind and water erosion of top soil, increased WUE through improved water infiltration and retention, increased nutrient-use efficiency through enhanced nutrient cycling and fertilizer placements adjacent to seed, reduced oscillation of surface soil temperatures, increased soil organic matter and diverse soil biology, reduced fuel, labour, and overall crop establishment costs, and more timely operations (Hobbs et al. 2008). In CA, mulch is important for intercepting rainfall energy and reducing erosion.

Ghosh et al. (2015) reported that the wheat equivalent yield was ~47% higher in plots under CA compared with conventional agriculture in a maize–wheat rotation under rainfed conditions of Uttarakhand. Mean runoff coefficients and soil loss with CA plots were ~45 and ~54% less, respectively, than conventional agriculture plots. Mean runoff during the growing seasons was 39.8% with conventional agriculture plots, whereas it was 21.9% with CA plots. Approximately 45% less runoff was observed with CA plots compared with conventional agriculture. Soil loss under conventional agriculture was 7.2 t/ha on average over a

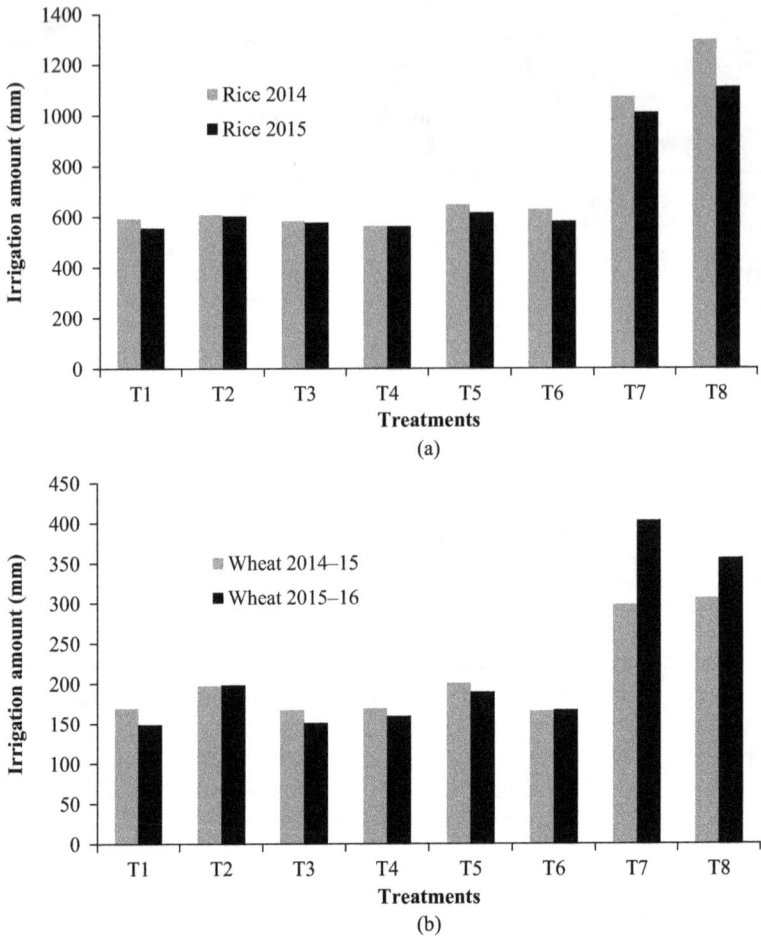

Figure 6.1 Effect of residue mulch, drip spacing, and flood irrigation on irrigation amount in (a) rice, (b) wheat.

T1: ZTRW+R+SDI$_{33.75}$ (zero till rice–wheat with residue; surface drip irrigation, laterals spaced at 33.75); T2: ZTRW-R+SSDI$_{33.75-15}$ (zero till rice–wheat without residue, sub-surface drip irrigation, laterals spaced at 33.75 at 15 cm depth); T3: ZTRW+R+SSDI$_{33.75-15;}$ T4: ZTRW+R+SSDI$_{33.75-20}$ (sub-surface drip irrigation, laterals spaced at 33.75 at 20 cm depth); T5: ZTRW-R+SSDI$_{67.5-15}$ (sub-surface drip irrigation, laterals spaced at 67.5 at 15 cm depth); T6: ZTRW+R+SSDI$_{67.5-15;}$ T7: ZTRW+R+FL (zero till flood with residue); T8: CTRW-R+FL (conventional till flood without residue).

Source: Sidhu et al. (2019).

five-year period, but soil loss in CA plots was 3.5 t/ha (Table 6.2). CA treatment reduced runoff significantly compared to conventional agriculture; probably due to the dense vegetation resulting in reduced runoff and silt deposition. The effect of cover cropping or residue mulch on soil water balance components has been

Table 6.2 Effect of CA on crop productivity and conservation efficiency on land with a 2% slope at Dehradun

Particulars	Conventional agriculture	Conservation agriculture
Water loss (% of rain)	39.8	21.9
Soil loss (t/ha/year)	7.2	3.5
Grain yield of maize (kg/ha)	1570	2000
Grain yield of wheat after maize (kg/ha)	950	1700
Wheat equivalent yield (kg/ha)	2320	3407
Dry grass yield (kg/ha/year)	–	610
Oil yield (kg/ha)	–	4.0
Weed biomass for mulching (kg/ha/year)	–	2100
Moisture conservation for wheat (mm) compared to fallow	28.1	58.5

Source: Ghosh et al. (2015).

documented in many locations around the world, but has not yet been explored in irrigated and rainfed crop production systems under CA in India.

Crop Water Use and Water Productivity Under CA

Soil evaporation (Es) is a non-beneficial loss of water (Jovanovic et al. 2020) and, from a ZT dry seeded RW system, Gupta et al. (2021) reported a soil evaporation loss of 600–700 mm. Studies (Gupta et al. 2019, 2021) have reported a significant loss from DSR (358–462 mm) and wheat (127–186 mm) under different tillage and mulch treatments. In some cases, the response to mulch in water-limited situations was also due to higher water uptake from the deeper soil profile due to a larger and deeper root system (Rahman et al. 2005). In a two-year study into rainfed wheat, Chakraborty et al. (2008) found that total water use (ET) was reduced by 79 mm under mulch during the higher and well-distributed rainfall year, and by only 14 mm in a comparatively dry year. Mulch increased WPET each year, but by more in the higher rainfall year due to both reduced ET and higher grain yield, while in the low rainfall year, the increase in WPET was mainly due to higher grain yield. However, in some studies ET of mulched crop remained unchanged due to transfer of water saved from suppressing Es used to increase transpiration. There are few studies on the influence of mulch on WPET of fully irrigated wheat, or whether mulch reduces irrigation requirements.

Straw mulching might reduce the irrigation requirement of wheat during the growing season due to reduced soil evaporation. Ram et al. (2013) reported that wheat yields were similar when three irrigations under rice straw mulch and four irrigations with no mulch were used, thereby saving one irrigation (75 mm). The effects of mulch in reducing irrigation requirement are well documented (Sidhu et al. 2015), however, its effects on other components of the water balance are less studied.

The effect of surface retention of residues on soil water status and irrigation requirement depends on the incidence and amount of rainfall. For example, in the RW system in NW India, mulching of wheat reduces the irrigation requirement in some years, but not in others (Yadvinder-Singh et al. 2008; Gupta et al. 2016). The mulching of wheat with rice residue generally delayed the need for irrigation of wheat, and reduced irrigation input to wheat by 50 mm in some years, but the residual mulch had a negligible effect on the irrigation requirement for DSR (Gupta et al. 2016). There is a need for in-depth research to develop IW scheduling based on SMP for ZT wheat with mulch conditions to reduce the IW requirement.

CA and Water-Use efficiency

Rice–Wheat System

Choudhary et al. (2018) compared conventional till rice–wheat–mungbean (CTRW + MB), zero-till rice–wheat along with residue retention and precise irrigation (ZTRW + R + PI), and ZTRW + MB + R + PI in a CA-based RW system. They reported that PTR consumed significantly more water compared to the rest of the treatments during three years of experimentation. Zero-till DSR irrigated using an SMP-based approach consumed 23–32% less irrigation water compared to PTR. However, ZTRW + MB + R + PI provided opportunities for a 79% saving of water. ZTRW+MB+R+PI recorded 24% and 41% higher system productivity and total (irrigation + rainfall) WP compared to CTRW, respectively. Kumar et al. (2018) compared the four scenarios, viz. conventional rice–wheat–fallow system in Scenario 1, the reduced-till rice–wheat–mungbean system in Scenario 2, full CA-based rice–wheat–mungbean system in Scenario 3, and full CA-based maize–wheat–mungbean system in Scenario 4. They reported the highest water inputs (irrigation + rainfall) in Scenario 1, followed by Scenarios 2 and 3, and the lowest in Scenario 4. Compared to the conventional RW–fallow system in Scenario 1, irrigation water savings in Scenarios 2 and 3 were 15% and 28%, respectively. In all five years, the irrigation input was 15–40% lower in direct-seeded rice–wheat–mungbean systems with CA (Scenario 3) than in the conventional rice–wheat system in Scenario 1. With the substitution of rice with maize in Scenario 4, irrigation water application decreased by 71%, 66%, and 59% compared with Scenarios 1, 2, and 3, respectively. In all five years, the WP of the system in Scenario 4 was higher (2.8–5.4 times) than in Scenarios 1 and 2. Gupta et al. (2016) reported inconsistent effects of tillage and mulch on grain WP over three years in a direct-seeded rice–wheat cropping system. Conventional tillage wheat decreased grain WP significantly (by 7–14%) in the first two years, and increased it by 9% in the third year, in comparison with ZT wheat. They further reported that the application of mulch significantly increased biomass WP in years 1 (from 6.2 to 8.4 kg/ m^3) and 3 (from 5.0 to 6.8 kg/m^3) in wheat.

Maize–Wheat and Other Systems

Permanent-bed (PB) planting facilitates the maintenance of uniform permanent soil cover for higher moisture/water capture and conservation. Jat et al. (2018) evaluated the effects of tillage and crop establishment methods and residue management options on crop yields and water productivity in an MW system at Karnal. Maize and wheat were irrigated at −60 and −50 kPa SMP, respectively. System crop productivity and WP under the MW system sown on PB and integrated with mungbean (PB+MB) were significantly increased by 28.2–30.7% and 36.8–40.5% compared to CT, respectively. The irrigation water use was 27% lower in PB plots compared to CT flat plots, irrespective of mungbean crop. Retention of crop residues in the PB system helped in reducing evaporation losses and conserving soil moisture (Jat et al. 2013; Parihar et al. 2016), which resulted in additional water saving (Jat et al., 2015) compared to the CT system. In a bed planting system, the furrow acts as an irrigation channel through which water moves quickly, thereby requiring less water for each irrigation compared to a conventional planting system (Jat et al. 2013, 2015). Higher WUE of maize, wheat, and a MW system was recorded with PB+MB and PB compared to CT and CT+ MB systems, which was due to the combined effect of less irrigation water applied and higher crop productivity. Higher WUE with PB has also been observed by many other researchers from the same agro-ecology (Jat et al. 2013; Singh et al. 2016; Parihar et al. 2016; Choudhary et al. 2016).

In a six-year study, Parihar et al. (2016) evaluated the performance of CA-based management practices (PB, ZT) and conventional till (CT) for four intensified irrigated maize-based systems (maize–wheat–mungbean, MWMb; maize–chickpea–*Sesbania* green manure, MCS; maize–mustard–mungbean, MMuMb; and maize–maize–*Sesbania*, MMS). In the initial two years, higher system productivity (maize equivalent yield) was recorded in PB (8.2–8.5 t/ha), while from the third year onwards, ZT registered maximum productivity (11.3–12.9 t/ha). Zero-till flat and zero-till PB practices reduced the irrigation water requirement by 40–65 ha-mm and 60–98 ha-mm compared to the CT system, respectively. The total water input (irrigation + rainfall) for all four maize-based cropping systems was significantly higher (2.5–10.1%) in CT compared to PB and ZT. In the initial two years, the highest WP was observed in PB plots (8.39–9.77 kg/ha- mm), while in succeeding years (3rd through 6th), the highest WP was recorded in ZT plots (7.37–12.64 kg/ha/mm) and the least was with CT planting (5.97–9.72 kg/ha/mm). The retention of residue at the surface in the ZT system helped in reducing evaporation losses and hence conserving more soil moisture. Higher soil moisture in the root zone not only helped with better crop growth but also increased the water productivity under ZT and PB compared to CT. Choudhary et al. (2018) compared conventional tillage fresh bed MW (FBMW) with CTMW + MB, PBMW + residue (R) + PI and PBMW + MB + R + PI in CA-based MW system. They reported that maize under PBMW (with or without MB) irrigated using an SMP-based approach consumed 87–92% less

Table 6.3 Amount of water applied and water productivity as influenced by different management scenarios in rice/maize–wheat–mungbean cropping systems

Treatment	Irrigation water applied (mm/ha)			System water productivity (kg grain/m³ of water applied)		
	2012–13	*2013–14*	*2014–15*	*2012–13*	*2013–14*	*2014–15*
CTRW	2508[b]	1798[b]	1710[b]	0.39[e]	0.47[f]	0.46[g]
CTRW+MB	2671[a]	1956[a]	1955[a]	0.41[e]	0.50[f]	0.49[g]
ZTRW+R	1828[d]	1238[d]	1328[d]	0.49[d]	0.63[e]	0.59[f]
ZTRW+MB+R	1940[c]	1348[c]	1501[c]	0.53[c]	0.68[d]	0.65[e]
FBMW	662[f]	435[f]	405[f]	0.90[b]	1.06[c]	1.07[c]
FBMW+MB	817[e]	588[e]	651[e]	0.87[b]	1.03[c]	1.01[d]
PBMW+R	402[h]	210[h]	205[h]	1.27[a]	1.52[b]	1.57[b]
PBMW+MB+R	467[g]	271[g]	273[g]	1.29[a]	1.57[a]	1.70[a]

Source: Choudhary et al. (2018).

Notes: CTRW, conventional till rice–wheat; CTRW+MB, conventional till rice–wheat with mungbean; ZTRW+ R+PI, zero till direct-seeded rice–wheat-residue retained; ZTRW+MB+R+PI, zero till direct-seeded rice–wheat with mungbean-residue retained; FBMW, conventional till fresh bed maize–wheat; FBMW+MB, conventional till fresh bed maize–wheat with mungbean; PBMW+ R+PI, permanent raised-bed maize–wheat-residue retained; PBMW+MB+R+PI, permanent raised-bed maize–wheat with mungbean-residue retained.

In a given column, the values with same superscript are at par and with different superscript are significantly different at $P \leq 0.05$.

irrigation water compared to PTR. Wheat needed 61% and 65% less water under PBMW irrigation based on an SMP approach compared with FBMW and CTRW (irrespective of MB and residue) irrigated at critical growth stages, respectively. System irrigation water consumption was significantly lower by 83–85% and 20–27% with PBMW+MB+R+PI and ZTRW+MB+R+PI, compared to CTRW (with or without MB), respectively (Table 6.3). They further reported that PBMW+MB+R+PI recorded 38% higher system productivity, saved 1660 mm of irrigation water, and increased irrigation + rainfall WP (WPI+R) by 270% compared to CTRW. As a result, CA-based sustainable intensification of MW systems (PBMW + MB + R + PI) was found to be a better alternative.

Parihar et al. (2017) reported that the amount of applied input water (irrigation + effective rainfall) was significantly lower for maize (5.1–9.5% in ZT and 7.6–14.2% in PB), wheat (10.5–11.9% in ZT and 15.8–17.7% in PB), and mungbean (7.9–10.7% in ZT and ZT and 11.9–16.1% in PB) compared to CT. They further reported that WUE was significantly higher with PB (1.75–1.97 kg/ha/m³) in 2013 and 2014, wheat (1.64–2.05 kg/ha/m³), mungbean (0.84–1.58 kg/ha/m³), and MWMb system (1.89–2.39 kg/ha/m³) in all three years of study compared to ZT and CT. In irrigated semi-arid tropics of India, the seven years' experiment of Parihar et al. (2018) showed that ZT and PB treatments in maize consumed 16.8–22.9% less irrigation water compared to CT. They reported that maize planted on PBB recorded ~11% lower water

Table 6.4 Impact of CA on irrigation water productivity of a maize–wheat system

Treatment	Maize		Wheat	
	Irrigation water applied (mm)	Water productivity (kg/ha/mm)	Irrigation water applied (mm)	Water productivity (kg/ha/mm)
CT	531.5[a]	6.94[d]	503.0[a]	9.75[bc]
PNB	497.0[ab]	8.09[c]	443.0[b]	10.48[b]
PNB + R	490.0[ab]	8.65[abc]	442.5[b]	11.01[ab]
PBB	483.0[b]	9.14[b]	429.5[b]	11.09[ab]
PBB + R	467.0[b]	10.12[a]	426.5[b]	11.60[a]
ZT +R	505.5[ab]	9.08[b]	452.5[b]	11.28[a]
ZT	516.0[ab]	8.43[abc]	512.5[a]	9.27[c]

Source: Das et al. (2018).

Notes: CT, conventional-till; PNB, permanent narrow bed; R, residue; PBB, permanent broad bed; ZT, zero-till.

In a given column, the values with same superscript are at par and with different superscript are significantly different at $P \leq 0.05$.

use and 16% higher WUE compared to CT. The PBB and ZT required 24.7% and 10.8% less irrigation water than the CTF system, respectively, with 11.5% higher system productivity.

Das et al. (2014) compared CT with different treatments of ZT, permanent broad-bed (PBB) and permanent narrow-bed (PNB) with and without residue (R) under a cotton–wheat system. They reported higher total water input (including rainwater) in CT plots, whereas PBB + R plots received the least water. Residue retention invariably reduced the amount of water applied. Further, there were about 3% and 10% water savings in the PBB plots compared to PNB and ZT plots, respectively, and about 13% and 14% water savings with PBB + R compared to ZT and CT plots. Das et al. (2018) compared conventional tillage with different treatment sets of ZT, permanent broad-bed (PBB), and permanent narrow-bed with and without residue under a maize–wheat cropping system. They reported that residue retention improved water productivity by 9% and reduced the irrigation water requirement by 62 mm/ha in two years in maize (Table 6.4). The partial residue retention (~40% for both maize and wheat) significantly improved two-year mean total water productivity in wheat crop under ZT plots.

Conclusion and Future Outlook

Agriculture in India is at risk due to continuously depleting aquifers and increasing pressures on surface and ground water resources. A single approach for irrigation management would not be able to tackle the forthcoming challenge of producing 'More Crop Per Drop' and also contributing to the 'Jal Shakti' (water power) mission. Integration of irrigation techniques (water-saving irrigation methods, irrigation scheduling approaches, etc.) with modern agronomic and soil manipulations (viz. direct-seeded rice, bed planting, crop diversification, zero tillage, and crop residue management, etc.) are essentially required to harness the full potential of available irrigation water for higher WP and profitability in dominant cereal-based systems on a sustainable basis. Improved irrigation management practices (timing and amounts) and methods (micro-irrigation, surface, sub-surface drip) based on real-time monitoring of water status in soil-crop systems are required to maximize the WP by efficiently managing water resources and allocation at regional scales. CA-based practices are gaining momentum in India and have helped to improve the resource-use efficiency, including WP. Limited studies on water management measures under CA have shown that coupling these practices conserves soil water by reducing evaporation while also improving crop yields, resulting in an increase in WP.

Integration of CA with drip irrigation methods helps in the saving of precious water used for irrigation in field crops and in increasing WP under different cropping systems. Crop management techniques like tillage, crop establishment, residue management, and fertilizer management should be integrated with real-time water application using modern methods and sensors to get higher water productivity and more water saving. In a rainfed ecosystem, deficit irrigation approaches are the key to get higher WP and profitability. In NW India, water resources are depleting at a very fast rate, and the adoption of SMP-based drip irrigation systems in CA-based RW systems are the need of the hour. Water balance studies are required to analyse rainfall capture, soil storage, and crop water use, including simple measurements of rainfall productivity of CA compared to conventional farming methods. Quantifying the benefits of residue retention (in different densities and types) and ZT practices on water balance are needed to reduce water stress and improve yield. Policy reforms are required to end the subsidy of methods and systems that lead to low water productivity on a systems basis.

References

Ali, M.H. and Talukder, M.S.U. 2008. Increasing water productivity in crop production – a synthesis. *Agricultural Water Management* 95: 1201–1213.

Aujla, M.S., Thind, H.S. and Buttar, G.S. 2005. Cotton yield and water use efficiency at various levels of water and N through drip irrigation under two methods of planting. *Agricultural Water Management* 71: 167–179.

Aujla, M.S., Thind, H.S. and Buttar, G.S. 2007. Fruit yield and water use efficiency of eggplant (*Solanum melongema* L.) as influenced by different quantities of nitrogen and water applied through drip and furrow irrigation. *Scientia Horticulture* 112: 142–148.

Balwinder-Singh, Eberbach, P.L., Humphreys, E., et al. 2011. The effect of rice straw mulch on evapotranspiration, transpiration and soil evaporation of irrigated wheat in Punjab, India. *Agricultural Water Management* 98: 1847–1855.

Bouman, B.A.M. and Tuong, T.P. 2001. Field water management to save water and increase its productivity in irrigated lowland rice. *Agricultural Water Management* 49:11–30.

Chakraborty, D., Nagarajan, S., Aggarwal, P. et al. 2008. Effect of mulching on soil and plant water status, and the growth and yield of wheat (*Triticum aestivum* L.) in a semi-arid environment. *Agricultural Water Management* 95: 1323–1334.

Choudhary, K.M., Jat, H.S., Nandal, D.P., et al. 2018. Evaluating alternatives to rice-wheat system in western Indo-Gangetic Plains: crop yields, water productivity and economic profitability. *Field Crops Research* 218: 1–10.

Choudhary, R., Singh, P., Sidhu, H.S., et al. 2016. Evaluation of tillage and crop establishment methods integrated with relay seeding of wheat and mungbean for sustainable intensification of cotton-wheat system in South Asia. *Field Crops Research* 199: 31–41.

Das, T.K., Bhattacharyya, R., Sudhishri, S., et al. 2014. Conservation agriculture in an irrigated cotton-wheat system of the western Indo-Gangetic Plains: crop and water productivity and economic profitability. *Field Crops Research* 158: 24–33.

Das, T.K., Saharawat, Y.S., Bhattacharyya, R., et al. 2018. Conservation agriculture effects on crop and water productivity, profitability and soil organic carbon accumulation under a maize-wheat cropping system in the North-western Indo-Gangetic Plains *Field Crops Research* 215: 222–231.

Erenstein, O. and Laxmi, V. 2008. Zero tillage in the rice-wheat systems of the Indo-Gangetic Plains: A review. *Soil and Tillage Research* 100(1–2): 1–11.

Gathala, M.K., Kumar, V., Sharma, P.C., et al. 2014. Optimizing intensive cereal-based cropping systems addressing current and future drivers of agricultural change in the Northwestern Indo-Gangetic Plains of India. *Agriculture, Ecosystems and Environment* 187: 33–46.

Ghosh, B.N., Dogra, P., Sharma, N.K., et al. 2015. Conservation agriculture impact on soil conservation in maize-wheat cropping system in the Indian sub-Himalayas. *International Journal of Soil Water Conservation Research* 3: 112–118.

Gupta, N., Eberbach, P.L., Humphreys, E., et al. 2019. Estimating soil evaporation in dry seeded rice and wheat fields after wetting events. *Agricultural Water Management* 217: 98–106.

Gupta, N., Humphreys, E., Eberbach, P.L., et al. 2021. Effects of tillage and mulch on soil evaporation in a dry seeded rice-wheat cropping system. *Soil and Tillage Research* 209: 104976.

Gupta, N., Yadav, S., Humphreys, E., et al. 2016. Effects of tillage and mulch on the growth, yield and irrigation water productivity of a dry seeded rice-wheat cropping system in north-west India. *Field Crops Research* 196: 219–236.

Hira, G.S. 2009. Water management in northern states and the food security in India. *Journal of Crop Improvement* 23: 136–157.

Hobbs, P.R., Sayre, K. and Gupta, R. 2008. The role of conservation agriculture in sustainable agriculture. *Philosophical Transactions of the Royal Society B* 363: 543–555.

Humphreys, E., Meisner, C., Gupta, R., et al. 2005. Water saving in rice–wheat systems. *Plant Production Science* 8: 242–258.

Jat, M.L., Gathala, M.K., Saharawat, Y.S., et al. 2013. Double no-till and permanent raised beds in maize–wheat rotation of northwestern Indo-Gangetic plains of India: effects on crop yields, water productivity, profitability and soil. *Field Crops Research* 149: 291–299.

Jat, R.D., Jat, H.S., Nanwal, R.K., et al. 2018. Conservation agriculture and precision nutrient management practices in maize-wheat system: Effects on crop and water productivity and economic profitability. *Field Crops Research* 222: 111–120.

Jat, H.S., Singh, G., Singh, R., et al. 2015. Management influence on maize–wheat system performance, water productivity and soil biology. *Soil Use and Management* 31: 534–543.

Jovanovic, N., Pereira, L.S., Paredes, P., et al. 2020. A review of strategies, methods and technologies to reduce non-beneficial consumptive water use on farms considering the FAO56 methods. *Agricultural Water Management* 239: 106267.

Kang, S.Z., Hu, X.T., Cai, H.J., et al. 2004. New ideas and development tendency of theory for water saving in modern agriculture and ecology. *Journal of Hydrological Engineering* 12: 1–7.

Kar, J., Bremer, H., Drummond, J.R., et al. 2004. Evidence of vertical transport of carbon monoxide from measurements of pollution in the troposphere (MOPITT). *Geophysical Research Letters* 31: 23–105.

Kukal, S.S. and Aggarwal, G.C. 2002. Percolation losses of water in relation to puddling intensity and depth in a sandy loam rice (*Oryza sativa*) field. *Agricultural Water Management* 57: 49–59.

Kumar, V., Hat, H.S., Sharma, P.C., et al. 2018. Can productivity and profitability be enhanced in intensively managed cereal systems while reducing the environmental footprint of production? Assessing sustainable intensification options in the bread-basket of India. *Agriculture, Ecosystems and Environment* 252: 132–147.

Kumar, V. and Ladha, J.K. 2011. Direct seeding of rice: recent developments and future research needs. *Advances in Agronomy* 111: 297–313.

Narayanamoorthy, A. 2006. Potential of Drip and Sprinkler Irrigation in India. Gokhale Institute of Politics and Economics India.

Parihar, C.M., Jat, S.L., Singh, A.K., et al. 2016. Conservation agriculture in irrigated intensive maize-based systems of north-western India: Effects on crop yields, water productivity and economic profitability. *Field Crops Research* 193: 104–116.

Parihar, C.M., Jat, S.L., Singh, A.K., et al. 2017. Bio-energy, biomass water-use efficiency and economics of maize-wheat-mungbean system under precision-conservation agriculture in semi-arid agroecosystem. *Energy* 119: 245–256.

Parihar, C.M., Jat, S.L., Singh, A.K., et al. 2018. Energy auditing of long-term conservation agriculture based irrigated intensive maize systems in semi-arid tropics of India. *Energy* 142: 289–302.

Pathak P., Sahrawat K.L., Wani S.P., et al. 2009. Opportunities for water harvesting and supplemental irrigation for improving rainfed agriculture in semi-arid areas, pp. 197–221. In: Wani, S.P., Rockström, J., Oweis, T. (Eds.), Rainfed Agriculture: Unlocking the Potential. CAB International, Wallingford, UK.

Prabhjyot-Kaur, Sandhu, S.S., Singh, S., et al. 2013. Climate Change – Punjab Scenario, pp. 16. Research Bulletin. AICRPAM, School of Climate Change and Agricultural Meteorology, Punjab Agricultural University, Ludhiana.

Prihar, S.S., Gajri, P.R. and Narang, R.S. 1974. Scheduling irrigation to wheat using open pan evaporation. *Indian Journal of Agricultural Sciences* 44: 567–571.

Rahman, M.A., Chikushi, J., Saifizzaman, M., et al. 2005. Rice straw mulching and nitrogen response of no-till wheat following rice in Bangladesh. *Field Crops Research* 91: 71–81.

Ram, H., Singh, Y., Saini, K.S., et al. 2013. Tillage and planting methods effects on yield, water use efficiency and profitability of soybean–wheat system on a loamy sand soil. *Experimental Agriculture* 49: 1–19.

Ram, H., Yadvinder-Singh, Kler, D.S., et al. 2011. Agronomic and economic evaluation of permanent raised beds, no tillage and straw mulching for an irrigated maize-wheat system on loamy sand in northwest India. *Experimental Agriculture* 48: 1–18.

Sandhu, O.P., Gupta, R.K., Thind, H.S., et al. 2019. Drip irrigation and nitrogen management for improving crop yields, nitrogen use efficiency and water productivity of maize-wheat system on permanent beds in north-west India. *Agricultural Water Management* 219: 19–26.

Sayre, K.D. and Hobbs, P.R. 2004. The raised-bed system of cultivation for irrigated production conditions, pp. 337–355. In: Lal R., Hobbs P.R., Uphoff N. and Hansen D.O. (Eds.). Sustainable Agriculture and the International Rice–Wheat System. Marcel Dekker, New York.

Sepaskhah, A.R. and Ahmadi, S.H. 2012. A review on partial root-zone drying irrigation. *International Journal of Plant Production* 4(4): 241–258.

Sharda, R., Mahajan, G., Siag, M., et al. 2017. Performance of drip irrigated dry-seeded rice (*Oryza sativa* L.) in South Asia. *Paddy Water Environment* 15: 93–100.

Sidhu, H.S., Jat, M.L., Yadvinder-Singh, et al. 2019. Sub-surface drip fertigation with conservation agriculture in a rice-wheat system: A breakthrough for addressing water and nitrogen use efficiency. *Agricultural Water Management* 216: 273–283.

Sidhu, H.S., Singh, M., Yadvinder-Singh, et al. 2015. Development and evaluation of the Turbo Happy Seeder for sowing wheat into heavy rice residues in NW India. *Field Crops Research* 184: 201–212.

Singh, V.K., Yadvinder-Singh, Dwivedi, B.S., et al. 2016. Soil physical properties, yield trends and economics after five years of conservation agriculture based rice-maize system in north-western India. *Soil and Tillage Research* 155: 133–148.

Yadvinder-Singh, Sidhu, H.S., Manpreet-Singh, et al. 2008. Straw mulch, irrigation water and fertilizer N management effects on yield, water use and N use efficiency of wheat sown after rice, pp. 171–181. In: Humphreys, E. and Roth, C.H. (Eds.). Permanent Beds and Rice-residue Management for Rice–Wheat Systems in the Indo-Gangetic Plain. ACIAR Proceedings No. 127.

Yadvinder-Singh, Thind, H.S. and Sidhu, H.S. 2014. Management options for rice residues for sustainable productivity of rice-wheat cropping system. *Journal of Research* (PAU) 51: 239–245.

7 Weed Management in Conservation Agriculture

T.K. Das, A.R. Sharma, N.T. Yaduraju, and Sourav Ghosh

Introduction

Modern intensive agriculture systems established on the doctrines of yield maximization have been afflicted with numerous constraints such as declining factor productivity, decreasing resource use efficiency, soil organic matter deterioration, salinization, soil structural degradation, water, and wind erosion, reduced water infiltration rates, surface sealing and crusting, soil compaction especially at the subsurface, declining groundwater table, pest and disease outbreak, weed resistance and weed shifts, etc. To address these overwhelming problems, conservation agriculture (CA) has been identified as a feasible solution and defined as 'a concept of resource-saving agricultural crop production which is based on enhancing the natural and biological processes above and below the ground' (FAO 2017). These are achieved through innovative CA practices combining the three principles of 'minimum soil disturbance', 'permanent soil cover', and 'crop rotation and diversifications'. Minimum soil disturbance is achieved through zero tillage and controlled tillage seeding systems that normally do not disturb more than 20–25% of the soil surface. In CA systems, surface retention of residues is the usual practice, as opposed to burning or incorporation. Recent advances in planting technology have made it possible to sow crops successfully into heavy residues and facilitate the use of residues as mulches for weed suppression. In India, the rotary disc drill and turbo happy seeder can seed wheat in heavy residue mulch of up to 8–10 t/ha without any adverse effects on crop establishment (Sharma et al. 2008; Kumar and Ladha 2011). Diversified crop rotations help moderate or mitigate possible weed, disease, and pest problems, utilize the beneficial effects of soil-building crops on the productivity of subsequent crops, and provide farmers with economically viable cropping options that minimize risk.

The primary reason for the promotion of CA is to reduce the cost of production, save water and nutrients, enhance yields, improve soil quality, and utilize resources efficiently. CA reverses soil degradation processes and improves soil and water quality through soil fertility and organic carbon enhancement, elevated microbial diversity in the rhizosphere, increase in water-holding capacity, and facilitating better infiltration of rainwater and higher ground-water

DOI: 10.4324/9781003292487-9

storage (Das et al. 2013; Das et al. 2020). Despite the numerous benefits that have been reported from CA systems, effective control of weeds is a major impediment to their productivity as weed infestations compromise crop yields (Kumar et al. 2008). The linear negative relationship between weed density and crop yields, especially during critical periods of crop–weed competition, makes weed management vital for achieving the potential yield gains under CA systems. Under CA, since mechanical weed control through tillage is restricted, growers mainly rely on selective and non-selective herbicides and alternative weed control strategies. Long-term CA-based weed management practices alter weed population and density over time, and a decline in overall weed intensity has been reported by Muoni et al. (2014).

Limited research work has been done on weed dynamics under CA systems in India. The few studies have mostly been on weed control in the short term, and the long-term effects on weed seed banks under CA are missing. The following sections present the information on weed management in CA with due references to the literature available elsewhere.

Weed Ecology and Distribution

Weeds are omnipresent, possessing a wide range of ecological amplitude which determines their adaptability. Weed shifts occur either due to disturbance in habitat (tillage, mulching, fire, flooding, drought, etc.) or alteration in the agronomic practices towards raising a crop such as submerged to aerobic rice, conventional tillage to zero tillage, etc. These practices have a tremendous influence on the composition of weed flora, seed distribution in soil, and dormancy. Tillage and inter-culture in crop fields are as responsible for the control of weeds as they are to their proliferation. Tillage systems also influence soil properties, such as organic matter, microbial populations, soil moisture, temperature, and pH, which can affect herbicide activity by influencing herbicide adsorption, movement, persistence, and efficacy.

Considering the lifecycle of annual weeds comprising seed bank, dormancy, viability, and seed erosion, Nichols et al. (2015) suggested four possible areas of intervention in a CA system as follows: (i) induction of weed seed dormancy and augmenting natural loss of seed viability in the soil, (ii) disrupting normal seedling establishment through cultural practices, (iii) curtailing seed production by mature weeds, and (iv) prevention of weed seed dispersal and re-establishment. In the case of perennial weeds like *Cyperus rotundus* which mainly proliferate through an underground vegetative network of rhizomes and tubers, a different set of interventions is necessary. CA practices can impact the vertical weed seed distribution in soil, soil moisture and temperature regime, light availability, and the population of seed predators and microbes. All these factors have a direct or indirect effect on weed seed dormancy, emergence, and seed mortality, thereby influencing weed recruitment in the field. The effect of the three CA principles on weed ecology, distribution, and dynamics are discussed below.

Distribution of Weed Seeds

Tillage affects weed seed distribution in the soil profile and the differential distribution of the seed in soil profile has the potential to change weed population dynamics. The conventional tillage system leads to burying the weed seeds in deeper soil layers. Burial increases seed survival, while seeds on or close to the soil surface can lose viability faster due to desiccation and harsh weather. In zero tillage (ZT) systems, owing to the absence of soil inversion, weed seeds (60–90%) accumulate on the top 0–5 cm of the soil profile (Nichols et al. 2015). These top-dwelling seeds are forced to germinate with pre-sowing irrigation and can be killed with any non-selective herbicide, such as glyphosate, glufosinate-ammonium, or paraquat. However, in the absence of herbicides, competition from weeds may increase as weeds germinate before the crop. Furthermore, depending on the extremity of the environment, the accumulation of seeds on untilled soil surfaces may increase the proportion of unviable weed seeds in the seed bank.

Germination and Viability of Weed Seeds

Tillage influences soil physical properties, namely bulk density, penetration, aggregation, and surface roughness, which can influence crop and weed emergence. Tillage provides germination stimulus for weeds through light flashes, scarification, temperature fluctuation, ambient CO_2 concentrations, and/or higher nitrate concentrations for breaking dormancy (Benech-Arnold et al. 2000). Better tilth, aeration, and exposure of the weed seeds to upper soil may be responsible for greater infestation of annual weeds under conventional tillage than zero tillage (Liebman et al. 2001). Owing to the distribution of weed seeds at different depths, weeds emerge in multiple flushes under conventionally tilled (CT) systems and therefore are difficult to control. In ZT systems, a higher proportion of the seed bank is concentrated in the top layer of the soil, which will germinate uniformly compared with CT seed banks (Gallandt et al. 2004).

Surface residues may reduce seed germination through the reduction in light and soil surface insulation. Insulation of the soil surface has implications for both soil temperature and moisture. Even under heavy crop residue loads, most seeds on the soil surface receive sufficient light to trigger germination. Residue mulching decreases the daytime maximum soil temperature but has a lesser influence on the daily minimum, thereby resulting in lower average soil temperatures and lower fluctuations. Therefore, residues delay germination in many weed species requiring certain threshold soil temperatures. In some weed species where germination is triggered by higher temperature fluctuations, the buffering of soil temperature and lower diurnal range under residue could therefore reduce germination rates and delay germination. In CA systems, residue decomposition or cover crops may release allelopathic substances and alter the chemical environment of the weed seed. Allelopathic effects from crop residue tend to have more pronounced effects on small-sized weed seeds compared to those of large-seeded crops.

Growth and Establishment of Germinated Weeds

Lower soil penetration resistance and bulk density under tilled soils lead to the successful emergence of germinated weed seeds. Further, soil inversion through tillage allows seedling emergence from deeper soil layers compared to untilled soils (Chhokar et al. 2007). Apart from emergence, tillage also impacts weed seedling establishment. Under the CA system, the radicle of germinated weed seeds may have difficultly penetrating ZT soil surfaces owing to higher penetration resistance, resulting in lethal germination.

Crop residues provide physical barriers that can prevent both light penetration and seedling emergence. The reduction in available light under surface residue has significant effects on seedling growth; the absence or poor supply of light will result in etiolated and weak seedlings which would be prone to competition and possibly could be more susceptible to certain types of herbicide damage. A low light environment will have a more profound effect on small-seeded annual weeds and crops, as they are initially more dependent on light compared to perennials and large-seeded species.

Composition and Diversity of Weed Species

The shift from CT to ZT alters the disturbance frequency of soil, often resulting in a change in weed species. CT systems with high soil disturbance have been shown to favour annual broadleaf species, while ZT systems favour perennial weeds like *Cyperus* and annual grasses. ZT having similarity with pasture or roadside situations have higher precedence of weed incursion from these environments. Comparison of CT and ZT soil seed banks showed higher weed species diversity and emerged weed communities under ZT (Erenstein and Laxmi 2008; Nichols et al. 2015). Under the rice–wheat system lower emergence of annual grassy weed *Phalaris minor* was reported under ZT than CT in wheat (Malik et al. 2002; Gupta and Seth 2007), while a reverse trend was observed in some broadleaf species, such as *Oxalis corniculata* (Chhokar et al. 2007). Lower emergence of *P. minor* under ZT was mainly attributed to higher soil strength (Chhokar et al. 2007), higher weed seed predation, lower soil temperature fluctuation (Gathala et al. 2011), and lower levels of light stimulation, N mineralization, or gas exchange. The higher weed density of *Oxalis corniculata* under ZT wheat following transplanted rice may be due to a higher concentration of their seeds on the soil surface (Chhokar et al. 2007, 2009).

Adoption of ZT in both rice and wheat has been reported to cause a shift in weed flora towards perennial species like Bermuda grass and purple nutsedge because tillage is not used to disrupt perennation. Moreover, poor management of perennial weeds before seeding crops under ZT also leads to a continuous build-up of stored food in the underground parts of perennial weeds.

Weed Seed Predation

ZT could augment weed seed predation and seed decay owing to a greater proportion of weed seeds situated on the soil surface where they are more prone to seed predation. Surface residue may indirectly encourage seed predation by providing foraging and nesting habitats for predators. Residue on the soil surface provides an insulated soil–atmosphere boundary that will decrease evaporative losses and maintain humidity. The increased micro-flora activity and biomass under residue would seem to encourage higher rates of seed losses under residue due to decay (Chauhan and Johnson 2010). Chauhan et al. (2010) reported a high rate (78–91%) of seed predation of grassy weed species, including *Echinochloa colona*, *Eleusine indica*, and *Digitaria* spp., from ZT soil surface in rice fields in the Philippines. Under ZT, the surface soil having a higher proportion of weed seeds, higher soil moisture, and microbial diversity, favours microbial seed decay.

Weed Management in CA Systems

With the absence of tillage, CA systems rely heavily on herbicides and cultural practices for weed control. Generally, low-persistence, non-selective herbicides, e.g. glyphosate, paraquat, or glufosinate, are usually applied before crop seeding, to render weed-free conditions to the emerging seedlings. Otherwise, a pre-emergence herbicide 1–2 days after sowing (DAS) of the crop may be taken. Selective post-emergence herbicides, if available, may be applied at recommended doses at an appropriate crop growth stage depending on the weed intensity. To reduce the herbicide residue build-up in CA systems, additional non-chemical options for weed control must be integrated with herbicides, with an emphasis on increasing the competitiveness of the crop against weeds. Such practices are discussed below.

Stale Seedbed Technique

The stale seedbed technique aims at eliminating initial flushes of germinating weeds before crop sowing. Weeds are stimulated to germinate and emerge before crop sowing by providing irrigation, which is then terminated through non-selective and low-residual herbicides like paraquat, glyphosate, or glufosinate. This technique is effective in reducing weed pressure during the crop season as well as reducing the weed seed bank (Rao et al. 2007; Singh et al. 2009; Kumar and Ladha 2011). This technique is most effective against: (i) weed seeds present in topsoil, (ii) weeds with low initial dormancy, and (iii) weed seeds requiring light to germinate. Susceptible weed species include *Cyperus iria*, *Digitaria ciliaris*, *Eclipta prostrata*, *Leptochloa chinensis*, and *Ludwigia hyssopifolia*.

Crop Establishment Methods

Crop establishment methods can play a significant role in weed control. Zero tillage, with or without residues, has been reported to reduce the population of

Phalaris minor (Malik et al. 2002; Chhokar et al. 2007; Gupta and Seth 2007). Franke et al. (2007) observed a lower emergence rate of all flushes of *Phalaris minor* in wheat under ZT compared with CT with about a 50% reduction of the initial flush. Chhokar et al. (2007) reported 39% lower dry matter of *Phalaris minor* under ZT compared with CT because of lower weed density. ZT, when combined with residue retention on the surface and early sowing, results in further suppression of *P. minor* and other weeds of wheat. Hajebi et al. (2016) reported ZT being superior to CT in terms of weed control, besides giving a 5.2% higher chilli yield.

Raised-bed planting of wheat reduces the weed infestation, including *P. minor*, due to burial of weed seeds at greater depths during bed preparation. The weed seeds situated on the top of the bed show poor germination and growth due to less availability of irrigation water on the top of the bed. Reshaping the bed before planting wheat can kill the first flush of *P. minor* seedlings.

Cultivar Selection

Certain varieties are more competitive with weeds than others. Additionally, the role of CA-specific cultivars for weed competitiveness under CA conditions is an active area of research (Mahajan and Chauhan 2013). Designing breeding programmes to select for competitive ability under CA is challenging due to the complexity of characteristics and large variation between locations and years. However, the development of such varieties would be highly beneficial not only for weed control but for other CA-specific characteristics also.

Sowing Time and Planting Geometry

The germination of many weed species follows a specific seasonal calendar owing to their innate dormancy and climatic requirements. Early planting of crops before the normal onset of weeds provides a competitive advantage to the crops. Likewise, planting can be delayed to allow weeds to germinate and be controlled before crop planting. In the case of *Phalaris minor* in wheat, adoption of ZT permitted early planting of wheat crops by 1–2 weeks, allowing the crop to establish before the emergence of the dormant *Phalaris minor* (Chhokar et al. 1999; Chauhan et al. 2012).

Higher seed rates and modified crop geometry, including row spacing and planting pattern, can impact crop–weed interference. Closer row spacing can tilt the competitive balance in favour of crops through faster canopy closure and smothering of weeds (Chauhan and Johnson 2011). Weed biomass was 25% lower under paired-row sowing (15–30–15 cm) of rice cultivar 'PR 115' compared with uniform row spacing of 23 cm (Mahajan and Chauhan 2011). Narrow row spacing (15 cm) reduced *P. minor* biomass by 16.5% compared with the normal spacing of 22.5 cm (Mahajan and Brar 2002). A higher seed rate of 150 kg/ha was found to be helpful in reducing populations of *Phalaris minor*, *Oxalis corniculata*, and *Melilotus alba* compared with a normal seed rate of 125 kg/ha.

Resource Management and Selective Stimulation

Weeds having more aggressive nutrient uptake capacity compared to crops may deprive the crop of essential nutrients. Therefore, alteration of timing, placement, and source of fertilizers should be planned in order to preferentially support nutrient uptake by the main crop rather than weeds (Das and Yaduraju 2007). Band placement of fertilizers can reduce weed biomass compared to broadcasting, with deep banding being more effective than surface banding. Under ZT, seed drills and happy seeders can place basal applications of fertilizer below the seeds, thereby denying/reducing access to nutrients to weeds compared to broadcasting fertilizers. In irrigated environments, spatial and temporal variation of soil moisture offers opportunities for weed control. When the top layer of soil is dry, planting large-seeded crops into the deep-soil moisture zone can provide crops with an initial advantage over weeds.

Residue Mulching

In situ residue mulching in CA systems provides an effective method of weed suppression. Residue mulching hinders weeds and reduces their recruitment and early growth by (i) a smothering effect on germinating weed seedlings which are sensitive to light and burial depth, (ii) imparting a physical barrier to emergent weeds, and (iii) release of allelo-chemicals in the soil. Chhokar et al. (2009) observed that 5.0 and 7.5 t/ha residue mulch reduced weed biomass by 26–46%, 17–55%, 22–43%, and 26–40% of *Phalaris minor*, *Oxalis corniculata*, *Medicago sativa*, and *Setaria glauca*, respectively, compared with ZT without residue. Under CA systems with a high demand for crop straw, residue mulch can be generated by cultivating short-duration crops such as mungbean or cowpea between wheat harvest and rice planting and retaining the entire residue of this crop as mulching material.

Cover and Intercropping

Another approach that has shown promise for suppressing weeds in ZT rice production involves sowing *Sesbania* seed at 25 kg/ha along with rice. *Sesbania* suppresses weeds with its rapid growth, which are then killed with 2,4-D ester at 25–30 days (Susha et al. 2018). Singh et al. (2007) reported 76–83% lower broadleaf weed densities and 20–33% lower grass densities with this practice. However, this practice may pose some risks, including (i) competition of *Sesbania* with rice if the 2,4-D application is ineffective or its application is delayed because of continuous rain, and (ii) additional costs associated with *Sesbania* seeds and management.

Crop Rotation and Diversification

Rotating crops with dissimilar life cycles or cultural conditions affect the growth cycle of weeds, hence they may prove to be one of the most effective methods of weed control. Every crop applies a unique set of biotic and abiotic constraints on the weed community; this will promote the growth of some weeds while inhibiting that of others. In this way, any given crop can be thought of as a filter, only allowing certain weeds to pass through its management regime. Monocultures often lead to weed simplification with only a few dominant weeds, potentially simplifying the choice of herbicide but potentially increasing selection pressure for herbicide-resistant weeds. Rotating crops will alter selection pressures, preventing one weed from being repeatedly successful, and thus preventing its establishment.

For containing infestations of *Phalaris minor* in wheat under the rice–wheat system, alternate systems like rice–fallow–sugarcane–ratoon and sugarcane–sunflower–rice–wheat–sugarcane rotations can be recommended. These rotations offer less scope for *P. minor* to proliferate. Other promising rotations include rice–potato–sunflower, rice–mustard–sugarcane, and rice–potato–onion (Chhokar et al. 2008). Inclusion of berseem in the rice–wheat cropping system reduced the seed bank of *P. minor* substantially as emerged plants of *P. minor* were also cut with each cutting of berseem before the weed had any opportunity to set seeds (Singh et al. 1999). Similarly, in potato-based rotations, earthing-up or digging operations in potatoes lead to the elimination of *P. minor* plants. In heavy soils, infestations of wild oats that predominated in maize–wheat systems were eliminated by growing rice instead of maize.

Prevention of Weed Seed Introduction

Depletion of the weed seed bank is of paramount importance to minimize weed pressure. Even after pre-emergence and post-emergence weed control in CA systems, some weeds escape and can produce a large number of persistent seeds, which add to future soil seed banks. Weeds should be prevented from setting seeds (annuals) or producing underground networks (perennials) during the off-season period and on bunds and channels as they can contribute significantly to the soil seed bank. Weed entry into fields through contaminated owner-saved seeds, manure, or compost, and irrigation water should also be restricted. Use of certified crop seeds, and well-decomposed and good-quality manure/compost free from weed seeds are good preventive methods worth practicing.

Chemical Weed Management

A CA system is dependent on herbicides and good agronomic practices for controlling weeds. Usually less persistent, non-selective herbicides like glyphosate, paraquat, or glufosinate are recommended before crop sowing to ensure weed-free conditions for crop germination. In addition, a pre-emergence herbicidal treatment is required to control flushes of annual weeds coming up with the germination of crops. Selective post-emergence herbicides, if available, may be applied at recommended doses at an appropriate crop growth stage depending on the weed intensity. In order to reduce the herbicide residue build-up in CA systems, non-chemical options must also be integrated with herbicides with an emphasis on increasing the competitiveness of crops against weeds. The use of herbicides for managing weeds is becoming popular because they are cheaper than traditional weeding methods, require less labour, tackle difficult-to-control weeds, and allow flexibility in weed management. Pre-emergence herbicides may not be efficient in controlling weeds in CA systems as the presence of crop residues intercepts and prevents herbicide spray from reaching the soil. The use of a high spray volume and at higher dosage levels, and use of granular formulations, are suggested to overcome such problems associated with pre-emergence weed control. However, post-emergence herbicides are preferred in the CA system. As a long-term strategy, the use of non-selective herbicides during the off-season to control perennial weeds and as a pre-plant treatment is suggested.

Weed management is a science as well as an art. The chemistry of herbicides and the art of application of herbicides play a role in achieving effective weed management. Like '4R Stewardship' in nutrients management, '5R Stewardship' should be followed for higher weed control efficacy with no phytotoxicity to crops and few implications to the environment. These are:

i   *Right choice of herbicide:* Herbicide should be selected based on the spectrum of prevalent weed to be controlled, time of application, and crop selectivity.

ii   *Right source of herbicide:* Herbicide should be procured from authentic sources/companies to ensure that the required active ingredient labelled is present and the herbicide is within the expiry date.

iii *Right dose of herbicide:* Herbicide should be applied at the doses recommended for crops considering the time of application, crop selectivity, and soil conditions.

iv   *Right time of application of herbicide:* Herbicide should be applied at the appropriate time, pre-planting, pre-emergence, or post-emergence, and the weed emergence and growth stages should be considered upon post-emergence applications.

v   *Right method of application of herbicide:* Herbicide should always be applied by using a sprayer, preferably a knapsack sprayer fitted with a flat fan or flood jet nozzle, proper pressure for delivery of spray droplets, and proper volume rate of water.

Several selective pre-emergence and post-emergence herbicides, some of which are low-dose and high-potency molecules, are now available to effectively manage weeds (Table 7.1).

Herbicide-Tolerant Crops

Crops have been subjected to genetic manipulation to withstand the application of non-selective herbicides such as glyphosate and glufosinate. This technology controls all weeds with a single application of herbicide and thus it has been readily adopted by farmers in many countries across the globe since its first commercialization in 1996. Since then, the adoption of CA has also been triggered

Table 7.1 Promising herbicides for weed control in different field crops under CA

Herbicide	Dose (g/ha)	Time of application	Remarks
Rice			
Pendimethalin	1000–1500	2–3 DAS/DAT	Broad-spectrum control of annual grasses and some broad-leaved weeds. Ensure sufficient moisture at the time of application. Does not control sedges, *Digera arvensis, Parthenium hysterophorus* effectively
Oxadiargyl	80-90	3–5 DAS/DAT or 15–20 DAT	Broad-spectrum control of weeds; do not spray if rain is expected within 6 hours. Could be used under surface-sown direct-seeded rice using a drum seeder
Pretilachlor (S)	750	3–5 DAS	Broad-spectrum control of weeds
Pyrazosulfuron-ethyl	25–30	20–25 DAS/DAT	Annual grasses and some broad-leaved weeds
Azimsulfuron	35	20 DAS/DAT	Annual grasses, sedges, and some broad-leaved weeds
Bispyribac-Na	25	15–25 DAS/DAT	Annual grasses and some broad-leaved weeds
Cyhalofop-butyl	100	20–30 DAS	Good control of grassy weeds
Chlorimuron-ethyl + metsulfuron-methyl	4	15–20 DAS/DAT	Annual broad-leaved weeds and sedges
2,4-D	750	20–25 DAS/DAT	Annual broad-leaved weeds and sedges
Ethoxysulfuron	20–22	3–5 DAT or 15–20 DAT	Broad-spectrum control but mostly broad-leaved weeds
Wheat			
Pendimethalin	1000–1250	0–3 DAS	Broad-spectrum control of annual grasses and some broad-leaved weeds. Ensure sufficient moisture at the time of application
Clodinafop-propargyl	60	30–35 DAS	Annual grasses, especially wild oat, and *Phalaris minor*
Sulfosulfuron	25	30–35 DAS	Broad-spectrum weed control; can control resistant *Phalaris minor*
Pinoxaden	30	30–35 DAS	Annual grassy weeds
Clodinafop-propargyl + metsulfuron-methyl	60 + 5	30–35 DAS	Ensures broad-spectrum weed control; clodinafop controls resistant *Phalaris minor*
Clodinafop-propargyl+ carfentrazone-ethyl	60 + 20	30–35 DAS	Ensures broad-spectrum weed control; clodinafop controls resistant *Phalaris minor*

(continued)

Table 7.1 Cont.

Herbicide	Dose (g/ha)	Time of application	Remarks
Sufosulfuron + metsulfuron	25 + 2	25–30 DAS	Annual grasses and broad-leaved weeds
Mesosulfuron+ idosulfuron	12 + 24	20–25 DAS	Annual grasses and broad-leaved weeds
Isoproturon + metsulfuron	1000 + 4	20–25 DAS	Annual grasses and broad-leaved weeds
2,4-D	500–750	20–25 DAS	Annual broad-leaved weeds and sedges
Maize			
Atrazine	1000–1500	1–2 DAS or 15–20 DAS	Broad-spectrum control of weeds. Does not control *Acrachne racemosa, Commelina benghalensis*, or sedges
Pendimethalin	1000–1500	1–2 DAS	Broad-spectrum control of annual grasses and some broad-leaved weeds. Ensure sufficient moisture at the time of application. Does not control sedges, *Digera arvensis, Parthenium hysterophorus* effectively
Metolachlor	1000–1500	1–2 DAS	Controls a broad spectrum of weeds, mainly grasses
Atrazine + pendimethalin (tank-mix)	750 + 750	1–2 DAS	Controls a broad spectrum of weeds and can render maize fields almost free of weeds, except sedges
Atrazine + metolachlor (tank-mix)	1000 + 1000	1–2 DAS	Controls a broad-spectrum of weeds and can render maize fields almost free of weeds, except sedges
Atrazine fb tembotrione	1000 and 100 (resp.)	1–2 DAS + 25–30 DAS	Controls a broad spectrum of weeds and can render maize fields almost free of weeds including sedges
Atrazine fb topramezone	1000 and 70 (resp.)	1–2 DAS) + 20–25 DAS	Controls a broad spectrum of weeds
2,4-D	750	30–35 DAS	Controls broad-leaved weeds and some sedges
Bentazon	1000–2000	15–20 DAS	Controls broad-leaved weeds effectively, moderate control of *Cyperus esculentus*
Soybean			
Pendimethalin	1000	1–2 DAS	Broad-spectrum control of annual grasses and some broad-leaved weeds. Ensure sufficient moisture at the time of application. Does not control sedges, *Digera arvensis, Parthenium hysterophorus* effectively

Table 7.1 Cont.

Herbicide	Dose (g/ha)	Time of application	Remarks
Metribuzin	300–400	1–2 DAS	Annual grasses and broad-leaved weeds
Imazethapyr	100	20–25 DAS	Annual grasses, broad-leaved weeds, and sedges
Pendimethalin+ imazethapyr (tank-mix)	750 + 100	1–2 DAS	Broad-spectrum control of weeds including sedges
Pendimethalin fb imazethapyr	750 and 100 (resp.)	1–2 DAS + 20–25 DAS	Broad-spectrum control of weeds including sedges
Pendimethalin fb quizalofop-ethyl	750 and 50 (resp.)	1–2 DAS + 20–25 DAS	Broad-spectrum control of weeds, but quizalofop should be applied as post-em where grassy weeds are a problem, otherwise not
Chlorimuron ethyl	9–12	15–20 DAS	Annual grasses, broad-leaved weeds, and sedges
Fenoxaprop	80–100	20–25 DAS	Annual grasses
Fenoxaprop + Chlorimuron	80 + 9	20–25 DAS	Annual grasses and broad-leaved weeds

Pigeonpea and mungbean

Herbicide	Dose (g/ha)	Time of application	Remarks
Pendimethalin	1000–1500	1–2 DAS	Broad-spectrum control of annual grasses and some broad-leaved weeds. Ensure sufficient moisture at the time of application. Does not control sedges, *Digera arvensis, Parthenium hysterophorus* effectively
Pendimethalin fb imazethapyr	750 and 100 (resp.)	1–2 DAS + 25–30 DAS	Broad-spectrum control of weeds including sedges
Pendimethalin fb quizalofop-ethyl	750 and 50 (resp.)	1–2 DAS + 25–30 DAS	Broad-spectrum control of weeds, but quizalofop should be applied as post-em where grassy weeds are a problem, otherwise not

Groundnut/ sunflower

Herbicide	Dose (g/ha)	Time of application	Remarks
Pendimethalin	1000	1–2 DAS	Broad-spectrum control of annual grasses and some broad-leaved weeds. Ensure sufficient moisture at the time of application. Does not control sedges, *Digera arvensis, Parthenium hysterophorus* effectively
Imazethapyr (for groundnut)	100	20–25 DAS	Broad-spectrum control of weeds including sedges
Oxadiazon (for sunflower)	500–1000	20–25 DAS	Soil moisture is a pre-requisite for its activity

(*continued*)

Table 7.1 Cont.

Herbicide	Dose (g/ha)	Time of application	Remarks
Sugarcane			
Atrazine	2000	1–2 DAT	Broad-spectrum control of weeds but a broad-leaved killer; less active against *Acrachne racemosa*, *Commelina benghalensis*, and perennial sedges
Metribuzin	1000–1500	1–2 DAT	Broad-spectrum control of weeds
Halosulfuron-ethyl	90	25–30 DAS	Controls *Cyperus* spp. along with other weeds
2,4-D	1000	35–40 DAT	Only for broad-leaved weed control
Chickpea/lentil/pea			
Pendimethalin	1000–1500	1–2 DAS	Broad-spectrum control of annual grasses and some broad-leaved weeds. Ensure sufficient moisture at the time of application
Pendimethalin fb quizalofop-ethyl	750 and 50 (resp.)	1–2 DAS + 20–25 DAS	Broad-spectrum control of weeds, but quizalofop should be applied as post-em where grassy weeds are a problem, otherwise not
Cotton			
Pendimethalin	1000–1500	1–2 DAS	Broad-spectrum control of annual grasses and some broad-leaved weeds. Ensure sufficient moisture at the time of application. Does not control sedges, *Digera arvensis*, *Parthenium hysterophorus* effectively
Pendimethalin fb quizalofop-ethyl	750 and 50 (resp.)	1–2 DAS + 25–30 DAS	Broad-spectrum control of weeds
Pendimethalin fb pyrithiobac-Na	750 and 62 (resp.)	1–2 DAS + 25–30 DAS	Broad-spectrum control of weeds
Rapeseed & mustard			
Pendimethalin	1000	1–2 DAS	Broad-spectrum control of annual grasses and some broad-leaved weeds. Ensure sufficient moisture at the time of application
Isoproturon	750–1000	1–2 DAS	Broad-spectrum weed control; does not control sedges
Oxadiazon	500	1–2 DAS	Broad-spectrum weed control
Oxyfluorfen	200–250	1–2 DAS	Broad-spectrum weed control

Source: Yaduraju et al. (2016).

Note: Non-selective herbicides like paraquat and glyphosate should be applied prior to sowing to kill existing weeds.

with the introduction of genetically modified crops, of which 47% of the area of which is under herbicide resistant and 41% with stacked traits (herbicide + insect resistant) (ISAAA 2019). These biotech crops are also considered to be the fastest adopted crop technology in the history of modern agriculture. These two technologies, i.e. CA and GM crops, have revolutionized world agriculture and grown hand in hand showing double-digit growth over the past two decades and covering 180 and 190 M ha in 2015–16 and 2019, respectively. Despite some apprehensions, this technology has had positive impacts on crop productivity and profitability.

Herbicide-tolerant crops (HTCs) facilitate the adoption of CA practices in selected crops but may restrict crop rotations. HTCs must be grown in conjunction with other weed control methods, particularly rotational use of herbicidal mode-of-actions to avoid resistance. Herbicides such as glyphosate and glufosinate, being non-selective, provide complete control of weeds and the crops/varieties being resistant remain completely safe from herbicidal effects even though herbicide is applied across their growth stages. Therefore, traditional pre-emergence herbicides selective to that crop can also be applied in HTCs, if the situation demands. However, the technology is associated with several concerns and controversies, and has not found favour in some countries, including India.

Integrated Weed Management

A single isolated approach cannot be a fool-proof strategy for achieving the desired level of weed control under CA systems. Integrated weed management (IWM) is not meant to replace selective, safe, and efficient herbicides but is a sound strategy to encourage judicious use of herbicides along with other effective, economical, and eco-friendly control measures. The three CA principles are interlinked and interactive enough to counter and reduce weed problems over the course of time. It would be wise if approaches such as stale seedbed, uniform and dense crop establishment, cover crops, crop residues as mulch, crop rotations, and practices for enhanced crop competitiveness with a combination of pre- and post-emergence herbicides are integrated to develop sustainable and effective weed management strategies under CA systems. Therefore, an IWM practice involving zero tillage, residue, cover crops, intercrops, and herbicides is needed to widen the weed control spectrum for efficient weed management and sustainable crop production under CA.

Practical Weed Management

Weeds are often considered to be the major constraint for adoption of CA. In fact, increased weed infestations are cited as the foremost reason for the poor performance or failure of crops under CA by many researchers. All of us have been taught from an early stage that one of the major purposes of tillage is weed control, besides its other benefits. Therefore, it is understandable that ZT will lead to increased weed proliferation, particularly of the perennial species like *Cyperus rotundus, Cynodon dactylon, Sorghum halepense,* and others if adequate control measures are not taken. Experience has shown that with ploughed systems, over the decades, weed problems become aggravated. In fact, with tillage operations, weed seeds in the soil obtain a sort of stimulation due to exposure to light and air, and the seeds lying in the lower layers are brought to the surface and vice versa. The result is that weed seed distribution becomes quite uniform in different soil layers, which keep germinating at different stages or flushes, and thus provide competition to crop plants throughout the growth period (Chauhan et al. 2012; Malik et al. 2014; Nichols et al. 2015).

Another dimension of increased weed infestation in CT systems is the general emphasis on control of weeds during the period of critical crop–weed competition. It is advocated that weeds should be controlled up to a certain stage only in the early part of the crop growth period to avoid economic loss, as the weeds emerging in the later stages do not cause much harm to the crop. However, these weeds in the later part of the season produce enough seeds by the time the crop is harvested. Also, these weeds keep growing even after the crop harvest during the fallow period and enrich the weed seed bank in the soil. The result is that the weed problem in the next season will either be the same or even greater than in the previous season, and the cycle goes on. Therefore, under CA systems, the focus is on minimization of the weed seed bank and season-long weed control.

A strategy of integrated weed management is needed for successful CA that may include herbicides other techniques. Residue retention on the soil surface helps in suppressing the emergence and growth of weeds to some extent but it cannot provide effective and full season control. Growing cover crops either as a sole or mixed/intercropping system with the objective of developing a full canopy within the shortest possible time is another way of checking weed growth. Crop rotation is also one of the essential requirements of CA systems as these help in controlling weeds and other pests. Certain weeds are associated with specific crops, and therefore, crop rotations are advocated to get rid of particular weed species. For example, if pulses and oilseeds are grown in a field infested with broadleaved weeds like *Chenopodium album, Rumex dentatus, Anagallis arvensis, Medicago sativa, Cichorium intybus, Vicia sativa,* and others, manual weeding is the only option because there are no selective post-emergence herbicides for such situations. Therefore, it is advisable to opt for cereal-based rotations where such broad-leaved weeds can be effectively controlled by the available herbicides like 2,4 D, metsulfuron, carfentrazone, bensulfuron, and others.

In the CA system, only the lines where the seed and fertilizer are placed are opened, and the inter-row spaces remain virtually undisturbed and covered with the mulch of crop residues. Weed seeds lying in the upper soil layer (0–5 cm) emerge in the first flush after sowing and there is relatively lower emergence of weeds in the subsequent flushes, by which time the crop develops adequate canopy to suppress weed growth. Further, sporadic manual weeding by uprooting or cutting plants from the base with a hand hoe is suggested to ensure weed-free conditions throughout the season. The key to weed management in CA is to prevent the weeds from flowering and setting seed.

It is essential that weeds growing in the field at the time of sowing are killed, otherwise these weeds will take an upper hand even before the crop emerges. Since the ploughing of land is not practiced for killing the existing weeds, the only alternative is to use a complete weedkiller and non-selective herbicide like paraquat or glyphosate to kill them. Paraquat is recommended for rapid desiccation and for killing of the annual not-so-hardy type of weeds, while glyphosate is a systemic herbicide that is used for the control of perennial and relatively hardy weed species. In fact, CA is a herbicide-driven technology requiring rational use of herbicides before sowing as well as after sowing, and during the crop growth period.

In conventional agriculture systems, we aim at controlling weed plants, while under CA, the focus is on minimization of the weed seed bank. For this constant watch is kept on weed plants throughout the crop season to prevent them from flowering and setting seeds. As the famous quote in weed science states 'One year seeding is seven years weeding'; therefore, the preventive measures are given equal or even more importance than the curative measures of weed control under CA. Weeds are to be controlled even after the crop harvest. In some countries, weeds seeds after crop harvest are either collected or burnt through flaming using special machines.

Tips for Efficient Weed Management Under CA

CA works well when all the required principles along with best management practices are followed. In fact, it is a controversial subject for some researchers and other stakeholders who discount it as not being relevant in Indian conditions, primarily because of weed infestation. The following tips will ensure effective weed control and lead to the success of CA.

Dos

- Season-long weed control is needed, not just during the critical period of crop–weed competition.
- Kill all the existing weeds before sowing with blanket spray of a non-selective herbicide.
- Ensure perfect levelling of the field for uniform sowing and irrigation.
- Ensure good initial crop stand and vigour, which in fact is the best weed management practice.
- Maintain optimum soil moisture at sowing – sow under residual moisture and apply light irrigation if needed.
- Uniform and adequate amounts of crop residues and other biomass as mulch to prevent light penetration.
- Placement of seed at the desired soil depth, and fertilizer close to the seed to make it less available to the weeds.
- Spray need-based herbicide as per the specific weed flora present in the field.

- Focus on minimization of the weed seed bank – prevent weed plants from flowering and setting weeds.
- Follow sporadic manual weeding for the control of left-over weeds after the main growth period of crop.
- Top dressing of N fertilizer should be made after weed control in cereal and other crops.
- Start CA with winter-season crops, when weed infestations are less and are also easy to control. After gaining experience, opt for summer- and rainy-season crops.
- Follow a cropping system approach, and include a cover crop, preferably a legume in the system.
- Follow CA in all crops in the sequence and dynamic rotations for maximum benefits.

Don'ts

- Do not follow ZT without mulch of crop residues or any other biomass.
- Do not broadcast basally applied fertilizer on the soil surface as it may benefit weeds.
- Do not use a conventional seed drill for sowing as it opens wide furrows and exposes the weed seeds.
- Do not keep land fallow or uncovered at any time of the year as weeds will proliferate.
- Do not allow the perennial weeds to establish – keep a check and nip them in the bud.
- Do not break the CA cycle as benefits multiply and weed infestations decrease over successive cropping cycles.

Conclusion and Future Outlook

The adoption of CA alters weed dynamics and communities, and thereby entails modifying weed control methods. ZT systems allow weed seeds to concentrate near the soil surface, where they have a greater probability of germinating but are also exposed to higher mortality risks through weather variability and predation. Considering no seed input into the system, germinable seed banks under ZT reduce more speedily than under CT. Reducing tillage causes the shifting of weed communities from annual dicots to grassy annuals and perennials. Surface residues moderate soil temperatures, thereby altering the emergence of both crops and weeds. Crop rotation affects weeds through allelopathy and altered timing of both crop management and resource demands. Crop rotations also aim at the use of different herbicides, preferably with different modes of action to overcome problems of herbicide resistance in weeds. The literature indicates that implementing ZT without crop rotation can result in severe weed problems and that greater rotational crop diversity results in easier weed management. Weed management in CA involves herbicide usage; however additional cultural practices must include: (i) selecting highly competitive cultivars, (ii) adjusting planting dates, (iii) farm hygiene to prevent the introduction of new weeds, (iv) adjusting planting geometry, densities, and fertilizer placement, and (v) microbial bio-controls. Further research is needed towards: (i) acquiring knowledge of the interactive effects of tillage and surface residue on weeds, (ii) the application of models and/or meta-analyses to anticipate weed responses, and to analyse intervention points in CA, and (iii) the weed-suppressive potential of longer crop rotations. Extensive research is needed concerning the interactions between CA practices for weed control, particularly tillage and residue retention.

CA offers a new paradigm for agricultural research and development, which is different from the conventional system. Weed management through the use of herbicides is promising but some tweaking is required in their selection and application, considering no-tillage and the presence of residues on the surface. HTCs complement CA but because of the anticipated concerns and controversies attached to them; this technology has not yet found approval in India. Even if it does gain approval in the near future, the development of herbicide resistance in weeds will be a major limitation and require integration of appropriate non-chemical methods of weed control for long-term sustainability.

Models are increasingly being used to explore cropping system scenarios and their predicted effects on weed populations. These could prove to be a valuable tool for investigating the effects of CA in various environments. Developing a standardized template for data collection could aid in performing meta-analyses, which could offer further insights into weed responses to CA adoption. Exploring the weed-suppressive potential of stacked and long-term crop rotations is another promising area that has received little attention in India.

References

Benech-Arnold, R.L., Sánchez, R.A., Forcella, F., et al. 2000. Environmental control of dormancy in weed seed banks in soil. *Field Crops Research* 67: 105–122.

Chauhan, B.S. and Johnson, D.E. 2010. The role of seed ecology in improving weed management strategies in the tropics. *Advances in Agronomy* 105: 221–262.

Chauhan, B. S. and Johnson, D.E. 2011. Row spacing and weed control timing affect yield of aerobic rice. *Field Crops Research* 121: 226–231.

Chauhan, B.S., Migo, T., Westerman, P.R., et al. 2010. Post-dispersal predation of weed seeds in rice fields. *Weed Research* 50(6): 553–560.

Chauhan, B.S., Singh, R.G. and Mahajan, G. 2012. Ecology and management of weeds under conservation agriculture: a review. *Crop Protection* 38: 57–65.

Chhokar, R.S., Malik, R.K. and Balyan, R.S. 1999. Effect of moisture stress and seeding depth on germination of littleseed canary grass (*Phalaris minor* Retz.). *Indian Journal of Weed Science* 31: 78–79.

Chhokar, R.S, Sharma, R.K., Jat, G.R., et al. 2007. Effect of tillage and herbicides on weeds and productivity of wheat under rice–wheat growing system. *Crop Protection* 26: 1689–1696.

Chhokar, R.S., Sharma, R.K., Singh, R.K., et al. 2008. Herbicide resistance in little seed canary grass (*Phalaris minor*) and its management, pp. 106. Proceedings of the 14th Australian Agronomy Conference. Adelaide, South Australia.

Chhokar, R.S., Singh, S., Sharma, R.K., et al. 2009. Influence of straw management on *Phalaris minor* control. *Indian Journal of Weed Science* 41: 150–156.

Das, T.K., Bhattacharyya, R., Sharma, A.R., et al. 2013. Impacts of conservation agriculture on total soil organic carbon retention potential under an irrigated agro-ecosystem of the western Indo-Gangetic Plains. *European Journal of Agronomy* 51: 34–42.

Das, T.K., Nath, C.P., Das, S., et al. 2020. Conservation agriculture in rice-mustard cropping system for five years: Impacts on crop productivity, profitability, water-use efficiency, and soil properties. *Field Crops Research* 250: 107781.

Das, T.K. and Yaduraju, N.T. 2007. Effect of irrigation and nitrogen levels on grassy weed competition in wheat and comparative eco-physiology of *Phalaris minor* Retz. and *Avena sterilis* ssp. *ludoviciana* Dur. in wheat. *Indian Journal of Weed Science* 39(3&4): 178–184.

Erenstein, O. and Laxmi, V. 2008. Zero tillage impacts in India's rice–wheat systems: a review. *Soil and Tillage Research* 100: 1–14.

FAO. 2017. FAO CA website at: www.fao.org/ag/ca.

Franke, A., Singh, S., McRoberts, N., et al. 2007. *Phalaris minor* seedbank studies: longevity, seedling emergence and seed production as affected by tillage regime. *Weed Research* 47: 73–83.

Gallandt, E.R., Fuerst, E.P. and Kennedy, A.C. 2004. Effect of tillage, fungicide seed treatment, and soil fumigation on seed bank dynamics of wild oat (*Avena fatua*). *Weed Science* 52: 597–604.

Gathala, M.K., Ladha, J.K., Kumar, V., et al. 2011. Tillage and crop establishment affects sustainability of South Asian rice-wheat system. *Agronomy Journal* 103: 661–672.

Gupta, R. and Seth, A. 2007. A review of resource conserving technologies for sustainable management of the rice–wheat cropping systems of the Indo-Gangetic Plains (IGP). *Crop Protection* 26: 436–447.

Hajebi, A., Das, T.K., Arora, A., et al. 2016. Herbicides tank-mixes effects on weeds and productivity and profitability of chilli (*Capsicum annuum* L.) under conventional and zero tillage. *Scientia Horticulturae* 198: 191–196.

ISAAA. 2019. *Global Status of Commercialized Biotech / GM Crops*. International Service for the Acquisition of Agri-biotech Applications (ISAAA). Brief No. 55. Ithaca, New York.

Kumar, V., Bellinder, R.R., Gupta, R.K., et al. 2008. Role of herbicide resistant rice in promoting resource conservation technologies in rice–wheat cropping systems in India: a review. *Crop Protection* 27: 290–301.

Kumar, V. and Ladha, J.K. 2011. Direct-seeding of rice: recent developments and future research needs. *Advances in Agronomy* 111: 297–413.

Liebman, M, Mohler, C.L; and Staver, C.P. (Eds.). 2001. Ecological Management of Agricultural Weeds. Cambridge University Press, U.K.

Mahajan, G. and Brar, L.S. 2002. Integrated management of *Phalaris minor* in wheat: rationale and approaches - a review. *Agricultural Reviews* 23: 241–251.

Mahajan, G. and Chauhan, B.S. 2011. Effects of planting pattern and cultivar on weed and crop growth in aerobic rice system. *Weed Technology* 25: 521–525.

Mahajan, G. and Chauhan, B.S. 2013. The role of cultivars in managing weeds in dry-seeded rice production systems. *Crop Protection* 49: 52–57.

Malik, R.K., Kumar, V., Yadav, A., et al. 2014. Conservation agriculture and weed management in south Asia: Perspective and development. *Indian Journal of Weed Science* 46(1): 31–36.

Malik, R., Yadav, A., Singh, S., et al. 2002. Herbicide resistance management and evolution of zero tillage - a success story. Research Bulletin, pp. 1–43. Haryana Agricultural University, Hissar, India.

Muoni, T., Rusinamhodzi, L., Rugare, J. T., et al. 2014. Effect of herbicide application on weed flora under conservation agriculture in Zimbabwe. *Crop Protection* 66: 1–7.

Nichols, V., Verhulst, N., Cox, R., et al. 2015. Weed dynamics and conservation agriculture principles: A review. *Field Crops Research* 183: 56–68.

Rao, A.N., Johnson, D.E., Sivaprasad, B., et al. 2007. Weed management in direct-seeded rice. *Advances in Agronomy* 93: 153–255.

Sharma, R.K., Chhokar, R.S., Jat, M.L., et al. 2008. Direct drilling of wheat into rice residues: experiences in Haryana and western Uttar Pradesh, pp. 147–158. In: Humphreys, E. and Roth, C.H. (Eds.) Permanent Beds and Rice-Residue Management for Rice–Wheat Systems in the Indo-Gangetic Plain. Canberra, Australia.

Singh, S., Chhokar, R.S., Gopal, R., et al. 2009. Integrated weed management: a key to success for direct-seeded rice in the Indo-Gangetic plains, pp. 261–278. In: Ladha, J.K. et al. (Eds.). Integrated Crop and Resource Management in the Rice–Wheat System of South Asia. IRRI, Los Banos, Manila, Philippines.

Singh, S., Kirkwood, R.C. and Marshall, G. 1999. Biology and control of *Phalaris minor* Retz. in wheat. *Crop Protection* 18(1): 1–16.

Singh, S., Ladha, J.K., Gupta, R.K., et al. 2007. Evaluation of mulching, intercropping with sesbania and herbicide use for weed management in dry-seeded rice (*Oryza sativa* L.). *Crop Protection* 26: 518–524.

Susha, V.S., Das, T.K., Nath, C.P., et al. 2018. Impacts of tillage and herbicide mixture on weed interference, agronomic productivity and profitability of a maize–wheat system in the north-western Indo-Gangetic Plains. *Field Crops Research* 219: 180–191.

Yaduraju, N.T., Sharma, A.R. and Das, T.K. (Eds.). 2016. *Weed Science and Management*, 402 p. Indian Society of Weed Science, Jabalpur and Indian Society of Agronomy, New Delhi.

8 Crop Residue Management in Conservation Agriculture

Yadvinder-Singh, H.S. Sidhu, M.L. Jat, Naveen Gupta, C.M. Parihar, and H.S. Jat

Introduction

India has achieved record foodgrain production of over 300 million tonnes (M t), and hence, crop residue production has also increased proportionately. Crop residues are those parts of the plants left in the field after crops have been harvested and threshed. Rice–wheat (RW), rice–rice (RR), pearlmillet–wheat (PmW), soybean–wheat (SW), maize–wheat (MW), and cotton–wheat (CW) are the major cereal-based production systems of India, occupying areas of over 10.5, 5.9, 2.3, 2.2, 1.9, and 1.5 million ha (M ha), respectively. The two northwestern (NW) states of Punjab and Haryana constitute a highly productive RW zone contributing about 69% of the total foodgrain output of the country (about 84% wheat and 54% rice).Unsurprisingly, this region is known as the 'food bowl of India'. Tillage-based agriculture and removal or burning of crop residues in India faces significant challenges to meet the food production requirement without significantly increasing the area under cultivation and degrading the environment. In recent years, another challenge to sustainable food production has remained in part due to climate change. There is a growing interest in conservation agriculture (CA)-based management practices, which can increase soil organic carbon (SOC) to enable sustainably increasing food production in South Asia (Jat et al. 2014).

Crop residue is a primary substrate for SOC accrual and improvement of soil productivity. Crop residue management can be considered as an intersection of two factors: the use of crop residues (retention or export) and the type of tillage applied. Crop residue management influences soil resilience, agronomic productivity, and greenhouse gases (GHG) emissions by: (i) aiding nutrient cycling, (ii) intercepting raindrops, thereby allowing water to gently percolate into the soil, (iii) lowering soil evaporation, (iv) improving soil structure, (v) reducing run-off and erosion, and (vi) increasing tolerance to biotic and biotic stresses (Jat et al. 2014; Yadvinder-Singh and Sidhu 2014).

The chosen residue management strategy affects crop yields and soil quality indicators (physical, chemical, and biological) including soil organic matter content (total quantity and distribution within the soil profile). Several excellent reviews have examined the effects of crop residue management (mainly

DOI: 10.4324/9781003292487-10

incorporation in soil) on crop production and soil quality, while little quantitative information exists regarding the response of crop yield to crop residue retention in CA and its influencing factors in India (Yadvinder-Singh et al. 2005; Bijay-Singh et al. 2008). The influence of crop residue management on crop production remains unclear; in some cases, crop production is enhanced by residue retention, but in others crop residues can reduce crop yield. This chapter summarizes the state of knowledge on the effects of residue management practices in relation to CA on crop yields and soil quality parameters in important cropping systems of India.

Crop Residue Availability in India

The quantity of crop residues produced depends on two main factors: crop yield and crop type. It is important to note that crop residues are not only the above-ground part not harvested for crop production, but also the below-ground parts. Root systems are crop residues consistently incorporated into the soil. According to an estimate, a gross quantity of 686 M t crop residues is available in India on an annual basis, generated by 26 crops. Of the total residue produced in India, cereals (rice, wheat, maize, pearlmillet, barley, sorghum, small millets) contribute the highest amount of 368 M t (54%), followed by sugarcane 111 M t (16%). At the individual crop level, rice contributes the highest amount of 154 M t residues, followed by wheat (131 M t). Gross residue potential is the total amount of residue produced, while surplus residue potential is the residue left after any competing uses such as cattle feed, animal bedding, heating and cooking fuel, and organic fertilizer. In NW India, about 23 M t of rice residues produced in rice-based cropping systems have been considered a nuisance by farmers and disposed through burning in fields (NAAS 2017). About 25% (a total of about 16 M t) of wheat residues left in the field after their collection for fodder are still burnt by the farmers for no convincing reason.

Crop Residues as a Source of Plant Nutrients

Crop residues are good sources of plant nutrients and are the primary source of organic matter (as C constitutes over 40% of the total dry biomass) added to the soil. About 30–40% of N, 25–35% of P, 70–85% of K, and 35–45% of S absorbed by cereals remain in the vegetative parts at maturity (Yadvinder-Singh et al. 2014). They supply essential plant nutrients upon mineralization and improve soil biophysical conditions. For example, rice straw contains 5–8 kg N, 0.7–1.2 kg P, 12–17 kg K, 0.5–1.0 kg S, 3–4 kg Ca, and 1–3 kg per 1000 kg straw on a dry weight basis. The potassium concentration is generally higher (up to 25 kg per 1000 kg) in rice straw of north-western Indo-Gangetic plains (IGP) than that from other regions of India or other parts of the world (Yadvinder-Singh et al.2014). According to Goswami et al. (2020), rice crop residues on average contain 0.7% N, 0.23% P, and 1.75% K. Therefore, the amount of NPK contained in rice crop residues produced is about 22.13×10^6 t/year in Asia. Maize stover contains more

N and K than wheat straw. Besides N, P, and K, 1 tonne of rice and wheat residues contains about 9–11 kg S, 100 g Zn, 777 g Fe, and 745 g Mn. Thus, the amount of NPK contained in rice and wheat residues produced (197 M t) is about 4.1 Mt. However, the nutrient concentration in residues depends on the soil conditions, nutrient management, variety, and season.

Burning of Crop Residues

A large portion of unused crop residues are burnt in the fields primarily to clear the left-over straw and stubble after the harvest, as they interfere with the tillage and seeding operation of the succeeding crop. On an overall basis, 18–30% of crop residues are burnt in India, but for the IGP, the figure is much higher at 30–40%. In the NW states comprising Punjab, Uttar Pradesh, and Haryana, nearly 23 M t of rice residues are burned *in situ* annually, leading to a loss of about 9.2 M t of C equivalent to a CO_2 load (NAAS 2017).

The problems of open-field burning of straw not only include air pollution (particulate matter and GHGs) and nutrient loss but it also has adverse effects on soil properties as well as on soil flora and fauna. One tonne of crop residue on burning releases 3 kg of particulate matter (PM2.5 and PM10), 1,515 kg CO_2, 92 kg CO, 3.83 kg NOx, 2 kg SO_2, 2.7 kg CH_4, 199 kg of ash, and 15.7 kg non-methane volatile organic compounds, causing severe atmospheric pollution (Gupta et al. 2004), which leads to health risks, aggravating asthma and chronic bronchitis, with decreased lung function, and corneal irritation and temporary blindness. Estimated emissions from open-field burning of crop residues assuming 25% of the available residue is burnt in the field in 2000 from open-field burning of rice and wheat straw in India were 110 Gg CH_4, 2306 Gg CO, 2.3 Gg N_2O, and 84 Gg NOx (Gupta et al. 2004). Every year, about 66,000 deaths attributable to particulate matter with an aerodynamic diameter of less than 2.5 micrometres (PM2.5) are associated with agricultural burning (HEI 2018).

Burning of rice residues causes loss of soil organic matter and plant nutrients, thereby adversely affecting soil health. About 90–100% of C and N, 60% of S, and 15–20% of P and K contained in rice residue are lost during burning. Burning of 23 M t of rice residues in NW India would lead to a loss of about 9.2 M t of C equivalent to a CO_2 load of about 34 M t per year and a loss of about 1.4×10^5 t of N annually.

Crop Residue Management with Reference to CA

Apart from burning of crop residues by farmers, the other options for crop residue management are: baling/removal for use as feed and bedding material for animals, *in situ* incorporation in the soil with tillage, and complete/partial retention on the surface as mulch using zero or reduced tillage systems. After baling, crop residues can also be used for paper and bioethanol production, mushroom cultivation, bioconversion, and engineering applications (Yadvinder-Singh et al. 2010a, 2014).

The use pattern of crop residues is not uniform amongst the different regions of India. In Uttarakhand and Himachal Pradesh, for example, the crop residues are mostly used as feed for livestock. The use of rice straw as animal feed is common among the rural households of Assam, West Bengal, Odisha, and Jharkhand, and to some extent in Bihar. However, rice straw is not preferred as a cattle feed in NW India due to its high silica content (Yadvinder-Singh et al. 2014).

The use of combines for harvesting rice and wheat in NW India is a common practice, leaving behind both standing as well as loose residues in the field, whose disposal or utilization in the short window of opportunity is very difficult. On-farm management through *in situ* incorporation, composting, and surface retention in the field are the best and most viable options to address the issue of burning as well as to maintain soil health and the long-term sustainability of the RW system. However, *in situ* incorporation and composting are energy- and cost-intensive. Furthermore, the time needed for decomposition of rice residue is a major limitation, because of the short turnaround time (20–25 days) available between rice harvesting and wheat sowing.

In recent times, the development of the 'turbo happy seeder' has provided the RW farmers of the IGP with the capability of direct drilling of wheat seeds into full rice residues. This technology for wheat sowing is a perfect climatic adaptation and mitigation strategy because it reduces GHG emissions, enables early sowing, reduces fuel consumption, weed population, and crop lodging, saves irrigation water, and increases the crop yields, particularly during high rainfall and/ or years with terminal heat stress (NAAS 2017). Keil et al. (2021) evaluated the profitability of residue management options for farmers to adopt no-burn technologies, especially the 'happy seeder' (HS), which is capable of sowing wheat directly into large amounts of crop residue. While we do not find any evidence of a yield penalty, our analysis reveals significant savings in wheat production costs, amounting to 136 USD per ha. Recent analysis by Keil et al. (2021) showed that the HS saves water and facilitates timely wheat sowing. They concluded that the economic benefits of HS use combined with its societal benefits of reducing air pollution and enhancing agricultural sustainability justify particular policy support for its large-scale diffusion, to be supplemented by a stricter enforcement of the ban on residue burning.

Crop Residue Decomposition and Nutrient Release

The degradation of crop residues varies depending on their lignin, polyphenol, and cellulose contents and their C:N ratio, which is crop dependent, but also on the environment and soil (texture, moisture) conditions (Yadvinder-Singh et al. 2005). Residues with a high C:N level (e.g., cereal residues having C:N ratios of 60:1 to 100:1) decompose slowly, resulting in the immobilization of soil N when incorporated in soil. This can have a positive influence in ZT systems, creating mulch that protects the soil from erosion and evaporation. Whether N is mineralized or immobilized upon incorporation of the residue depends on the

C:N ratio of the organic matter being decomposed by soil microorganisms. It has been suggested that the net immobilization phase during the initial years of ZT adoption is transitory, and the immobilization of N under CA systems in the longer term reduces the possibility of leaching and denitrification losses of soil mineral N. Some researchers argue that 15–25% extra available moisture during the growing season with ZT as compared to CT also favours N losses from the system through leaching and denitrification.

A number of approaches have been used to predict the C and N release from the organic materials in soil. A first-order kinetic equation has been extensively employed to describe crop residue decomposition. Using litter bags, Yadvinder-Singh et al. (2010b) showed that rice residue decomposition with time (temporal scale as growing degree days, GDD) can be described by a first-order exponential model. Decomposition was significantly affected by the method of residue placement and soil type. The buried rice residue lost about 80% of its initial mass by wheat harvest, leading to a decomposition rate (k) that was about 2.5 times as fast as that in the surface-placed residue. The effects of soil moisture and temperature on decomposition rate constants need to be considered to predict the decomposition of residue C under field conditions.

Crop Residue Management and Soil Quality

Depletion of the inherent capacity of soil in sustaining crop production is the greatest challenge for improving food security and sustainability of agriculture in India. Crop residues when applied to soil have a significant effect on SOM, and the physical, chemical, and biological properties of soil (Bijay-Singh et al. 2008; Yadvinder-Singh and Sidhu 2014).

Good soil quality for crop growth also depends on the presence of water-stable aggregates while soil organic matter sustains many key soil functions by providing the energy, substrates, and biological diversity to support biological activity, which affects soil aggregation and water infiltration.

Soil Organic Matter and C Sequestration

Soil organic C (SOC) has long been recognized as a key component of soil quality. Understanding SOC and N dynamics in soil leading to changes in their forms and contents under different management practices and agro-ecologies has always been and remains a subject of great interest. The magnitude of increase in SOC in CA depends on the extent of the reduction in tillage (e.g. ZT) and also on the amount of crop residues retained on the soil surface (Yadvinder-Singh et al. 2014). On the basis of changes in soil C values and the amount of C applied, 12–25% of applied rice straw-C (incorporated into the soil) was sequestered by the loamy sand soil after 3–7 years (Yadvinder-Singh and Sidhu 2014). Rates of C sequestration are highly influenced by soil type and climate. Numerous calculations have been made of the amount of residues needed to maintain organic matter at a particular level. Since most of the agricultural soils of India are low in SOC, significant potential for C sequestration is expected.

In the eastern IGP of India, SOC concentrations were measured after seven years of RW rotation under various combinations of tillage, crop residue management, and crop establishment methods (Sapkota et al. 2017). About 50% rice and 25% wheat residues were recycled in +R treatments. The total carbon input from above-ground residues was ca. 14.5 t/ha in ZTDSR–ZTW+R, which is almost six-fold greater than in the ZTDSR–ZTW-R system. After seven years, ZTDSR/ZTW+R increased SOC at 0–60 cm depth by 4.7 t C/ha, whereas the CTR–CTW system resulted in a decrease in SOC of 0.9 t C/ha (Sapkota et al. 2017). Over the same soil depth, ZT without residue retention (ZTDSR–ZTW–R) only increased SOC by 1.1 t C/ha. Variations in crops, climate, and soils are probably the major reasons for the range of values reported in different studies. The increase in SOC in ZT systems compared with the other treatments could be due to: (i) surface retention of crop residues (or stubbles in the case of no residue), (ii) higher plant biomass production leading to large amounts of root residues left in the system, and (iii) a lower rate of organic matter decomposition due to minimum soil disturbance. The increase in SOC content in ZT compared to CT is also attributed to lower disruption of macro-aggregates, which protected SOC against oxidation (Gupta-Choudhury et al. 2014). Surface placement of crop residues coupled with ZT reduced the rate of decomposition and carbon loss due to limited contact of residue with soil and sub-optimal moisture content, thereby increasing SOC content (Yadvinder-Singh et al. 2010b).

Marked stratification of SOC in the surface layer under ZT compared with CT can be attributed to the application of crop residues and other biomass on the surface in ZT and incorporation within the plough layer in CT systems. Residue retention under ZT+R and PB+R increased SOC stocks (by about three-fold) compared with no residue retention, which increased total SOC by 3.0–4.7 at 0–60 cm depth (Sapkota et al. 2017).

Choudhary et al. (2018a) reported that SOC was increased by 83% and 72% with CA-based rice–wheat–mungbean and maize–wheat–mungbean systems, respectively, compared to a conventional RW system (4.6 g/kg). In another study, Choudhary et al. (2018b) observed an increase in SOC stock by 2.0 t/ha and 1.2 t/ha in PB+R compared to PB in RW and MW systems after three years of cropping, respectively. Singh et al.(2016) reported that the SOC content after five years of using the rice-maize (RM) system was increased by 22% and 55% at 0–15 cm, and 0.29% and 15–30 cm in no residue (0.45% at 0–15 cm and 0.29% at 15–30 cm) treatment, respectively. Chatterjee et al. (2018) reported that the application of crop residue mulch (wheat straw 10 t/ha) significantly increased the SOC concentration by 14.9% at 0–5 cm soil depth compared to the no-mulch treatment. However, the effect of crop residue mulch was not significant on the SOC concentration at 5–15 cm soil depth.

Many other researchers have also recorded a significant increase in SOC with residue mulch under CA due to less soil disturbance and retention of crop residues after 2–5 years of study (Bera et al. 2018a). The effect of CA is more spectacular on the labile pools of SOC. In an irrigated RW system of eastern IGP, six years of double ZT with cereal residue retention alone (ZT-ZT+R) or along with brown manuring (ZT-ZT+R+BM) resulted in a significantly greater increase in labile SOC compared to the total SOC content in a 0–15 cm soil layer (Dey et al. 2018). From a review of the literature, Dhaliwal et al. (2020) concluded that different fractions of SOC, namely total SOC, particulate organic carbon, soil microbial biomass carbon, and potentially mineralizable carbon, were significantly increased by tillage and straw management systems. Mondal et al. (2021) reported a beneficial effect of CA on crop yields and 46 and 40% increases in SOC concentration and stock, respectively, under full CA over a conventional system in the 0–7.5 cm soil layer after 11 years of an RW system. There was enrichment in SOC content of aggregates under CA irrespective of the size class.

Effect on Soil Physical Properties

Crop residues as mulch in ZT systems play an important role in maintaining/ improving soil physical conditions, including soil moisture conservation and suppression of weeds. In most climates, the removal or burning of crop residues leads to a deterioration in the soil physical properties (Yadvinder-Singh et al. 2005; Yadvinder-Singh and Sidhu 2014).

Soil Erosion and Runoff Losses

Crop residues provide soil cover, which decreases runoff and soil loss, especially on low-incline slopes. Surface mulch is important for intercepting rainfall energy, maintaining structural integrity, increasing infiltration rates, and reducing runoff, thereby increasing water storage in the soil.

Infiltration Rate

Crop residues affect the hydraulic conductivity and infiltration rate (IR) by modifying soil structure, proportion of macropores, and aggregate stability. The retention of rice residue in wheat may help reduce the adverse effects of hard pan in the RW system and benefit the wheat crop (Yadvinder-Singh and Sidhu 2014). Marked increases in hydraulic conductivity and IR have been reported in treatments where crop residues were retained on the surface or incorporated by conventional tillage over treatments where residues were either burned or removed (Singh et al. 2016). After five years of an RM system, Singh et al. (2016) reported that crop residues increased mean steady-state IR by 24% compared to no residue treatment. The increase in IR in ZT+R plots was possibly due to an improvement in the aggregate stability, and pore size distribution and continuity. Some researchers (Gathala et al. 2011; Jat et al. 2013) have reported higher IR (initial as well as steady state) under ZT and residue retention compared to CT and removal of residues.

Soil Aggregation

The soil structure quality greatly depends on the SOC content, especially on the fraction of labile SOC (also called the 'particulate organic matter') because this fraction cycles relatively quickly in the soil. After five years of the RM system, residue retention increased the water stable aggregates (WSA) of >2 mm and 0.25–2.0 mm size by 23% and 10.1% over residue removal, respectively (Singh et al., 2016). The proportion of smaller size (< 0.053 mm) WSA was, however, lower for +R compared to –R treatment. ZT with residue retention increased WSA as well as mean weight diameter (MWD) compared to ZT-R. Gupta-Choudhury et al. (2014) showed that ZT with residue cover had higher aggregate stability, aggregate size values, and total organic carbon in soil aggregates than CT. Mondal et al. (2021) reported improvements in the macro-aggregate content, MWD and GMD of aggregates, and aggregation ratio in a CA system compared to a CT RW system. The impact of CA was mostly limited to the 0–7.5 cm soil layer and a maximum of up to 15 cm soil depth.

Soil Moisture, Temperature, and Hydraulic Conductivity

Crop residues placed on the soil surface reduce water loss by evaporation due to their mulching effect. Gupta et al. (2021) reported that rice straw mulch in a ZT rice–wheat system in the IGP of India reduced soil evaporation by around 45 mm. This is the one of the major advantages of CA, especially in low rainfall years. Residues insulate the soil surface and increase resistance to heat and vapour transfer, leading to increased available soil water. The beneficial effect of residue mulch on soil moisture and temperature changes can affect seed germination, seedling emergence, root growth, and N_2 fixation, which ultimately determine the growth and yield of crops. Studies have demonstrated that mulching in wheat has a significant positive effect on soil water conservation in ZT systems and the effect was more pronounced in dry periods (Bijay-Singh et al., 2008; Yadvinder-Singh et al., 2014). CA-based scenarios (ZT with surface residue retention) significantly influenced the soil water content in comparison to CT at both soil depths (Jat et al. 2018a). The volumetric water content at 0–15 cm soil depth was significantly increased by 7.0–8.9% under CA-based RW and MW systems. Gupta et al. (2016) reported that mulching of wheat with rice residue generally delayed the need for irrigation of wheat, and reduced irrigation input to wheat by 50 mm in some years, but the residual mulch had a negligible effect on the irrigation requirement for rice in a direct-seeded RW cropping system. The mulched soil retains higher water during unsaturated conditions. An increase in water infiltration can result in an increase in plant water availability.

Crop residues have low thermal conductivity and act as an insulating layer over the soil surface. Residue mulch is known to lower the maximum soil temperature but increases the minimum soil temperature due to the increased solar reflection and the insulating effect of residue cover. Furthermore, higher soil moisture observed in the surface layers under +R compared to –R also reduced the soil thermal regime (Singh et al. 2016). In an RM system, Singh et al. (2016) reported that in winter maize, the minimum soil temperature in December to February was 1.0–3.0°C higher in +R plots but the maximum temperature was lower by 2.1–7.1°C compared to the –R plots. In March and April, both minimum and maximum soil temperatures were lower by 1.2–4.7°C and 0.5–6.6°C, respectively, under +R compared to –R. An increase in minimum temperatures, when temperatures are below optimal, is very important and may allow better conditions for seedling emergence, and root and plant growth early in the maize cycle. Hydraulic conductivity at the soil surface generally decreases with residue removal or incorporation into the soil because of the destabilization of soil aggregates. There have been few studies on the influence of crop residue management on soil hydraulic properties under CA systems.

Soil Bulk Density, Compaction, and Penetration Resistance

Generally, soil compaction is quantified by one of four indicators: total porosity, pore size distribution, bulk density (BD), and penetration resistance. ZT with residue helps in improving soil aggregation and reducing BD (Gathala et al. 2011). Singh et al. (2016) reported a significant reduction in soil BD at 0–15 cm and 15–30 cm depths under crop residues mulch compared to residue removal after five years of an RM system. Similarly, Choudhary et al. (2018c) recorded a significant decrease in soil BD at 0–15 cm depth with crop residues recycling compared to residue removal after three years of RW and MW systems on a sandy clay loam soil. Lower BD with ZT with residue might be due to loose soil and more pore space created by residues.

Singh et al. (2016) reported a significant reduction in penetrometer resistance in the surface soil layer compared to no residue evaluated after five years of an irrigated RM system. Soil penetrometer resistance (SPR) followed a trend similar to that of BD. At 0–5 cm depth, mean SPR under +R compared to –R was reduced by 56% in conventional RM system plots (Singh et al., 2016). At 5–10 cm depth, +R decreased the SPR by 23–31% in conventional RM and ZTDSR/ZTM over –R. Choudhary et al. (2018c) reported lower soil BD in a CA-based permanent-bed maize–wheat system due to loose soil and more pore space created in the beds by accumulations of organic carbon on the top of raised beds due to huge quantities of residue retention for six crop seasons.

Root Growth

Root mass density was 6–49% greater in rice and 21–53% greater in maize under +R plots compared to –R in different soil layers down to 60 cm depth (Singh et al. 2016). Higher maize root growth under ZT +R was indicative of moderated mechanical resistance due to improved soil moisture content induced by residue mulch (Chakraborty et al. 2008). Higher root growth of crops under +R was indicative of reduced soil penetrometer resistance and improved soil aggregation.

Weed Growth

Surface mulch of previous crop residue along with ZT helps reduce weed infest-ation through its physical presence on the soil surface, preventing penetration of light and/or excluding certain wavelengths of light that are needed for weed seedlings to grow, reducing photosynthesis and causing allelopathy (Gupta-Choudhury et al. 2014; Yadvinder-Singh and Sidhu 2014). Sidhu et al. (2007) found that weed biomass was reduced by 60% with 7 t/ha of rice straw mulch 45 days after sowing wheat. However, levels of residue may influence the effect-iveness of herbicides by affecting the distribution of the herbicide and its contact with either weeds or the soil surface.

Crop Residues and Soil Chemical Properties

Electrical Conductivity and pH

Soil pH plays a major role in controlling nutrient availability to plants. the effects of crop residues on soil pH and EC are inconsistent. A three-year study by Choudhary et al. (2018c) showed that soil pH remained unchanged due to residue management in both RW and MW systems but the EC was significantly reduced by ZT and residue retention. The residue incorporation in CT had a greater influ-ence on EC than residue retention in ZT to maintain lower EC. Several other studies (Singh et al. 2016; Bera et al. 2018a) have shown no effect of crop residue mulch on pH and EC of soil. However, in a partially reclaimed sodic soil, Jat et al. (2018a) recorded a significant decrease in soil pH due to straw management in an RW system. Choudhary et al. (2018c) reported that lower soil pH was associated with treatments where higher amount of residues were recycled.

Nutrient Availability in Soil

Crop residues increase nutrient availability after their decomposition in cropping systems. *In situ* retention of crop residues, particularly cereal residues with high C:N ratios, may result in microbial N immobilization and a temporary decrease in crop-available N. However, surface application of crop residue is less likely to cause N immobilization, a common problem in soil that incorporates crop residues with high C:N ratio, rather improving soil water conservation and weed suppression (Yadvinder-Singh et al. 2014). Using a litter bag technique, Yadvinder-Singh et al. (2010b) calculated total N release from buried rice residue by the maximum tillering stage was about 6 kg N/ha (15% of initial) in sandy loam soil. The amount of N released from the buried residue increased to 12 kg/ha by the booting stage and to 26–28 kg/ha by maturity. In contrast, there was no release of N (rather N was immobilized) from the residues on the soil surface throughout the wheat growing season, with no N benefit to the growing wheat crop. With such small amounts of N released from the cereal residue, a benefit of savings in fertilizer N is unlikely over a short period. The above discussion suggests that N recommendations should be higher in CA systems than conventional systems, at least at the initial phases of establishment of a continuous CA system until new steady-state equilibrium between immobilization and mineralization is reached at a later phase and supply of N from the labile organic pool increases.

Goswami et al. (2020) reported that in a rice–rice system early residue incorporation improved the congruence between soil N supply and crop demand, although the size of this effect was influenced by the amount and quality of incorporated residue. Grain yields were 13–20% greater with early compared to late residue incorporation system without applied N or with moderate rates of N. There was 11–12% more C sequestration and 5–12% more N accumulation in a rice–rice system than rice–maize rotation and a greater amount was sequestered within N-fertilized treatments.

The phosphorus content of crop residues and its availability to subsequent crops can range from agronomically insignificant, to quantities in excess of crop P requirement. Crop residues with low P concentration, such as cereal stubbles, due to re-translocation of a large proportion of stubble P into grain, will not make an agronomically significant contribution to soil P availability, but may reduce P availability due to assimilation in the microbial biomass.

The contribution of crop residues to the P nutrition of subsequent crops has not been widely studied in India. However, some studies showed higher P availability in the surface soil layers under ZT with residue retention as compared to CT (Gupta et al. 2007; Singh et al. 2016). From a four-year study on an RW system, Gupta et al. (2007) reported that the retention of residues of both rice and wheat crops increased plant P uptake after three years. Crop residues increased both inorganic and organic P content in soil, reduced P sorption, and increased P release over straw that was burned.

Cereal residues are rich sources of K and release about 70% of K within 10 days after incorporation into the soil (Yadvinder-Singh et al. 2010b). Some authors have reported an increase in available K in soil continuously amended with rice residue but the increase was relatively small considering the total K added through the residues (Yadvinder-Singh et al. 2017). Adoption of CA in a rice–maize cropping system with residue management can be effective for improvement of system productivity, K use efficiency, and apparent K balance (Singh et al. 2018). An improved understanding of how crop residue management affects K cycling and different pools of K in the soil is required. To obtain a more meaningful picture of the changes taking place in the different fractions of soil K after long-term recycling of organic materials in the root zone, soil layers deeper than 0–15 cm should be taken into account (Yadvinder-Singh et al. 2017).

About 50–80% of micronutrient cations (Zn, Fe, Cu, and Mn) taken up by rice and wheat crops can be recycled through incorporated residue. Crop residues influence the availability of micronutrients such as Zn and Fe in rice (Yadvinder-Singh et al. 2005; Gupta et al. 2007). Jat et al. (2018a) assessed the effect of CA on soil chemical (N, P, K, and micronutrients) properties of a partially reclaimed sodic soil after four years in NW India (Table 8.1). Available N was

Table 8.1 Effect of CA practices on chemical soil properties (0–15 cm) after four years on partially reclaimed soil

Treatment	SOC (g/kg)	Total N (%)	Available nutrients (kg/ha)		
			N	P	K
Conventional RW	4.5[b]	0.14[b]	117[c]	16[b]	183[c]
CA–RW–mungbean	7.5[a]	0.19[a]	156[b]	22[a]	236[b]
CA–MW–mungbean	7.7[a]	0.19[a]	197[a]	20[a]	318[a]

	pH	Available micronutrients (mg/kg)			
		Zn	Cu	Fe	Mn
Conventional RW	8.06[a]	4.8	2.7[a]	132[b]	813[c]
CA–RW–mungbean	7.84[ab]	9.2	2.7[a]	136[a]	986[a]
CA–MW–mungbean	7.60[b]	7.2	2.6[a]	88[c]	873[b]

Source: Jat et al. (2018a).

Note: Values in the column followed by same letter do not differ significantly at P<0.05.

33% and 68% higher at 0–15 cm depth in CA-based RW–mungbean and maize–wheat–mungbean systems, respectively, than a conventional RW system. DTPA extractable Zn and Mn were significantly higher under CA-based cereal systems compared to a conventional RW system. The CA improved soil properties and nutrient availability and has the potential to reduce external fertilizer inputs over the long term.

Soil Biological Properties

The role of soil organisms as primary agents of decomposition, energy flow, and nutrient cycling has become a subject of increased interest. Crop residues affect biomass activity and the composition of micro- and macro-organisms and their functions by affecting the supply of carbon and other nutrients, and the physico-chemical characteristics of the soil environment.

Soil Microbial Biomass

Soil microbial biomass (SMB) reflects the soil's ability to store and cycle nutrients (C, N, P, and S) and organic matter, and plays an important role in the physical stabilization of aggregates. Crop residues retention/incorporation promotes, while removal of residues decreases, SMB (Bera et al. 2018b; Choudhary et al. 2018c). Significant amounts of P and S are also cycled by the microbial biomass. Both SMB-C and SMB-N were highly influenced by tillage and residue treatments at the end of the third MW and RW cropping cycles (Table 8.2). Compared to CT and residue removal, ZT and residue mulch increased SMB-C by 29% and 56%, whereas SMB-N increased by 27% and 84%, respectively.

Table 8.2 Effect of cropping system, tillage, and residue management on soil biological properties after three years

Treatment	SMB-C (µg/g)	SMB-N (µg/g)	DHA	Total microbial population		
				Bacteria (cfu × 10^4 per g soil)	Fungi (cfu × 10^2 per g soil)	Actinomycetes (cfu × 10^4 per g soil)
Rice–wheat system						
CTRW-R	646^e	201^d	180^e	74.7^f	45.3^f	38.5^f
CTRW+R	1113^c	243^c	256^d	84.0^{cde}	56.8^{cd}	48.2^{cd}
ZTRW-R	890^d	219^d	196^e	79.3^{ef}	52.0^e	41.2^e
ZTRW+R	1182^c	364^c	260^d	86.7^{cd}	64.3^{bc}	50.8^{bc}
Maize–wheat system						
CTMW-R	895^d	244^c	219^e	81.6^{def}	54.3^{de}	45.8^d
CTMW+R	1500^b	590^b	453^b	94.5^{ab}	73.2^a	69.3^a
ZTMW-R	1278^c	416^c	313^c	88.8^{bc}	66.3^b	54.2^b
ZTMW+R	1990^a	729^a	558^a	96.2^a	77.3^a	71.0^a

Source: Choudhary et al. (2018c).

Note: Values in the column followed by same letter do not differ significantly at P<0.05.

Higher levels of microbial biomass under ZT with residue mulch can be explained by greater availability of substrate to sustain the microbial biomass. Higher levels of soil SOC, SMB-C, SMB-N, and DHA were directly related to the surface accumulation of crop residues promoted by CA, residue quality, and greater C input via crop roots. Choudhary et al. (2018c) reported that SMB-C and SMB-N were 213% and 293% higher with CA over a conventional system (646 and 201 µg/g dry soil), respectively.

Microbial Population and Activity

Nutrient availability, soil aggregation, soil tilth, and decomposition of plant residues are governed by organic matter transformations. Residue incorporation into the soil always leads to increased bacterial and fungal activities. The addition of wheat straw significantly enhanced the population of asymbiotic N_2 fixers and nitrifying bacteria in both rhizosphere and non-rhizosphere soils. A recent study showed more soil fauna in ZT, residue-retained treatments compared with conventional tillage plots (Choudhary et al. 2018a). The mean numbers of bacteria, fungi, and actinomycetes were increased by 28%, 68%, and 98% respectively, under ZTMW+MB+R compared to CTRW-R. A higher microbial population is likely to be the result of improved food source availability supplied by residue amendment. Crop residues are also known to enhance N_2 fixation in soil by asymbiotic bacteria (*Azotobacter chrococcum* and *A. agilis*). Behera et al. (2021) recorded increases in DHA (54.7%), FDA (6.50%), alkaline phosphatases (4%), urease (4.80%), SOC (8.71%), and SMB-C (13.4%) with the addition of crop residues in ZT compared to conventional practice. They concluded that microbial population and diversity, and productivity of soil, could be maintained or improved with zero tillage and crop residue recycling.

Earthworms and micro-arthropods play a dominant role in organic matter decomposition and nutrient cycling associated with different crop residue management systems. ZT and residue retention improved the growth and multiplication of micro-arthropods, thereby protecting them from soil desiccation during summer (Choudhary et al. 2018b). Higher densities of micro-arthropods were reported under rice straw mulching than maize stover and other mulching.

Enzyme Activities

Different enzymatic activities such as nitrogenase, dehydrogenase, and phosphatase significantly increased (by 5–18%) under ZTW+R compared with ZTW-R and CTW-R after five years of the RW system (Table 8.3). Results from a three-year study by Choudhary et al. (2018c) showed that DHA and APA were improved by 210% and 49% under ZTMW-MB+R and 140% and 42% under ZTDSR/ZTW+R compared to a CTRW system, respectively. DHA and APA were associated with higher microbial activities including MBC and MBN through the release of organic substances thereby also creating a positive 'rhizosphere effect' on enzyme secretion in soil. Improvement of SOC, enzyme activity, MBC, and MBN as well as microbial population with CA-based management practices resulted in higher SQI values under maize-based systems in western IGP (Choudhary et al. 2018c).

Crop Residue Management and Productivity

The influence of residue management on crop productivity is complex and variable, and it results from direct and indirect effects and their interactions. Yield reduction is sometimes caused by unfavourable weather conditions (wet years) and the opposite is true in drier years. However, when crop residue retention enhances soil water conservation, it ultimately will produce better yields. The growth and yield of crops depend upon the weather. The results of a 2.5-year

Table 8.3 Soil biological properties in 0–7.5 cm layer after five years of a rice–wheat system

Treatment	MBC (µg/g)	Microbial respiration (µg CO_2/g/d)	DHA (µg/g/h)	APA (µg pNP/g/h)
CTW	159[c]	15.1[c]	6.85[c]	32.6[c]
ZTW	212[b]	17.9[b]	7.19[b]	38.3[b]
ZTW+R	224[a]	22.2[a]	8.97[a]	42.4[a]

Source: Bera et al. (2018a, b).

Note: Values in the column followed by same letter do not differ significantly at $P<0.05$.

study reported that growth and yields of both wheat and direct-seeded rice declined over successive seasons due to difference in seasonal conditions (Gupta et al. 2016). However, surface retention of rice residues improved the growth of ZT wheat right from the first crop, but the residual mulch did not help in improving the growth and yield of direct-seeded rice. A significant yield reduction in ZT wheat was observed when residues were removed (–R) over CT but the surface retention of 100% rice residues significantly increased the yield of ZT wheat by 11–30% over ZTW in a RW system on a sandy loam soil (Table 8.4). Furthermore, ZTW+R out-yielded CT wheat in the third year of the study and thereafter by 6.4–10.3%. Another study on the RW system in eastern India (Jat et al. 2014) showed that the benefits of ZTW + R in wheat were clear right from the second year and these benefits continued increasing over time during the seven years of the study and were possibly due to the cumulative effects of crop residues in reducing the adverse impact of terminal heat stress during the reproductive phase (February and March), providing optimum soil moisture and thermal regimes and improving soil quality. Choudhary et al. (2018a) recorded a 39% higher system yield of CA-based (with residue retention) rice–wheat–mungbean and maize–wheat–mungbean compared to a conventional RW system.

Rashid et al. (2019) reported a significant increase in yields of all the crops in a rice–maize–mungbean system with residue retention compared with no residue retention in the eastern IGP. The results from another three-year study from the eastern IGP of South Asia showed that strip tillage with high residue load is a potential crop management approach for the seasonally flooded rice–lentil/wheat–jute cropping systems to enhance soil nutrient status, crop yield, and farm economy (Salahin et al. 2021). Mondal et al. (2021) observed that the yield of rice in CA was comparable to or higher than that in CT, whereas the system rice equivalent yield was always higher (38–53%) under CA than under the conventional practices in an RW system. Therefore, a CA-based cropping system must be encouraged, to increase SOC status, improve aggregation stability and, consequently, sustain or increase system productivity, in order to achieve food and nutritional security in the eastern IGP of India.

Jat et al. (2018b) reported that residue mulch in PB provided favourable soil moisture and temperature conditions for better crop growth, resulting in higher grain yields of maize and wheat. This is consistent with observations made by other researchers (Jat et al. 2013; Parihar et al. 2016, 2017) who showed yield advantages of residue mulch on maize and wheat under irrigated conditions. The highest system grain yield in PB+MB was due to the combined effect of higher maize and wheat yields supplemented with additional yield provided by mungbean (Parihar et al. 2016, 2017). Mean MW system productivity (averaged across three years) under permanent beds + residue retention was significantly increased by 7.2% compared to CT.

The results of a six-year study clearly showed the positive effects of ZT and PB, and residue retention on grain yields of different maize-based cropping systems (Parihar et al. 2016). In an MW system on a sandy clay loam soil, maize yield

Table 8.4 Effect of crop establishment methods in wheat on grain yield of wheat (t/ha) in a rice–wheat system

Treatment	2010–11	2011–12	2012–13	2013–14
CTW	4.87[a]	5.32[a]	4.75[b]	5.30[b]
ZTW	3.71[b]	4.33[c]	4.46[c]	5.09[c]
ZTW+R	4.82[a]	5.51[a]	5.24[a]	5.64[a]

Source: Yadvinder-Singh and Sidhu (2014).

Note: Values in the column followed by same letter do not differ significantly at P<0.05.

in the permanent broad-bed plots with residue (PBB+R) and narrow-bed with residue (PNB +R) were 28% and 15% higher than in CT plots (Chakraborty et al. 2010).

Based on the results of a five-year study on the RM system, Singh et al. (2016) reported downward trends in yield of DSR in all crop establishment methods with a greater decline in –R compared to +R treatments. However, ZT maize yields showed a significant increasing trend after ZTDSR when the crop residues were retained. On average, residue retention increased the total yields of an RM system by 4.3% compared to when the residues were removed. Positive yield trends for ZT and residue retention were due to the improvement in soil quality parameters over time. Jat et al. (2014) also reported similar yield trends in wheat under CA-based management (ZT+R) in RWS. Choudhary et al. (2017) evaluated CA-based practices on a pearlmillet–mustard production system. ZT with 4 t/ha crop residue increased the grain yields of pearlmillet and mustard by 22.3% and 24.5%, respectively, in comparison to CT without residue. Although the positive effects of crop residue management on yield have been reported in several studies, some studies have reported no or a small effect of crop residue management on crop production over a short period of time depending on the residue load and soil type (Sidhu et al. 2015).

Crop Residue Management and GHG Emission

The short-term effects of cereal residues (wheat straw) incorporation into rice field include stimulation of CH_4 emissions (Bijay-Singh et al. 2008). However, incorporation of wheat straw before transplanting of rice may reduce N_2O emissions due to immobilization of mineral N. Applying crop residues on the soil surface may have no significant effect on the seasonal CH_4 and N_2O emissions. From the perspective of mitigating GHG emissions from upland crops (e.g. wheat), residues are not the primary crop management concern. When soil is at or near field capacity, there would be little CH_4 formation and N_2O emission, and the effect of crop residue mulch would be negligible. The global warming potential of conventional till wheat with *ad hoc* nutrient management was significantly higher than in ZT+residue mulch with precision nutrient management (Sapkota et al. 2014).

Crop residues under CA have the potential to slow/reverse the rate of emissions of CO_2 and other greenhouse gases such as methane and nitrous oxides by agriculture, by reducing tillage and residue burning and improving N-use efficiency. The effectiveness of these practices depends on factors such as the climate, soil type, input resources, and farming system. Understanding the fluxes of GHG under different tillage, residue, and nutrient management practices, their carbon equivalent, and net carbon storage are important in policy initiatives and climate change mitigation.

Constraints in Crop Residue Management

A comprehensive appraisal of the benefits and constraints related to crop residue management has been explored (NAAS 2017). Major constraints to successful crop residue management in CA systems are related to the appropriate machinery and other alternative economic uses of crop residues such as livestock feed, fuel, bedding during the rainy and winter seasons, and thatching for some farming households. Crop residues are needed to provide livestock feed during the dry season when feed is severely limited while manure is needed for crop production. The yield benefits derived from manure, whose quantity and quality partly depend on crop harvest residues, suggest that farmers face tradeoffs in crop residue management and it might be beneficial for them to follow the manure production pathway rather than apply crop residues as mulch. Livestock play a crucial role in many smallholder farming systems where animals contribute to food security, provide draft power, and add to capital. The importance of livestock in mixed farming systems and the competing use of residues for livestock feed are a major disincentive for CA adoption by farmers. Therefore, it is necessary to establish a compromise between crop residues used for retention on the field and for feed. Substituting cereal residues with nutrient-rich feed for animal diets or increasing total residue production so that a portion of the residue can be returned to the field without compromising livestock feed may incentivize CA adoption.

Conclusion and Future Outlook

Large amounts of surplus residues of different crops are available for *in situ* recycling for CA in India. Recycling crop residues containing considerable quantities of nutrients can help reduce the use of fertilizers. The importance of crop residue retention to the sustainability of crop production is widely acknowledged. Crop residue management practices, and the quality and quantity of crop residues returned, affect soil fertility through a series of chemical, physical, and biological changes in the soil. Retention of crop residues on the soil surface as in CA systems not only reduces runoff and soil erosion but also improves soil physical characteristics (such as hydraulic properties and soil aggregation) and increases SOM content, especially in the surface layer. Additionally, increases in soil microbial biomass and activity following crop residues addition improve the nutrient-supplying capacity of soil and reduce nutrient losses. Decomposition of crop residue is slower for surface retention compared with incorporation into the soil by CT, thereby making nutrients less prone to losses (via leaching, volatilization, and denitrification).

Crop residues provide elemental C and nutrient recycling in soil. On-farm management of crop residues offers several environmental and ecological benefits for the soil–water–plant system and in meeting the nutrient requirements of crops. The use of crop residue as mulch has been found to be beneficial as it reduces the maximum soil temperature and conserves water. Recent developments in machinery (turbo happy seeder) allow ZT sowing of wheat with rice residue as surface mulch while maintaining yield, reducing tillage costs, saving time, and avoiding the need for burning. Crop residue management helps in maintaining the soil moisture content by protecting the soil surface and increasing irrigation efficiency. Soils with higher organic matter content tend to have higher aggregate stability and therefore less risk of compaction and soil erosion. The microbial and macro-faunal (earthworm) populations and their activity increase in mulched plots. Generally, retention of crop residues increases soil porosity and reduces soil bulk density, regardless of tillage operations. In addition, crop residues provide multiple ecosystem services to the environment when retained in the field after harvest.

There is a need to develop a complete package of practices (fertilizer, irrigation, weed control, pest management, etc.) under crop residue management in major cropping systems. Support of on-farm adaptation of crop residue management technologies in both large and small fields, and developing focused institutional and policy support including appropriate incentives are needed for the widespread dissemination and adoption of crop residue management practices. In view of the often sub-optimal soil moisture levels in rainfed environments, it would be particularly instructive to quantify the effects of CA components on soil moisture availability to crops. The long-term impact of a variable quantity of rice straw in wheat and wheat straw in dry-seeded rice on the productivity, root mass density, and weed density needs to be studied in a systematic manner. Systematic long-term studies on the impact of tillage and residue mulch on different quality parameters, and the development of soil quality index under contrasting cropping systems, soil types, and agro-ecological environment need to be carried out. Evaluating SOC dynamics of different cereal-based systems under present and projected climate change scenarios, crop residue management practices, and their potential impacts on agricultural system sustainability would substantially benefit producers and policy makers.

References

Behera, U.K., Singh, G., Kumar, A., et al. 2021. Long-term effects of tillage, crop residue and crop rotations on soil microbial parameters under the wheat (*Triticum aestivum*) based cropping systems in semi-arid Northern India, *Indian Journal of Pure and Applied Bioscience* 9: 219–235.

Bera, T., Sharma, S., Thind, H.S., et al. 2018a. Soil biochemical changes at different wheat growth stages in response to conservation agriculture practices in a rice-wheat system of north-western India. *Soil Research* 56: 91–104.

Bera, T., Sharma, S., Thind, H.S., et al. 2018b. Changes in soil biochemical indicators at different wheat growth stages under conservation-based sustainable intensification of rice-wheat system. *Journal of Integrative Agriculture* 17: 1871–1880.

Bijay-Singh, Shan, Y.H., Johnson-Beebout, S.E., et al. 2008. Crop residue management for lowland rice-based cropping systems in Asia. *Advances in Agronomy* 98: 117–199.

Chakraborty, D., Garg, R.N., Tomar, R.K., et al. 2010. Synthetic and organic mulching and N effect on winter wheat (*Triticum aestivum* L.) in a semi-arid environment. *Agricultural Water Management* 97: 738–748.

Chakraborty, D., Nagarajan, S., Aggarwal, P. et al. 2008. Effect of mulching on soil and plant water status, and the growth and yield of wheat (*Triticum aestivum* L.) in a semi-arid environment. *Agricultural Water Management* 95: 1323–1334.

Chatterjee, S., Bandyopadhyay, K.K., Pradhan, S., et al. 2018. Effects of irrigation, crop residue mulch and nitrogen management in maize (*Zea mays* L.) on soil carbon pools in a sandy loam soil of Indo-Gangetic plain region.*Catena* 165: 207–216.

Choudhary, M., Datta, A. and Jat, H.S. 2018b. Changes in soil biology under conservation agriculture based sustainable intensification of cereal systems in Indo-Gangetic Plains. *Geoderma* 313: 193–204.

Choudhary, M., Jat, H.S., Datta, A., et al. 2018c. Sustainable intensification influences soil quality, biota, and productivity in cereal-based agroecosystems. *Applied Soil Ecology* 126: 189–198.

Choudhary, K.M., Jat, H.S., Nandal, D.P., et al. 2018a. Evaluating alternatives to rice-wheat system in western Indo-Gangetic Plains: crop yields, water productivity and economic profitability. *Field Crops Research* 218: 1–10.

Choudhary, M., Rana, K.S., Bana, R.S., et al. 2017. Energy budgeting and carbon footprint of pearl millet and mustard cropping system under conventional and conservation agriculture in rainfed semi-arid agro-ecosystem. *Energy* 141: 1052–1068.

Dey, A., Dwivedi, B.S., Meena, M.C., et al. 2018. Dynamics of soil carbon and nitrogen under conservation agriculture in rice-wheat cropping system. *Indian Journal of Fertilisers* 14: 12–26

Dhaliwal, S.S., Naresh. R. K., Gupta. R.K., et al. 2020. Effect of tillage and straw return on carbon footprints, soil organic carbon fractions and soil microbial community in different textured soils under rice–wheat rotation: a review. *Reviews in Environmental Science and Biotechnology* 19:103–115.

Gathala, M.K., Ladha, J.K., Saharawat, Y.S., et al. 2011. Effect of tillage and crop establishment methods on physical properties of a medium-textured soil under a seven-year rice–wheat rotation. *Soil Science Society of America Journal* 75: 1851–1862.

Goswami, S.B., Mondal, R. and Mandi, S.K. 2020. Crop residue management options in rice-rice system: A review. *Archives of Agronomy and Soil Science* 66:1218–1234.

Gupta, N., Humphreys, E., Eberbach, P.L., et al. 2021. Effects of tillage and mulch on soil evaporation in a dry seeded rice-wheat cropping system. *Soil and Tillage Research* 209:104976.

Gupta, P.K., Sahai, S., Singh, N., et al. 2004. Residue burning in rice-wheat cropping system: Causes and implications. *Current Science* 87 :1713–1715.

Gupta, N., Yadav, S., Humphreys, E., et al. 2016. Effects of tillage and mulch on the growth, yield and irrigation water productivity of a dry seeded rice-wheat cropping system in north-west India. *Field Crops Research* 196: 219–236.

Gupta, R.K., Yadvinder-Singh, Ladha, J.K., et al. 2007. Yield and phosphorus transformations in a rice-wheat system with crop residue and phosphorus management. *Soil Science Society of America Journal* 71: 1500–1507.

Gupta-Choudhury, S., Srivastava, S., Singh, R., et al. 2014. Tillage and residue management effects on soil aggregation, organic carbon dynamics and yield attribute in rice–wheat cropping system under reclaimed sodic soil. *Soil and Tillage Research* 136: 76–83.

HEI. 2018. Burden of disease attributable to major air pollution sources in India. *GBD MAPS Working Group Special Report* 21. Health Effects Institute (HEI).

Jat, H.S., Datta, A., Sharma, P.C., et al. 2018a. Assessing soil properties and nutrient availability under conservation agriculture practices in a reclaimed sodic soil in cereal-based systems of North-West India. *Archives of Agronomy and Soil Science* 64: 531–545.

Jat, M.L., Gathala, M.K. and Saharawat, Y.S. 2013. Double no-till and permanent raised beds in maize–wheat rotation of northwestern Indo-Gangetic plains of India: effects on crop yields, water productivity, profitability and soil. *Field Crops Research* 149: 291–299.

Jat, R.D., Jat, H.S., Nanwal, R.K., et al. 2018b. Conservation agriculture and precision nutrient management practices in maize-wheat system: Effects on crop and water productivity and economic profitability. *Field Crops Research* 222: 111–120.

Jat, R.K., Sapkota, T.B., Singh, R.G., et al. 2014. Seven years of conservation agriculture in a rice–wheat rotation of Eastern Gangetic Plains of South Asia: yield trends and economic profitability. *Field Crops Research* 164: 199–210.

Keil, A. Krishnapriya, P.P., Mitra, A., et al. 2021. Changing agricultural stubble burning practices in the Indo-Gangetic plains: is the Happy Seeder a profitable alternative? *International Journal of Agricultural Sustainability* 19: 128–151.

Mondal, S., Mishra, J.S., Poonia, S.P., et al. 2021. Can yield, soil C and aggregation be improved under long-term conservation agriculture in the eastern Indo-Gangetic plain of India? *European Journal of Soil Science* 72: 1742–1761.

NAAS. 2017. Innovative Viable Solution to Rice Residue Burning in Rice-Wheat Cropping System through Concurrent Use of Super Straw Management System-fitted Combines and Turbo Happy Seeder, 16 p. Policy Brief No. 2. National Academy of Agricultural Sciences, New Delhi.

Parihar, C.M., Jat, S.L., Singh, A.K., et al. 2016. Conservation agriculture in irrigated intensive maize-based systems of north-western India: Effects on crop yields, water productivity and economic profitability. *Field Crops Research* 193: 104–116.

Parihar, C.M., Jat, S.L. and Singh, A.K. 2017. Bio-energy, biomass water-use efficiency and economics of maize-wheat-mungbean system under precision-conservation agriculture in semi-arid agroecosystem. *Energy* 119:245–256.

Rashid, M.H.,Timsina, J., Islam, N. et al. 2019. Tillage and residue management effects on productivity, profitability and soil properties in a rice-maize-mungbeansystem in the Eastern Gangetic Plains. *Journal of Crop Improvement* 33: 683–710.

Salahin, N., Jahiruddin, M. Islam, M.R., et al. 2021. Establishment of crops under minimal soil disturbance and crop residue retention in rice-based cropping system: yield advantage, soil health improvement, and economic benefit. *Land* 10: 581.

Sapkota, T.B., Jat, R.K., Singh, R.G., et al. 2017. Soil organic carbon changes after seven years of conservation agriculture in a rice–wheat system of the eastern Indo-Gangetic Plains. *Soil Use and Management* 33: 81–89.

Sapkota, T.B., Majumdar, K., Jat, M.L., et al. 2014. Precision nutrient management in conservation agriculture based wheat production of northwest India: Profitability, nutrient use efficiency and environmental footprint. *Field Crops Research* 155: 233–244.

Sidhu, H.S., Manpreet-Singh, Humphreys, E., et al. 2007. The Happy Seeder enables direct drilling of wheat into rice stubble. *Australian Journal of Experimental Agriculture* 47: 844–854.

Sidhu, HS, Singh, M, Yadvinder-Singh, et al. 2015. Development and evaluation of the Turbo Happy Seeder for sowing wheat into heavy rice residues in NW India. *Field Crops Research* 184: 201–212.

Singh, V.K., Dwivedi, B.S., Yadvinder-Singh, et al. 2018. Effect of tillage and crop establishment, residue management and K fertilization on yield, K use efficiency and apparent K balance under rice-maize system in north-western India. *Field Crops Research* 224: 1–12.

Singh, V.K., Yadvinder-Singh, Dwivedi, B.S., et al. 2016. Soil physical properties, yield trends and economics after five years of conservation agriculture based rice-maize system in north-western India. *Soil and Tillage Research* 155: 133–148.

Yadvinder-Singh, Gupta, R.K., Singh, J., et al. 2010b. Placement effects on rice residue decomposition and nutrient dynamics on two soil types during wheat cropping in rice–wheat system in northwestern India. *Nutrient Cycling in Agroecosystems* 88: 471–480.

Yadvinder-Singh and Sidhu, H.S. 2014. Management of cereal crop residues for sustainable rice-wheat production system in the Indo-Gangetic Plains of India. *Proceedings of Indian National Science Academy* 80: 95–114.

Yadvinder-Singh, Sidhu, H.S., Jat, M.L., et al. 2017. Can recycling of potassium from crops and other organic residues be integrated into potassium rate recommendations? *Indian Journal of Fertilizers* 13: 60–66.

Yadvinder-Singh, Singh, M., Sidhu, H.S., et al. 2010a. Options for effective utilization of crop residues, p. 32. Directorate of Research, Punjab Agricultural University, Ludhiana, India.

Yadvinder-Singh, Singh, B. and Timsina, J. 2005. Crop residue management for nutrient cycling and improving soil productivity in rice-based cropping systems in the tropics. *Advances in Agronomy* 85: 269–407.

Yadvinder-Singh, Thind, H.S. and Sidhu, H.S. 2014. Management options for rice residues for sustainable productivity of rice-wheat cropping system. *Journal of Research* (PAU) 51: 239–245.

9 Machinery Development for Conservation Agriculture

Manpreet Singh, H.S. Sidhu, and Shiv Kumar Lohan

Introduction

Rice–rice, rice–wheat, cotton–wheat, rice–maize, maize–wheat, and sugarcane–wheat are the major cropping systems in irrigated ecologies of India. The rice–wheat (RW) system is responsible for phenomenal agricultural growth in north-west (NW) India. It is practiced on around 10 M ha across the Indo-Gangetic alluvial plains of NW India. With the increase in crop production in India, crop residue production has also increased proportionately. However, the sustainability of the conventional RW system is threatened by scarcities of water, energy, and labour, increasing costs of production, and air pollution due to the burning of surplus crop residues. In recent years, there has been increasing interest in the development of climate-resilient CA practices for important cropping systems of India to improve soil quality, save irrigation water, and increase nutrient-use efficiency, profitability, and crop productivity. CA is a method of crop production and soil management based on minimal tillage, leaving crop residues on the soil surface, and crop rotation. A much-needed revamping of whole-farm operations (e.g. seeding, irrigation water, fertilizer application, residue management, and herbicide application) is needed to effectively develop and promote technologies for CA in relation to site- and climate-specific conditions. The development of scale-appropriate machinery for different field operations (seeding, fertilizer management, residue management, irrigation, chemical applications) is urgently required for successful implementation of CA. Although efforts have been made in developing and promoting machinery for seeding wheat in zero-tillage systems, machinery for CA is yet to be developed and evaluated for a range of crops and cropping sequences.

Combine harvesting of rice and wheat predominates in NW India, with large amounts of loose and anchored crop residues left in the fields after harvest. Rice straw is considered to be a poor feed due to its high silica content and has no other economic use. The loose rice residues generated during combine harvesting hamper tillage and seeding operations for the subsequent wheat crop. Therefore, open-field burning of rice residue is the general practice in NW India. Substantial losses of plant nutrients (especially N and S) and organic C occur during this burning, with important implications for soil quality and human health (Singh

DOI: 10.4324/9781003292487-11

et al. 2010). *In situ* incorporation of rice residues is energy- and time-intensive and delays wheat sowing, thereby adversely affecting wheat yield (Singh et al. 2010). Therefore, in-field retention of crop residues could play an important role in replenishing soil nutrient stocks and organic matter, thus contributing to sustainable RW production systems under CA (Singh et al. 2003, 2005, 2014). Cost-effective management of crop residues is thus both a major challenge and a major opportunity for increasing the sustainability of the intensive RW system.

Relay seeding of different crops in wheat-based systems offers an excellent opportunity to improve crop productivity and farmers' income in NW India. In the cotton–wheat system, yields of wheat are markedly lower after cotton due to a delay in sowing compared to that after rice and maize. New machinery is needed to plant wheat into standing cotton for timely seeding of wheat. Similarly, integration of short-duration mungbean in wheat-based cropping systems will provide much needed protein for poorly nourished population, and enhance farmers' profits and improve soil health. However, sowing of mungbean after the harvesting of wheat gets delayed, leading to crop failure due to the overlapping of its maturity period with the onset of monsoons. Therefore, relay seeding of mungbean in standing wheat will help advance sowing so that it is harvested before the monsoon season.

Maize, an important crop for food and nutritional security in India, is grown in diverse ecologies and seasons. To meet the rising demand, a quantum jump in maize production is the need of the hour. Resource conservation technologies (RCTs) that include several practices, namely zero tillage (ZT), minimum tillage (MT), and surface seeding came into practice in various maize-based cropping systems and these are cost effective and environment friendly. A commonly available, inclined plate-metering mechanism can be attached to any ZT planter/ happy seeder to sow maize, whereas the happy seeder can be used to sow wheat into maize residues.

The degradation of natural resources, climate change, and sustainability of agriculture are the main driving forces for the adoption of CA in India. However, the small size of land holdings, poor economic condition of farmers, low seasonal use of machinery, irregular size and shape of fields, competition among machines and labour, and the mindset of farmers towards zero-till sowing of crops are the major constraints to the adoption of CA machinery in India.

Machinery for CA

Laser Land Leveller

Laser land levelling is a pre-requisite before the adoption of CA in any cropping system. For agricultural production in India, irrigation is the most crucial input because about 40% of cropland is irrigated, which accounts for 60–80% of food production (Singh et al. 2014). Water management by proper irrigation scheduling in combination with better crop management techniques (i.e. laser-assisted precision land levelling, CA-based management practices) are potential options for saving water and increasing water productivity. Results from a number of studies have indicated that laser land levelling (LLL) saved irrigation water to the tune of 20–25% and irrigation time by 30% and also improved the productivity of rice, wheat, and sugarcane by 10–15% (Jat et al. 2006; Sidhu et al. 2007; Jat et al. 2015). Latif et al. (2013) recorded lower weed intensity and about 11% lower herbicide cost in wheat on laser-levelled fields compared with TLL. The LLL system is also likely to enhance the cultivable area by 1.5–10.0% due to the removal of extra bunds and channels in the field over TLL (Sidhu et al. 2007; Jat et al. 2009b). Therefore, the use of a laser leveller is of utmost importance for the success of CA-based production systems.

Uneven fields and unlined irrigation channels cause the losses of a huge amount of irrigation water. Laser-assisted precision land levelling is the main step in the judicious use of irrigation water and enhancing water productivity. Precision land levelling is considered to be a precursor technology for CA and has been reported to improve crop yields and input use efficiency, including water and nutrients (Jat et al. 2006; Lohan et al. 2014). It is a laser-guided (light amplification by stimulated emission of radiation) precision levelling technique used for achieving very fine levelling with the desired grade on the field to within ±2 cm of its average micro-elevation. An automatic survey system has been found to be time- and resource-saving while also reducing cost, and thus proved more economical compared with a manual survey method (Table 9.1).

This study also showed that ASS reduced the earth work (EW) and LI by 72.1% and 72.7% compared with 63.3% and 63.6% with the manual survey method (MSM). ASS improved the LUC by 62.2% compared with 45.4% for the MSM. The differential indices (differential earth work – DEW; differential land uniformity coefficient – DLUC; differential levelling index – DLI) per unit fuel (l) and time (h) consumed were significantly ($P \leq 0.05$) better for ASS compared with MSM. The system field capacity for ASS was 53.0% higher and the total cost of the operation was 30.1% lower compared with MSM. In conclusion, both

Table 9.1 Comparative economics of an automatic survey system (ASS) and manual survey method (MSM)

Parameter	ASS	MSM	% change over MSM	P value
Survey diesel (l/ha)	0.3	–	–	–
Survey time (min/ha)	12.1±0.9	28.7±1.1	−57.9	<0.0001
Survey labour cost (Rs/ha)	7.5±1.1	36.2±1.3	−79.2	<0.0001
Total survey cost (Rs/ha)	22.4±1.6	36.2±1.3	−38.2	<0.0001
System capacity (ha/hr)	0.3±0.0	0.2±0.0	53.0	0.04
Laser unit cost (Rs/hr)	657.8	612.6	7.4	–
Laser cost (Rs/ha)	2121.1±196.7	3028.4±318.73	−30.0	0.03
Total system cost (Rs/ha)	2143.5±197.5	3064.6±319.12	−30.1	0.03

Source: Lohan et al. (2014).

quantitative and qualitative performance indicators were better for ASS compared to MSM. The laser leveller being a custom-operated machine, with the saving in time and labour benefiting the custom operator, while precision in levelling benefits the farmer. The use of less fuel for levelling also contributes towards less environmental pollution due to reduced greenhouse gases emissions. LLL has huge potential to improve and sustain agriculture production, ensure food security, save water, and reduce the environmental footprints of irrigated production systems in India. The laser leveller has been successfully adopted on an area of 1.75 M ha in Punjab state alone and more than 7850 laser units are in operation.

The cost of buying a laser leveller is quite high, and its limited use prevents individual farmers from owning the equipment. However, the use of laser levellers has become economically feasible and accessible through custom hiring services, even to lower-income farmers. The farm size affects the potential for achieving operational efficiency with the laser leveller to manifest its full potential.

Two-Wheel Tractor-Driven Laser Leveller

A normal laser leveller requires a 50-hp tractor for smooth operation in the field. Moreover, small-holding sizes and irregular shapes of the fields are hindrances to the economic use of four-wheel tractor-driven laser levellers in eastern parts of the IGP. Efforts have been made to design and develop laser levelling technology applicable for small plot sizes (e.g. units that can be mounted onto smaller or two-wheel tractors) in the eastern IGP of India. Keeping this in mind, a prototype of a two-wheel tractor laser leveller was developed by Borlaug Institute for South Asia (BISA), Ladhowal, Punjab, for marginal and small farmers in Asia.

CA Machinery for Rice–Wheat Rotation

Cost-effective management of heavy loads of crop residues is both a major challenge and opportunity for increasing the sustainability of the intensive RW system.

Planter for Direct Dry Seeding of Rice

Dry seeding of rice (DSR) refers to the process of establishing a rice crop from seeds sown under aerobic conditions rather than by transplanting seedlings from the nursery in puddled fields under waterlogged conditions.

A shift in rice production systems from puddled transplanted rice (PTR) to DSR is testimony of the resource conservation technologies (Gupta et al. 2006). The adoption of DSR cultivation as an alternative to PTR significantly decreases the cost of rice production (Jat et al. 2009a). Uncertainty of the availability of water/electricity early in the season is another reason for adoption of DSR for timely planting of rice in Punjab and other parts of India. In the absence of suitable seeding equipment, farmers often use a very high seeding rate for manual seeding of DSR or using inappropriate seed drills. This practice leads to poor yields of DSR because of the greater competition for sunlight and poor trans-location of photosynthates to reproductive parts. Thus, there was a dire need to develop a planter for DSR and other crops. A planter for DSR has been developed and is being promoted for adoption by farmers.

Functional Requirements of the DSR Planter

The machine for DSR should be able to maintain optimum plant to plant and row to row distances without mechanical seed injury at a seed rate of 15–20 kg/ha and a seeding depth of 2–3 cm. There are different seed metering inclined plates for different crops. The plates vary in the size of groove, number of grooves, and shape of the grooves. The size, number, and shape of the grooves are designed to suit specific crops. To change the plates, the nut in the centre of the plate is opened and then, after changing the plate, it is tightened again.

An additional inclined plate box can also be attached to the existing zero till drill as an alternative to buying a separate machine. This machine can be operated with a 35-hp tractor. Efforts are being made to develop an inclined plate metering mechanism attachment for a two-wheel tractor to increase its use and make it a multicrop and multifunctional machine.

The seed metering and delivery system of the planter consists of: (i) a seed box used to store the seed in the planter, and (ii) inclined rotary metering plates, which have grooves to guide the seed and drop it into the cups.

In India, the yields have been significantly lower (9.2–28.5%) in DSR than in PTR. Yields of bed-dry-DSR were lower by 29%, whereas those of DSR were lower by 9.2–10.3%. In the north-western IGP, there is a tendency for a yield penalty with DSR (Gathala et al. 2011a) but not in the eastern IGP (Singh et al. 2009b). A possible reason for this differential performance in north-western versus eastern IGP is the lower rainfall in the former (400–750 mm/year) than in the latter area (1000–1500 mm/year) (Gupta and Seth 2007). Flooding of rice after successful establishment can alleviate nutrient deficiencies (i.e. Fe and Zn) and soil-borne diseases (i.e. nematodes). Also in the eastern IGP, current yields of PTR are much lower than that in the north-western IGP; therefore, it is easier to achieve equivalent yield with DSR.

The causes of lower yield in DSR reported by researchers in different production zones may include (i) uneven or poor crop establishment, (ii) inadequate weed control, (iii) higher spikelet sterility than in puddled transplanting, (iv) higher crop lodging, especially in wet seeding and broadcasting, and (v) insufficient knowledge of water and nutrient management (micronutrient deficiencies).

Machinery for Direct Seeding of Wheat into Crop Residues

Turbo Happy Seeder

Minimum and zero-till (ZT) technologies for wheat seeding after removal or burning of rice residues are beneficial in terms of economics, irrigation water saving, and improved timeliness of wheat sowing in comparison with conventional tillage (Malik et al. 2004; Singh et al. 2008; Erenstein and Laxmi 2008; Lohan et al. 2018). However, there are problems with direct drilling of wheat into combine-harvested rice fields using the standard ZT seed drill due to: (i) straw accumulation in the seed drill furrow openers, (ii) poor traction of the seed metering drive wheel due to the presence of loose straw, and (iii) the need for frequent lifting of the implement under heavy residue conditions, resulting in uneven seed depth and thus crop establishment. In view of the serious problems associated with the burning of paddy residues, significant efforts are needed to find ways and means to efficiently utilize the huge amount of surplus rice residue produced in NW India for maintaining soil, human and animal health, and increasing farmers' profits. Management of rice residue is a serious problem, because there is very little turn-around time between rice harvesting and wheat sowing. Until recently, the availability of suitable machinery was a major constraint to direct drilling into heavy rice stubbles. The development of a new line of machines for seeding into rice residues commenced in 2002 with the development of the happy seeder (Sidhu et al. 2007; Singh et al. 2009a). Through a series of further modifications, an improved version of the latest happy seeder known as the turbo happy seeder was developed and evaluated for direct seeding of wheat into heavy loads of rice residue by Punjab Agricultural University, Ludhiana, in 2012 (Sidhu et al. 2015).

Development of the Turbo Happy Seeder

The latest version of the turbo happy seeder (THS) consists of a rotor for managing the paddy residues and a zero till drill for sowing of wheat. Flail-type blades are mounted on the straw management rotor which cuts (hits/shear) the standing stubble/loose straw coming in front of the sowing tine and cleans each tine twice in one rotation of the rotor for proper placement of seed in soil. The rotor blades/flails guide/push the residues as surface mulch between the seeded rows. The THS leaves the seeded rows exposed and clearly visible, enabling accurate lining up of adjacent sowing passes. This power takeoff (PTO)-driven machine can be operated with a 45-hp tractor and can cover 0.3–0.4 ha/hour. The THS can be used to sow wheat in rice residues, mungbean, and maize in wheat residues, and direct seeding of rice into wheat residues.

Cutting and shredding are achieved with hinged J-type flails mounted on a high-speed (1000–1300 rpm) rotor inside the straw management drum. The flails cut (shear) the anchored residues close to the soil surface and smash them and the loose residues against serrated blades fixed on the internal walls of the straw management drum, and against the seeding tines which are also inside the straw management drum, thus chopping and shredding them into small pieces. At the same time, flails sweep past each sowing tine twice per rotation, clearing the residues away and enabling the tines to pass freely through the residues. This technology also results in much less straw deposition on the seed rows than between the rows, and greatly reduces the generation of dust. The THS thus leaves the seeded rows exposed and clearly visible, enabling accurate lining up of adjacent sowing passes.

Field Evaluation of the Turbo Happy Seeder

The turbo happy seeder performed well in farmers' fields across Punjab, with a mean wheat yield increase of 3.2% in 2008–10 over farmers' usual practice (Table 9.2). The turbo happy seeder produced significantly higher yields compared to farmer practice in six out of 15 districts and the yields were similar in the other districts. Other studies on farmers' fields (Sidhu et al. 2007; Gathala et al. 2011b) and on-station experiments (Singh et al. 2011a) showed similar or higher yields of THS-seeded wheat compared to conventional practice. The increases in wheat yield for THS could be related to increased soil water availability due to reduced soil evaporation and a better soil thermal regime with surface residue retention (Singh et al. 2011b). The lower canopy temperature in the THS plots in mid-March was lower compared to farmers' fields, suggesting higher crop transpiration and improved soil water availability (Singh et al. 2011a).

Sowing wheat into rice residues with the THS has many benefits, both economic and environmental. By removing the need to burn the rice residues, significant air pollution is avoided, soil nutrients are retained, and soil organic matter is increased (NAAS 2017). Sowing of wheat using the THS can save as much as 83% of the energy use compared to CT, and also reduces emissions

Table 9.2 Wheat yield under the turbo happy seeder (THS) and conventional-till (CT) at different locations in Punjab

| Location (no. of sites) | Grain yield of wheat (t/ha) | | | | % change over CT |
| | Range | | Mean | | |
	CT	THS	CT	THS	
Fatehgarh Sahib (14)	3.18–5.43	3.50–5.53	4.34	4.54	4.6
Sangrur (8)	3.90–4.80	4.00–5.25	4.51	4.87	8.0
Barnala (3)	4.50–5.00	4.50–5.15	4.83	4.71	–2.5
Fatehagarh (4)	3.63–5.00	3.75–5.38	4.47	4.91	9.6
Ludhiana (23)	3.75–5.00	3.50–5.38	4.37	4.48	2.7
Kapurthala (3)	3.00–5.00	3.00–5.00	4.17	4.17	0.0
Jalandhar (16)	3.75–5.50	3.25–6.00	4.38	4.48	1.7
Ferozepur (4)	3.50–3.75	3.50–4.00	3.66	3.73	1.8
Nawanshahr (6)	4.88–6.00	4.38–5.50	5.17	5.13	–0.6
Tarantaran (3)	4.00–4.00	3.75–4.13	4.00	3.92	–2.1
Amritsar (22)	3.50–5.00	3.50–5.50	4.15	4.35	4.7
Overall mean (106)			**4.36**	**4.50**	**3.2**

Source: Sidhu et al. (2007).

of CO_2 through reduced fuel consumption. Singh et al. (2008) showed that the happy seeder technology is more profitable (Rs. 6757 per ha) than conventional cultivation when similar wheat yields were considered under the two practices. The net financial benefits almost doubled over conventional practice with a grain yield increase of 5% using the happy seeder. Reduced water, fertilizer, and herbicide inputs are also expected. In the majority of experiments, sowing wheat using the THS did not need pre-sowing irrigation due to sufficient residual soil moisture in the rice fields. Thus, 75–100 mm of irrigation water could possibly be saved with the adoption of THS technology. Singh et al. (2011a) found that the surface retention of rice residues reduced soil moisture loss through evaporation by about 40 mm, and thereby saved one irrigation of wheat when irrigation scheduling was based on soil moisture status. There was also less weed growth (saving in weedicide cost). Rice straw mulch has the potential to control weed growth by more than 65% compared with that without mulch (Sidhu et al. 2007) and thereby will help to reduce the use of weedicides, making the operation more economical and eco-friendly.

A two-wheel tractor-driven smaller version of THS with five seeding rows has also been developed for direct seeding of wheat into rice residue for farmers with smaller land holdings. The smaller version of THS can be mounted on a two-wheel tractor by removing the tiller attachment. The THS can also be used for direct seeding of summer mungbean or maize fodder immediately after wheat harvesting, thus providing additional income to farmers.

Happy Seeder for Sowing Wheat After Maize

The happy seeder can be used to sow two rows of wheat in maize residues on beds. A knife roller is attached to the front of the tractor to cut the standing maize residues after combine harvesting of maize. The knife roller cuts the maize residues into 10–15 cm pieces by shearing the maize stalks between knife roller blades and the soil surface. The rotor of the turbo happy seeder chops the maize residues using flail blades and tine interaction. The maize residues on the surface help to reduce weed emergence among the furrows and tops of beds.

Adoption of the Turbo Happy Seeder for Sowing of Wheat

The pace of adoption of the THS has been slow to date, despite large subsidies (up to 60% of the purchase price). Barriers to adoption, apart from the capital cost, include substantial manual input for uniform distribution of residues, the low window of operation of the exiting machine (25 days/year, 8–9 hr/d) and the lower seeding capacity (0.3 ha/hr) compared with 0.5 hr/ha with conventional seed drills. Other factors include high subsidization of diesel and electricity for pumping groundwater. Implementation of legislation banning in-field straw burning would greatly accelerate large-scale adoption of technology for drilling wheat into rice residues.

Above all, combine harvesting of rice leaves the cut residues in windrows, as the machines used for harvesting rice in India do not have straw spreaders. Uniform spreading of the loose residues is a pre-requisite for successful use of the turbo happy seeder, to avoid choking and the creation of patches of thick deposits of mulch which suppress crop establishment. A prototype straw-spreading system, which can be readily attached onto existing combine harvesters, has already been developed (Singh et al. 2010) and is now commercially available. Uniform straw spreading will also ensure more uniform soil moisture content throughout the field at the time of wheat sowing.

Straw Management System (SMS)

The combine harvester harvests cereal crops in a width equal to its cutter bar width, and throws straw residues from straw walkers into the centre of the harvested area. The width of the straw walkers is usually one-third of the cutter bar width of the combine. This forms lines of loose residues (as wide as the straw walker width) in the field parallel to the combine operation. This uneven residue load hinders the operation of straw management machines. Uniform spreading of loose straw is a precondition for the smooth operation of all second-generation drills, and this operation takes around 8–13 man-hr/ha for spreading of loose straw. It is very difficult to spread the entangled dry loose rice straw due to its light weight. Therefore, for smooth operation of the THS, it was felt that the existing combine harvester should be modified to achieve chopping and uniform spreading of loose straw.

The straw management system (SMS, spinning disc) is an attachment to the existing combine for managing and spreading the straw in the harvested area, which has been jointly developed and recommended by PAU, Ludhiana, and CIMMYT-BISA, Ludhiana. The straw spreader is attached to the rear side of the combine harvester just below the straw walkers and behind the chaffer sieves. The loose residues falling from the harvester straw walker are spread behind the harvester by the spinning discs. The SMS consists of two counter-rotating spinner discs made from an MS sheet. Three vertical vanes are riveted on each spinner disc at an angle of 120° and these divide the discs into three sections. The spinner discs are counter-rotated by a bevel gear arrangement. The counter-rotating discs rotate at 450–500 rpm and distribute the loose residues discharged from straw walkers evenly in the cutter bar width of the harvester. The SMS includes stationery housing for easy attachment/detachment to the rear of the combine. The spinner discs can be driven either by v-belts and a pulley arrangement or by hydraulic motors.

Super SMS

A new version of SMS called the Super SMS was developed and evaluated jointly by PAU and CIMMYT-BISA in 2016. The Super SMS can be mounted at the rear of the self-propelled combine harvester with a 4.27 m cutter bar and engine power of 110 hp. The straw coming out of the straw walkers of the combine harvester is fed into the unit from one side and is discharged from the outlet of the housing. The chopped material is blown off tangentially and deflected using a deflector for uniform spreading of the residues over the entire width of the combine harvester. PAU, Ludhiana, has already recommended the Super SMS attachment for self-propelled combine harvesters to chop and uniformly spread the loose straw coming out of the harvester straw walkers for farmers in Punjab and other states in India.

The Super SMS includes the stationary housing for attachment to the rear end of the combine harvester. The straw coming out of the straw walkers of the combine harvester is fed to the unit from one side and is discharged from the outlet of the housing. Inside the housing, a rotor is mounted with six lugs in a row, and four such equally spaced rows along the entire periphery of the rotor. The rotor operates at a speed in the range of 1400–1500 rpm and driven through a v-belt pulley arrangement. There are 24 stationary serrated blades fixed on the concave portion of the rotor housing. Each pair of flails during rotation passes over the stationary serrated blades and cuts the straw into pieces. The chopped straw is blown off tangentially and uniform spread is achieved with the help of a deflector attached at its outlet. The deflector spreads the chopped straw over the full width of the combine harvester.

Uniformity of spread of rice residues was measured by attaching the SMS behind the self-propelled combine harvester with a 4.27 m cutter bar width at PAU seed farm, Ladhowal, Ludhiana, and BISA, Ludhiana. The Super SMS attachment was operated at 1450 rpm. The uniformity of residue spread was measured by dividing the width of cut into four equal sections (left, middle, and right) of 1.06 m each. After the operation, loose residue dropped in these sections for 1 meter length of travel was collected to determine the coefficient of variation. The size reduction of straw was measured by segregating the collected samples after operation of the combine harvester with and without the Super SMS into various sizes ranging from <2.5 cm to >20 cm. Performance evaluation of the combine harvester in terms of cleaning efficiency, straw walker losses, sieve losses, fuel consumption, etc .was carried out with and without SMS operation (Manpreet-Singh et al. 2020a; Manpreet-Singh et al. 2020b). The weighted mean chop size was 13.7 and 27.0 cm for the Super SMS attachment and normal combine harvester operation (Table 9.3).

The fuel consumption and field capacity were 23.4% and 8% higher for the combine harvester with Super SMS attachment compared to without the SMS (Table 9.4). The cleaning efficiency was 96% for the combine harvester with and without the Super SMS, keeping the height of cut at 26 cm. The grain losses from the straw walker and sieves of the combine harvester were under the permissible limit of 2.5% as per IS 8122 (1994) for combine harvesters both with and without the SMS.

The additional cost of straw spreading with the Super SMS combine harvester over manual spreading was Rs. 475 per ha. However, the overall system saving

Table 9.3 Comparative chop size range of paddy residues harvested by a combine with and without the PAU Super SMS attachment

Range (cm)	Average range	Percent chop size	
		Without SMS	*Super SMS*
<2.5	1.25	8.01	16.20
2.5–7.5	5.00	6.33	13.78
7.5–12.5	10.00	5.34	15.98
12.5–20	16.25	11.25	22.67
>20	35.00	69.07	31.37
Average chop size (cm)	–	26.95	13.72

Source: Manpreet-Singh et al. (2020b).

Table 9.4 Performance parameters of a combine harvester with and without the PAU Super SMS attachment

Particular	Super SMS	Without Super SMS
Speed of operation (km/hr)	3.40	3.64
Field capacity (ha/hr)	1.16	1.24
Fuel consumption (l/ha)	11.98	8.92
Straw walker loss (%)	0.69	0.26
Sieve loss (%)	0.18	0.23
Cleaning efficiency (%)	0.96	0.96

Source: Manpreet-Singh et al. (2020b).

was Rs. 315 per ha when considering the increased field capacity of the happy seeder by 0.6 ha/day. The chopping of rice residue with the SMS attachment improved the working of THS used for *in situ* residue management. The Super SMS has an option of switching on or off using a small metal sheet. This enables the combine harvester with Super SMS attachment to be used as a traditional combine harvester without dismantling the SMS attachment (for collection of residues from the field).

The custom hiring cost of the Super SMS to chop and uniformly spread rice residues is around Rs. 750–875 per ha. The cost of hiring a stubble shaver for cutting and spreading the rice residues before burning is about the same. Therefore, there is no additional cost involved for farmers. The increased capacity of THS (due to uniform chopping and spreading of rice residues) will restrain the practice of residue burning and minimize environmental pollution. The 'PAU Super SMS' attachment for the self-propelled combine harvester will play a key role in tackling the problem of residue burning in states such as Punjab, Haryana, and western Uttar Pradesh.

Precautions for Super SMS mounting on a combine harvester

i   The rotating flail blades, bushes, plates, and nut bolts of the rotor should be of exactly the same dimensions, otherwise this may cause vibrations in the Super SMS unit. The rotor of the Super SMS should be dynamically balanced to minimize vibrations in the attachment as it works at 1600–1800 rpm.

ii   The Super SMS system should be supported from the combined chassis of the combine harvester body to minimize overhang and vibrations.

iii   The overlapping of the stationery blades and flails on the rotor can be adjusted by changing the angle and position of stationery blades assembly. The size of straw chop and load on the harvester is dependent on the overlap of the stationery blades and flails on the rotor. This overlap can be adjusted depending on the field conditions.

iv   The v-belt and pulley section should be so selected so that belt slippage is within acceptable limits.

v   The rotor rpm indicator may be attached to the Super SMS as an additional safety feature.

vi   The end sheet of the combine harvester may be modified to automatically open if the straw walkers of the harvester clog during harvesting as an additional safety feature.

CA Machinery for Cotton–Wheat Rotation

Cotton–wheat (CW) rotation covering an area of about 3.5 M ha is one of the potential candidates for major gains in future wheat production in India. The optimum time of wheat sowing in NW India is the last week of October to the first fortnight of November. Wheat planting after cotton is often delayed due to late picking of cotton and the time involved in its seed bed preparation. The sowing of wheat after 20 November reduces its productivity at the rate of 1.0–1.5% per day of delay (Nasrullah et al., 2010). The average productivity of wheat in the CW system is lower (about 3.2 t/ha) compared to productivity in the RW system (about 4.7 t/ha) of Punjab (Buttar et al. 2013). A delay in wheat sowing in the CW system can be avoided by relay seeding using a self-propelled walk-behind-type relay seeder (RS) in standing cotton. A two-wheel self-propelled relay seeder was developed in 2009 by the Cereal Systems Initiative for South Asia (CSISA)/ CIMMYT team in collaboration with Amar Agro Industries, Ludhiana, Punjab. Relay seeding of wheat increased cotton yield by creating the opportunity for one additional picking, which was made possible due to the extended growing period of the cotton for about 30 days. This extra growing period helped in full opening of the majority of the immature bolls at the time of pulling out of cotton stalks, leading to an 11–14% increase in seed cotton yield over CTW. The wheat yield was increased by 25% under relay seeding compared to conventional sowing.

The wheat yield gains with a self-propelled walk-behind-type relay seeder (RS) were 12–41% compared with CT wheat after cotton harvest. However, farmers' adoption of this three-row walk-behind-type RS for planting wheat in the CW system is very limited due its low capacity (<0.6 ha/day) and heavy work. Hence, there was a need for a four-wheel tractor-operated RS, which can sow wheat in standing cotton crop with different row geometries.

A four-wheel tractor was mounted on a high clearance platform, which increased the ground clearance from 45 to 115 cm and facilitated easy movement of the tractor above the standing cotton crop (Mahal et al. 2016). The working clearance (from the ground) of the tractor was 110 cm.

Development of a High-Clearance Platform for Tractors

The traditional four-wheel tractors with ground clearance of around 45–50 cm cannot move in a standing cotton field as the plants are about 100–130 cm tall. To address this issue, a high-clearance platform attachment for a four-wheel tractor was developed in collaboration with BISA Punjab, India, PAU Ludhiana, India, and Rajar Agricultural Works, Mullanpur, Ludhiana (Punjab). This platform increased the ground clearance of the tractor to 115 cm to make the tractor move easily above the standing cotton. Brief specifications of the platform for the high-clearance tractor are included in Table 9.5.

The track width of the mounted tractor was increased by 1.5 times the standard one (from 135 cm to 202.5 cm), which enables the high-clearance tractor to move in both 67.5 and 101 cm row geometries of cotton, and also increases the stability of the tractor. Any traditional tractor (ground clearance ~45 cm) can be converted to a high-clearance tractor by mounting on a high-clearance platform in 4–6 hours.

Four-Wheel Tractor-Operated Relay Seeder

Relay seeders (suitable for 67.5 and 101 cm cotton row spacing) fitted with three types of furrow openers [zero-till inverted T-type (ZTT), zero-till double disc (ZTDD) and strip rotor (SR)] were used for relay seeding of wheat in cotton crop.

Table 9.5 Brief specifications of a normal tractor and tractor mounted on a high-clearance platform

Specifications	Normal tractor	Tractor mounted on high-clearance platform
Ground clearance (mm)	450	1100
Turning radius (mm)	2700	3020
Weight (kg)	2005	2550
Speed reduction (over normal)	–	17%
Track width (mm)	1350	2025
Centre of gravity (x, y) from rear tyre	923.6, 802.5	903.4, 993.7

Early sowing of relay wheat by 31 days compared to CTW increased all three yield parameters, thereby increasing the grain yield by 19% (Manpreet-Singh et al. 2017). The average gross returns from the relay CW system were 15.5% higher compared with the CTW system due to lower tillage costs and higher yields of seed cotton, grain, and straw of wheat. Net returns were Rs 20,837–27,625 per ha more (an increase of 27–37%) under relay seeding of wheat using a high-clearance tractor compared with the CTW system. The relay seeding of wheat using different furrow openers included a single operation, whereas CTW needed five–six tillage operations.

Relay Seeding of Mungbean in Wheat

The high-clearance four-wheel tractor-driven relay seeder was also evaluated for relay seeding of mungbean into standing wheat during March. Wheat was planted in a paired row system and mungbean was planted in the wider space in the paired rows. Early relay sowing of mungbean by 20–25 days ensured a mungbean yield of about 1.0 t/ha, escaping the risks from early onset of monsoon rains obstructing the harvesting of the crop.

CA Machinery for Maize–Wheat Rotation

The maize–wheat rotation is the third most important cropping system followed in NW India after rice–wheat and cotton–wheat crop rotation. Maize is considered a promising option for diversifying agriculture in the upland areas of India. Traditionally, seed bed preparation for maize involves several tillage operations. However, maize can be grown without any preparatory tillage with zero till/happy seeder having an inclined plate planter attachment. Zero tillage has many benefits such as savings in diesel and time, reduced environmental pollution, and saving of irrigation water in the first irrigation, thus resulting in reduced cost of production. This also helps in timely planting of maize over large areas.

Tractor-Operated Double-Disc Planter

The planter uses double-disc openers to cut the residues coming in front of the seed row. The passive anti-blocking double-disc planter is powered by four-wheel tractors usually operated on medium-size farms for planting four rows of maize in wheat residues in a maize–wheat rotation. The press wheels are attached behind the double-disc openers to close the seed row for better soil seed contact to enhance the germination. The planter is equipped with an inclined plate seed metering mechanism of the planter.

Bed Former

Maize can be sown on ridges made by a tractor-drawn ridger to facilitate easy and economical irrigation during dry and hot weather conditions. A seed drill attachment mounted on the ridger can also be used for sowing by adjusting the position of the tines. The ridger consists of a frame mounted with staggered tines having miniature furrow openers. These furrow openers make furrows at a spacing of 33.5 cm and the remaining portion of the soil is converted into a seed bed. This tractor-operated machine is used in maize–wheat rotation to make and reshape permanent beds. The bed former can be operated with a 45-hp tractor and the field capacity of this machine is 0.4 ha/h. The same equipment can be used for mechanical weeding, shaping of ridges, and earthing-up of maize.

Raised-Bed Planting

Raised-bed planting is considered to be the best planting method for maize during monsoon and winter seasons both under excess moisture as well as limited water availability/rainfed conditions. Sowing/planting should be done on the southern side of the east–west ridges/beds, which helps in good germination. The raised bed planter having inclined plate seed metering systems facilitates the placement of seed and fertilizers at the proper place in one operation, which helps in getting a good crop stand, and higher productivity and resource-use efficiency. Using raised-bed planting technology, 20–30% of irrigation water can be saved with higher productivity. Moreover, under temporary excess soil moisture/water logging due to heavy rains, the furrows will act as drainage channels and crops can be saved from oxygen stress. Maize CA-based management practices (PB and ZT) in maize-based cropping systems led to higher system productivity and economic benefits compared to CT (Parihar et al. 2016).

Double-Disc Bed Planter for Maize

To realize the full potential of the bed planting technology, maize sowing on permanent beds can be done in a single pass without any preparatory tillage. The permanent bed planter consists of double-disc furrow openers along with a bed shaper. The double-disc planter sows one maize row at the centre of the bed and the seed-to-seed spacing is 67.5 × 20.0 cm. The double-disc furrow openers carry out the sowing of maize in a narrow furrow, thereby causing less damage to permanent beds compared to ZT tine openers.

Tractor Front-Mounted Knife Roller for Managing Crop Residues

A knife roller consists of two rollers having straight knives mounted on the entire periphery of the rollers. The knife roller rotates with passive power from the soil surface. The knives cut the maize or any other residue by shearing the residues between the soil surface and knife edge. The knife roller is a mounted on the tractor front and on the rear of the tractor the seeding machine can be used so that the direct drilling of next crop (wheat) can be carried out in a single pass of the tractor. The knife roller can be operated with a 45-hp tractor and the field capacity of this machine is 0.8 ha/hr.

CA Planters for Small Farmers

Manual and animal traction seeders are usually small, light-weight, simple in design, and easily manufactured, utilized, and maintained. These seeders are invariably used on small farms and hilly areas. The use of manual equipment is the basic level of mechanization, and improved equipment designs are intended to enhance productivity in respect to energy efficiency and ergonomics. A demand-driven development and an innovation system are necessary for the adoption of improved human- and animal-pulled equipment. The field capacity of the manually operated planters is around 0.06 ha/h. Some typical seeders in India are reviewed below.

Single-Row Punch Planter

The punch planter can be used for planting maize, soybean, sunflower, and cowpea under the CA system. It has two containers, for seed and fertilizer, a punching unit to dibble the seeds in residues, a furrow opener to place the fertilizer, and a press wheel to control the planting depth. The operator walks behind the planter and controls the operation through handlebars. The total weight of this single-row planter is 30 kg.

Manual Direct Planter

This is a multi-crop planter that generally consists of a zero till tine to open a small seed row. Seed and fertilizer are held in two separate hoppers and delivered into the slot by individual drop tubes. Mounted behind the tine is a seed and metering device drive wheel which may act as a seed-covering device and press wheel. The operator walks behind the seed drill and controls the operation through handlebars.

Manual Double-Disc Planter

A multi-crop manual planter generally consists of a double-disc opener to cut the residue and open a small seed row. Seed and fertilizer are held in two separate hoppers and delivered into the slot by individual drop tubes. Mounted behind the double-disc opener is a seed and metering device drive wheel which may act as a seed-covering device and press wheel. The operator walks behind the double-disc planter and controls operation through handle bars.

Suggestions for Large-Scale Adoption of CA Machinery in India

Currently there are small- to medium-sized zero/minimum-till seeders which can promote the growth of CA in India. However, their development is slow and their uptake is limited. In order to develop high-performance no/minimum-till seeders, the following recommendations can be made:

i Policy support is crucial for the rapid development of seeders, including provision of adequate research funding for the development of new machinery for the implementation of CA.

ii Although a variety of no-till seeders have been developed and fabricated, further improvements/modifications are needed to suit a range of crops, field sizes, and soils in different geographical regions for different power sources available.

iii For marginal farmers, with limited resources and farm size, the purchase of large machinery (built locally or imported) is a constraint. Therefore, the no-till seeders need to be light-weight, simple, affordable, and suited to low-horse-power tractors.

iv For heavy clayey soils, no-till seeders should use tines instead of discs, to facilitate planting and simplicity in design and operation.

v In no-till fields, especially in hilly areas, agricultural production is hampered by natural obstacles (rough/sloppy surfaces), therefore all no-till seeders should have a depth-control mechanism and contour-following capability with individual soil-covering devices and pressing wheels.

vi Anti-blocking mechanisms play an important role in the no-till seeder, especially in high-residue conditions. The development of no-till seeder anti-blocking technology is necessary to ensure good performance and adoption of the seeders.

vii Research institutions, enterprises, and farmers should have closer cooperation in designing suitable no/minimum-till seeders for different cropping areas.

Conclusion and Future Outlook

CA has been proposed as an alternative to conventional agriculture to sustainably improve agricultural production, increase nitrogen-use efficiency, and reduce emissions of GHGs, making the systems more resilient to climate change. The development of suitable agricultural machinery is the key to the adoption of CA in the region. Machines for seeding and managing residue simultaneously are required for practicing and adoption of CA in different cropping systems. Satisfactory progress has already been made in developing a number of new machines, such as the laser land leveller, multi-crop planters, happy seeder, and relay seeders for ZT seeding, straw management system, and relay seeding in CA-based cropping systems. The residue management machines such as the turbo happy seeder and super SMS reduce residue burning and improve soil health for long-term sustainability of the rice–wheat system. Relay seeding of wheat in cotton helps in timely sowing, capturing the residual soil moisture of the last irrigation to cotton, and increasing the productivity and profitability of the cotton–wheat system. Appropriate machinery is needed for fertilizer placement to increase nutrient-use efficiency in different CA-based cropping systems. To accelerate the pace of adoption of CA and diversification in the region, the development and evaluation of multi-crop, multi-utility machines for CA and human resource development need immediate action as '*Unsustainability cannot be an option in modern agriculture*'.

References

Buttar, G.S., Sidhu, H.S., Singh, V., et al. 2013. Relay planting of wheat in cotton: an innovative technology for enhancing productivity and profitability of wheat in cotton–wheat production system of South Asia. *Experimental Agriculture* 49: 19–30.

Erenstein, O. and Laxmi, V. 2008. Zero tillage impacts in India's rice–wheat systems. *Soil and Tillage Research* 100: 1–14.

Gathala, M.K., Kumar, V., Saharawat, Y.S., et al. 2011b. Happy Seeder technology: a solution for residue management for the sustainability and improved production of the rice–wheat system of the Indo-Gangetic Plains. In: Resilient Food Systems for a Changing World. Proceedings of the 5th World Congress on Conservation Agriculture. Brisbane, Australia. 25–29 September 2011.

Gathala, M.K., Ladha, J.K., Kumar, V., et al. 2011a. Tillage and crop establishment affects sustainability of South Asian rice-wheat system. *Agronomy Journal* 103(4): 961–971.

Gupta, R., Jat, M.L., Singh, S., et al. 2006. Resource conservation technologies for rice production. *Indian Farming* 56(7): 42–45.

Gupta, R. and Seth, A. 2007. A review of resource conserving technologies for sustainable management of the rice–wheat cropping systems of the Indo-Gangetic Plains. *Crop Protection* 26: 436–447.

Jat, M.L., Chandna, P., Gupta, R.K., et al. 2006. Laser Land Leveling: A Precursor Technology for Resource Conservation. Technical Bulletin Series 7. Rice–Wheat Consortium for the Indo-Gangetic Plains, New Delhi, India.

Jat, M.L., Gathala, M.K., Ladha, J.K., et al. 2009a. Evaluation of precision land leveling and double zero-till systems in the rice–wheat rotation: Water use, productivity, profitability and soil physical properties. *Soil and Tillage Research* 105: 112–121.

Jat, M.L., Gupta, R., Ramasundaram, P., et al. 2009b. Laser-assisted precision land leveling: A potential technology for resource conservation in irrigated intensive production systems of the Indo-Gangetic plains, pp. 223–238. In: Ladha J.K., et al. (Eds.). Integrated Crop and Resource Management in the Rice-Wheat System in South Asia. International Rice Research Institute, Los Banos, Philippines.

Jat, M.L., Singh, Y., Gill, G., et al. 2015. Laser-assisted precision land leveling impacts in irrigated intensive production systems of South Asia. *Advances in Soil Science* 22: 323–352.

Latif, A., Shakir, A.S. and Rashid, M.U. 2013. Appraisal of economic impact of zero tillage, laser land leveling and bed-furrow interventions in Punjab, Pakistan. *Pakistan Journal of Engineering and Applied Science* 1: 65–81.

Lohan, S.K., Jat, H.S., Yadav, A.K., et al. 2018. Burning issues of paddy residue management in north-west states of India. *Renewable and Sustainable Energy Reviews* 81: 693–706.

Lohan, S.K., Singh, M., Sidhu, H.S. 2014. Laser Guided Land Leveling and Grading for Precision Farming, pp. 148–158. In: Ram, T., Lohan, S.K., Singh, R. and Singh, P. Precision Farming – A New Approach. Astral International Pvt. Ltd. New Delhi.

Mahal, J.S., Sidhu, H.S., Manes, G.S., et al. 2016. Development and feasibility of innovative relay seeders for seeding wheat into standing cotton using a high clearance tractor in cotton-wheat system. *Applied Engineering in Agriculture* 32(4): 341–352.

Malik, R.K., Yadav, A., Gill, G.S., et al. 2004. Evolution and acceleration of no-till farming in rice–wheat cropping system of the Indo-Gangetic Plains. In: Proceedings of the 4th International Crop Science Congress, Brisbane, 29 September–3 October, 2004.

Manpreet-Singh, Sidhu, H.S., Goyal, R., et al. 2020b. Development and evaluation of straw management system for combine harvester for *in-situ* management of rice residue in rice-wheat system of South Asia. *Field Crops Research* (in press).

Manpreet-Singh, Sidhu, H.S., Mahal, J.S., et al. 2017. Relay sowing of wheat in the cotton–wheat cropping system in north-west India: Technical and economic aspects. *Experimental Agriculture* 53(4): 539–552.

Manpreet-Singh, Sidhu, H.S., Yadvinder-Singh, et al. 2020a. Performance evaluation of automatic vis-à-vis manual topographic survey for precision land levelling. *Precision Agriculture* 21: 300–310.

NAAS. 2017. Innovative Viable Solution to Rice Residue Burning in Rice-Wheat Cropping System through Concurrent Use of Super Straw Management System-fitted Combines and Turbo Happy Seeder, 16 pp. Policy Brief No. 2. National Academy of Agricultural Sciences, New Delhi.

Nasrullah, M.H., Cheema, S.M. and Akhtar, M. 2010. Efficacy of different dry sowing methods to enhance wheat yield under cotton-wheat cropping system. *Crop and Environment* 1: 27–30.

Parihar, C.M., Jat, S.L., Singh, A.K. et al. 2016. Conservation agriculture in irrigated intensive maize-based systems of north-western India: Effects on crop yields, water productivity and economic profitability. *Field Crops Research* 193: 104–116.

Sidhu, H.S., Mahal, J.S., Dhaliwal, I.S., et al. 2007. Laser Land Leveling–A Boon for Sustaining Punjab Agriculture. Farm Machinery Bulletin-2007/01:13. Punjab Agricultural University, Ludhiana, India.

Sidhu, H.S., Singh, S., Singh, Y., et al. 2015. Development and evaluation of the Turbo Happy Seeder for sowing wheat into heavy rice residues in NW India. *Field Crops Research* 184: 201–212.

Singh, R.P., Dhaliwal, H.S., Humphreys, E., et al. 2008. Economic evaluation of the Happy Seeder for rice–wheat systems in Punjab, India. In: Proceedings of the 52nd Australian Agricultural and Resource Economics Society Conference Canberra, Australia, 5–8 February, 2008.

Singh, B., Eberbach, P.L., Humphreys, E., et al. 2011a. The effect of rice straw mulch on evapotranspiration, transpiration and soil evaporation of irrigated wheat in Punjab, India. Field Crops Research 98: 1847–1855.

Singh, B., Humphreys, E., Eberbach, P.L., et al. 2011b. Growth, yield and water productivity of zero till wheat as affected by rice straw mulch and irrigation schedule. *Field Crops Research* 121: 209–225.

Singh, Y., Kukal, S.S., Jat, M.L., et al. 2014. Improving water productivity of wheat-based cropping systems in South Asia for sustained productivity. *Advances in Agronomy* 127: 157–258.

Singh, K.K., Lohan, S.K., Jat, A.S., et al. 2003. Influence of different planting methods on wheat production after harvest of rice. *Agriculture Mechanization in Asia, Africa and Latin America* 34(4): 18–19.

Singh, Y., Sidhu, H.S., Singh, M., et al. 2009a. Happy seeder: A conservation agriculture technology for managing rice residues. Technical Bulletin No. 01. Department of Soils, Punjab Agricultural University, Ludhiana.

Singh, U.P., Singh, Y., Kumar, V., et al. 2009b. Valuation and promotion of resource-conserving tillage and crop establishment techniques in the rice-wheat system of eastern

India, pp. 151–176. In: Ladha, J.K., Singh, Y., Erenstein, O., Hardy, B. Integrated Crop and Resource Management in the Rice-Wheat System of South Asia. International Rice Research Institute, Los Banos, Philippines.

Singh, Y., Singh, M., Sidhu, H.S., et al. 2010. Options for effective utilization of crop residues, 32 p. Research Bulletin 3/2010. Punjab Agricultural University, Ludhiana, India.

Singh, Y., Singh, B. and Timsina, J. 2005. Crop residue management for nutrient cycling and improving soil productivity in rice-based cropping systems in the tropics. *Advances in Agronomy* 85: 269–407.

10 Energy Use in Conservation Agriculture

U.K. Behera and P.K. Sahoo

Introduction

Traditional agriculture based on tillage has been accused of being responsible for soil erosion problems, surface and underground water pollution, and more energy and water consumption. Moreover, it is implicated in land resources degradation, wildlife and biodiversity reduction, low energy efficiency, and contribution to the global warming problems. Conservation agriculture (CA) is an ecosystem approach to sustainable agriculture and land management based on the practical application of three locally adapted interlinked principles: (i) continuous no or minimum mechanical soil disturbance (no-till seeding/planting and weeding, and minimum soil disturbance with all other farm operations including harvesting); (ii) permanent maintenance of soil mulch cover (crop biomass, stubble, and cover crops); and (iii) diversification of the cropping system (environmentally and socially adapted rotations and/or sequences and/or associations involving annuals and perennials, including legumes and cover crops), along with other complementary good agricultural production and land management practices. Conservation agriculture systems are present in all continents, involving rainfed and irrigated systems including annual cropland systems, perennial systems, orchards and plantation systems, agroforestry systems, crop–livestock systems, organic production systems, and rice-based systems.

Conservation agriculture is a way to cultivate annual and perennial crops, based on no vertical perturbation of the soil (conservation tillage), with crop residues management and cover crops, in order to offer a permanent soil cover and a natural increase of the organic matter content in surface horizons. The main environmental consequences of this method have been investigated worldwide with the objective of presenting a synthesis of the available studies and documents to farmers and scientific communities. It stresses the very beneficial impacts of a conservative method of cultivation on the global environment (soil, air, water, and biodiversity), compared to traditional agriculture. CA allows most soils to have a richer bioactivity and biodiversity, a better structure and cohesion, and very high natural physical protection against weather (raindrops, wind, dry or wet periods). Soil erosion is therefore highly reduced, soil agronomic inputs transport is slightly reduced, while pesticide bio-degradation is

DOI: 10.4324/9781003292487-12

enhanced. It protects surface and ground water resources from pollution and also mitigates negative climate effects. Hence, CA provides excellent soil fertility and also saves money, time, and fossil-fuel use. It is an efficient alternative to traditional agriculture.

Agriculture is basically an energy conversion process. It is both a producer as well as a consumer of energy. The process involves the conversion of solar energy into biomass by way of photosynthesis. The upper limit of maximum photosynthetic efficiency of terrestrial plants has been calculated to be of the order of 6%, but most cereals and oilseed plants achieve only 0.15–0.20%. Modern agriculture enhances the production of biomass by increasing photosynthesis through improved plant types and the use of irrigation water, fertilizer, pesticides, and other growth-promoting inputs. Countries with high agriculture productivity use very high levels of non-renewable commercial energies as compared to developing counties which usually have low levels of agricultural productivity.

There is a direct relation between energy consumption and per capita gross national product (GNP). Currently, commercial energy sources such as coal, oil, natural gas, and electricity have been the foundation inputs for the modernization of agriculture and the development of industries. The developing nations have made strides, based on the experiences of the developed nations, in boosting their agricultural, industrial, and national productivity. However, rapid cost escalations around the mid-1970s in commercial energy inputs to every sector of the economy and living hampered these aspirations. Feasible substitution and supplementation of commercial energy by new and renewable sources have become very important.

Energy in Agriculture

Technically, sources of energy are classified into commercial and non-commercial, and renewable and non-renewable. Commercial sources of energy are direct, such as coal, oil, natural gas, and electricity; and indirect, such as chemical fertilizers, plant protection chemicals, and machinery. Non-commercial energies could also be direct, such as human labour, draft animals, vegetative fuels, as well as indirect like seeds and organic manures. Coal oil, natural gas, and fossil fuels are non-renewable, whereas solar energy, biomass, wind energy, and draft animals are renewable.

Another aspect of the energy consumption pattern in agriculture is of its peculiar nature. Agriculture is a seasonal industry where the energy demand fluctuates throughout the year. There are certain months of the year when agriculture demands more energy to meet its requirement to complete crucial operations like sowing, transplanting, harvesting, and threshing in time. No doubt, some resources are surplus, such as human resources, but they are also rendered unemployed in the lean months, whereas animal resources do not have work throughout the year. However, in the peak periods, animals draft power is insufficient to cope with the heavy load. These fluctuations are essential for planning the resources to meet the energy demands of agriculture fully.

Traditionally, men and animals have been the main sources of energy for Indian agriculture and this continues to be the case despite substantial agricultural mechanization. India has about 200 million agricultural workers, providing 43,000 TJ of human energy, and 80 million draught animals, providing 81,000 TJ of animal energy. For a desirable level of agricultural activity (200% cropping intensity), farm power requirement is considered at 0.77 kW/ha. India, with limited land mass and increasing human population, cannot afford to have more draught animals. Therefore, the desired level of farm energy availability has to be met from electro-mechanical sources.

Energy Identification and Energy Equivalents

For a clear understanding of energy consumption for different farm operations, energy identification and accounting are carried out in such a way that the energy input is classified on the basis of the source and use as direct and indirect energy. Direct energy is energy which is released directly from power sources for crop production, while indirect energy is dissipated during various conversion processes such as energy consumed in manufacturing, storage, distribution, and related activities. Both direct energy and indirect energy inputs are considered as energy in farm operations, except sequestered energy of mechanical power sources and implements. However, for the purpose of computation and analysis, three groups of energy resources are considered, namely physical, chemical, and biological energy inputs. The chemical and biological energy inputs are considered as indirect energy inputs, whereas physical energy inputs are considered as both indirect and direct energy inputs (Sahoo and Behera 2011).

Energy Inputs in Crop production

Direct Energy Inputs

Direct energy input is the energy consumption for physical work during field operations. Field operations consume significant energy in agricultural production, with most of the usage being fuel consumption. Physical energy input, such as human labour, and draft animal and mechanical power sources are considered as direct energy inputs. Energy equivalents of these power sources are as follows:

a   **Human labour:** Human muscle power is inputs for physical work in field operation activities in crop production. A power equivalent of 74.6 W (0.1 hp) for human labour is considered appropriate.

b   **Draft animals:** A power equivalent of 746 W (1.0 hp) for a pair of bullocks is considered appropriate.

c   **Mechanical power:** Energy consumed during farm operations is affected by many factors, including weather, soil type, and depth of tillage. Therefore, information on fuel consumption and the working hours of mechanical power sources for different farm operations are used for calculation of mechanical

Table 10.1 Energy coefficients used in energy calculations in crop production

Particulars	Units	Equivalent energy (MJ)
Human labour	Man-hour	1.9
Bullocks	Pair-hour	10.1
Diesel	Litre	56.3
Electricity	Kwh	11.9
Fertilizers		
N	kg	60.0
P	kg	11.1
K	kg	6.7
FYM	kg (dry)	0.3
Chemicals		
Superior chemicals	kg	120.0
Zinc sulphate	kg	20.9
Inferior chemicals	kg	10.0
Wheat		
Seed/grain	kg	14.7
Straw	kg	12.5
Machineries		
Electric motor	kg	64.8
Prime movers other than electric motor	kg	68.4
Farm machinery	kg	62.7

Source: Panesar and Bhatnagar (1994).

energy inputs. These data are gathered from field surveys from individual farmers at the farm level. In the case of a farmer using a hired machine or no information on fuel consumption, the average and estimated value based on the type and size of mechanical power source gathered from the field survey are adopted. The energy equivalents are given in Table 10.1.

Indirect Energy Inputs

Physical energy input in terms of mechanical power source, and chemical and biological energy inputs are considered as indirect energy input. Chemical fertilizers and pesticides are considered as chemical energy inputs, while seeds and hormones are considered as biological energy inputs. Energy equivalents of these power sources are as follows:

a   **Physical energy input:** Only the indirect energy of a mechanical power source is accounted for. Energy for manufacturing, repair, and maintenance as well as transportation and distribution of machinery and equipment is considered as energy sequestered or indirect energy input for mechanical power sources. The energy sequestered in manufacturing is energy used in producing the raw materials and energy required in the manufacturing process. The standard unit for energy in manufacturing is MJ/kg of the final product.

b   **Chemical energy input:** Fertilizers and pesticides are the main sources of chemical energy inputs. The energy equivalent values of 60.0, 11.1, and 6.7 MJ/kg for N, P_2O_5, and K_2O, respectively, are used for the calculation of total fertilizer and energy inputs. Energy equivalents of 120 and 10 MJ/kg for undiluted and diluted pesticides, respectively, are used to calculate energy.

c   **Biological energy inputs:** Seeds and hormones are included as biological energy inputs.

Energy Output in Crop Production

Crop production output consists of the main product and by-products. Straw and bagasse are considered as by-products.

Energy Ratio

The ratio of energy output of the production to input energy is described as the energy ratio or energy efficiency. This expression is extensively used to measure the energy efficiency in agricultural and food systems. The energy ratio can be defined as the quotient of energy value of outputs and energy value of the sum of all direct and indirect inputs. The highest energy ratios are achieved in those systems having only human effort without fossil fuel input.

Computation of Energy Inputs and Outputs

The following formulae are used for calculation of the energy inputs and outputs as well as the energy ratio for each crop (Chamsing et al. 2006).

Total Energy Input

$$\text{Total energy input} (MJ/ha) = E_f + E_s \tag{10.1}$$

where,

E_f = Energy input in farm operations (MJ/ha)
E_s = Energy sequestered of machinery (MJ/ha)

Energy input in farm operations (E_f)

$$\text{Energy input in farm operation} (MJ/ha) = \sum_{k=1}^{k=r} (Phy + chem + Bio)_k \tag{10.2}$$

where,

Phy = physical energy input in farm operation k^{th} (MJ/ha)

Chem = chemical energy input in farm operation k^{th} (MJ/ha)
Bio = biological energy input in farm operation k^{th} (MJ/ha)
k = farm operation k^{th}

(i) Physical energy input

The total physical energy input for each farm operation is calculated as the sum of energy inputs from human labour, draft animal, and mechanical power sources.

Labour Energy Input

$$\text{Labour energy input (MJ/ha)} = \sum_{l=1}^{l=s} \frac{\left[\left(0.268L_f \cdot wd_{lf} \cdot wh_{lf}\right) + \left(0.268L_h \cdot wd_{lh} wh_{lh}\right)\right]_l}{A_p}$$

(10.3)

where,

L_f and L_h = number of family labour and hired labour (persons)
wd_{lf} and wd_{lh} = number of working days for family labour and hired labour (day)
wh_{lf} and wh_{lh} = number of working hours for family labour and hired labour (h/day)
A_p = planted area (ha)
l = time for applying input for time l^{th}

Mechanical Energy Input

$$\text{Mechanical energy input in field operations ((MJ/ha)}$$
$$= \sum_{l=1}^{l=s} \frac{\left[\left(MF_f N_{mf} \cdot wd_{mf} \cdot wh_{mf} F_{eq}\right) + \left(MF_h N_{mh} \cdot wd_{mh} wh_{mh} F_{eq}\right)\right]_l}{A_p}$$

(10.4)

where,

MF_f and MF_h = fuel consumption of power source machines (L/h) for owned and hired machines
N_{mf} and N_{mh} = number of owned farm machines and hired machines
wd_{mf} and wd_{mh} = working days of owned farm machines and hired machines (day)
wh_{mf} and wh_{mh} = working hours for owned farm machines and hired machines (h/day)
F_{eq} = energy equivalent of fuel (MJ/L)
A_p = planted area (ha)

(ii) Biological energy input

Seeds and other hormones are considered as biological energy resources. Hormones are not much used in crop production; therefore, only seeds are considered to calculate biological energy input.

Total seed input (MJ/ha) = Amount of seed applied (kg/ha) ×
Energy equivalent of seed (MJ/kg)   (10.5)

(iii) Chemical energy input
(a) Fertilizer (MJ/ha)

Total fertilizer input (MJ/ha)

$$= \sum_{l=1}^{l=s} \left[\sum_{n=1}^{n=u} \frac{\left[N.N_{eqv} \right]_n}{A_p} + \sum_{n=1}^{n=u} \frac{\left[P_2O_5.P_{eqv} \right]_n}{A_p} + \sum_{n=1}^{n=u} \frac{\left[K_2O.K_{eqv} \right]_n}{A_p} \right]_l \qquad (10.6)$$

where,

N_{eqv} = energy equivalent value of N = 60.0 MJ/kg
P_{eqv} = energy equivalent values of P_2O_5 = 11.1MJ / kg
K_{eqv} = energy equivalent values of K_2O = 6.7 MJ/kg
N = compound fertilizer rate applied × percentage of N ingredient (kg)
P_2O_5 = compound fertilizer rate applied × percentage of P_2O_5 ingredient (kg)
K_2O = compound fertilizer rate applied × percentage of K_2O ingredient (kg)
n = compound fertilizer for applied time 1th

(b) Herbicide energy input (MJ/ha)

$$\text{Total herbicide input (MJ/ha)} = \sum_{l=1}^{l=s} \sum_{n=1}^{n=u} (Her.Heqv)_n \qquad (10.7)$$

where,

H_{er} = applied rate (kg or lit/ha) of herbicide 1th for applied time k^t
H_{eqv} = Energy equivalent (MJ/kg or lit) of herbicide 1th

(c) Pesticide energy input (MJ/ha)

$$\text{Total pesticide input (MJ/ha)} = \sum_{l=1}^{l=s} \sum_{n=1}^{n=u} (Pes.Peqv)_n \qquad (10.8)$$

where,

P_{es} = application rate (kg or litre/ha) of herbicide with applied time k^{th}
P_{eqv} = energy equivalent (MJ/kg or lit) of pesticide 1th

(iv) Energy sequestered in mechanical power sources and their equipment

Energy sequestered in machinery for each farm was calculated using the following formula:

$$\text{Total energy sequestered (MJ/ha)} = \frac{\sum_{m=1}^{m=t}\left[\frac{(M+T+R)}{L}\right]_m}{Aa} \qquad (10.9)$$

where,

M = energy sequestered in manufacturing for machinery m^{th} (MJ) = weight of machinery m^{th} (kg) × 62.7 (MJ/kg)
T = energy sequestered in transportation or distribution for machinery m^{th} (MJ) = weight of machine or equipment (kg) × 8.8 (MJ/kg)
R = energy for repair and maintenance for machinery m^{th} (MJ) = energy in manufacturing (MJ) × conversion factor
Aa = Annual planted area (ha)
L = Economic life of machinery m^{th}

Energy Output

Energy output is considered for the main product and by-product.

$$\text{Total energy output (MJ/ha)} = (\text{yield} \times E_{eq}) + (\text{by-product} \times E_{eq}) \qquad (10.10)$$

where, E_{eq} = energy equivalent value of main product or by-product.

Energy Ratio

$$\text{Energy ratio} = \text{Total energy output (MJ/ha)/Total energy input (MJ/ha)} \qquad (10.11)$$

Mechanization Index and Machinery Energy Ratio

The mechanization index and machinery energy ratio can be computed using the equations given below to develop the relationship between mechanization level and energy usage.

(i) Mechanical index = MI = $\sum_{i=1}^{n} \left(\left(\frac{M_{e(a,i)}}{M_{av}} \right) \left(\frac{L_{(a,i)}}{TL_{(a)}} \right) \right)$ 　　　　(10.12)

where,

MI = mechanization index for the production unit 'a'

$M_{e(a,i)}$ = overall input energy due to machinery for crop 'i' in the production unit 'a'

M_{av} = regional average energy due to machinery

$L_{(a,i)}$ = land area cultivated with crop 'i' in the production unit 'a'

$TL_{(a)}$ = total farm land ownership of the production unit 'a'

(ii) Machinery energy ratio = MER = $\sum_{i=1}^{n} \left(\frac{M_{e(a,i)}}{T_{e(a,i)}} \right)$ 　　　　(10.13)

where,

MER = ratio between machinery energy and total input energy

$T_{e(a,i)}$ = total input energy (from labour, machine, seed, fertilizers, agrochemicals, animals) for the production of crop 'i' in the production unit 'a'.

Optimization of Energy Usage

High energy ratios correspond to high efficiency in the use of energy and low mechanization level. India has made considerable progress in increasing agricultural production and productivity due to the introduction of high-yielding varieties, intensive cropping systems, increased usage of chemicals and fertilizers, and a high level of mechanization. In contrast, agriculture production has become quite energy intensive. Keeping in mind the scarce and expensive energy resources, an attempt has been made to develop a functional relationship between energy input for different farm operations, seed, fertilizer, and chemicals, and crop productivity to optimize crop production systems.

Various models (linear, semi-log, log linear ,and polynomial) have been tried to develop the functional relationship between energy input and crop productivity to the crop production system. The data on energy used in performing various farm operations and from various sources for growing wheat, rice, maize, and cotton crops have been collected. Although all important variables are taken into account for developing the functional relationship except energy input from solar radiation, weather, and soil, there is the possibility that in some cases the intercept value may be negative, because it will purely be a mathematical relationship based upon data. The selection of the model is made on the basis of the R value. The general equation fitted for different crops is of the following type (Sidhu et al. 2004):

$$Y = A_{o} + \sum_{i=1}^{m} \sum_{j=1}^{n} A_{ij} (x_j)^i \qquad (10.14)$$

where,

m is equal to 1 for a linear relationship and 2 for a quadratic relationship
Y is yield (kg/ha)
A_0 is intercept
n is number of independent variables
A_{ij} is regression coefficients, and
x_j is the independent variables.

Machinery for Energy Conservation

Farm mechanization has played a pivotal role in improving agricultural production as well as productivity through the timeliness of field operations and by enabling proper use of critical inputs. A large number of farm equipment types have been developed which save energy and are highly efficient. This includes equipment for seedbed preparation, sowing, transplanting, interculturing, chemical and fertilizer application, harvesting, and threshing. Selection of the most appropriate farm machines for field preparation can save time and energy, and enable farmers to sow their crops in time and thus obtain a higher yield. The use of appropriate harvesting and threshing machinery helps farmers to overcome the vagaries of weather during the harvesting season and also to save energy.

Farm mechanization also improves quality of life by reducing drudgery and by providing respectability to farm operators. Energy-efficient machines to suit the requirements can be adopted, if care is taken of the following factors: (i) selection of proper machinery, (ii) matching the size of the implement with the power source, (iii) proper load matching, and (iv) proper maintenance and adjustments.

Equipment for Seedbed Preparation, Sowing, and Planting

Tillage is defined as the mechanical manipulation of soil to provide conditions favourable for crop production. It is an energy-intensive operation. Surface tillage and no tillage are collectively referred to as reduced tillage. Surface tillage is breaking, tearing, cutting, or otherwise loosening of the surface layers of soil with variously shaped disks or sweep or chisel cultivators prior to seed placement.

Rotavator

Roto-tilling is done by a rotavator which combines primary and secondary tillage in one operation. The rotavator is powered by the tractor PTO. When the rotavator works, it exerts a forward push on the tractor and thus facilitates traction. The advantage of rotary cultivation has been found to be more significant in seedbed preparation during the turn-around time between the *kharif* harvest and *rabi* sowing. Savings of 60–70% in operational time and 55–65% in fuel consumption with single rotavation compared to the conventional method of seedbed preparation with separate ploughing and harrowing operations have been observed, in addition to conservation of moisture due to destruction of capillaries.

Strip-Till Drill

In strip-till drilling, seeds are drilled directly in a narrow tilled strip in a single pass. Strip tillage technology has the distinct advantage that the soil is loosened in a band to allow effective root development. It avoids undesirable compaction of the soil in the crop rows. It is more suitable for subsequent crops after rice/soybean for timely planting in a single operation. The system is comprised of a seed-cum-fertilizer drill with a rotary attachment fixed ahead of the furrow openers for soil manipulation in strips. The rotary attachment consists of a rotor with flanges in which tines (blades) are remounted. The spacing between the flanges is the same as the row spacing of the strips to be planted. Power to the rotor is provided from the tractor PTO with the required speed reduction.

Zero-Till Drill

Zero-till drilling refers to when seeds are drilled directly into uncultivated seedbeds. This drill is similar to that of the conventional seed-cum-fertilizer drill except that the furrow openers are of a specially designed inverted 'T' type, which create furrow grooves with reduced surface exposure and thereby help to maintain the groove humidity in a reasonably wet soil for better germination of seeds and emergence of seedlings. This system offers the apparent advantage of timely planting at reduced time, fuel, and labour costs. Investigations have revealed that compared to conventional sowing of wheat, zero-till drilling saved 70% sowing time and 64% cost of the seeding operation with an overall increase in cost:benefit of 20–25%. The system is more effective in situations where late harvesting of rice compels a delay in the planting of wheat. The first irrigation is given within a week, which helps in providing better seed–soil contact and also moisture.

Inclined Plate Planter

An inclined plate planter facilitates uniform placement of a single seed in soil; and therefore, helps in saving of costly seeds. It avoids the problem of thinning of an over-planted population and helps to improve crop yield as the individual plant gets the required nutrients, water, and sunlight. The unit consists of individual seed boxes, furrow openers, and a transmission system with a ground drive wheel. The seed boxes are of modular design with an independent inclined plate-type seed-metering mechanism. The seed boxes are bolted to furrow openers. The furrow opener assemblies are adjustable for crop row spacing and work as a modular unit for sowing in each row. The planter is suitable for small- and bold sized seeds such as maize, sorghum, pigeonpea, soybean, chickpea, rapeseed, and mustard by use of different seed-metering plates. For a six-row tractor-mounted unit, the field capacity varies between 0.50–0.65 ha/h at an overall field efficiency of 70–75%.

Pneumatic Precision Planter

Precision planting is practiced to save seeds and place them in the proper soil environment for better seed distribution and uniformity at pre-determined seed and row spacing. It also has provision for drilling fertilizer simultaneously with seed. It consists of a pneumatic disc and suction-type seed-metering mechanism. It can be used for different crops by changing the disc. The planter is suitable for small to bold seeds and also for intercropping since each row has a separate planting mechanism (Sahoo and Rajan, 2016).

Sugarcane Sett Cutter Planter

This equipment consists of a carriage with an operator seat, sett cutter, two seed boxes, a rectangular box for fertilizer, a chute for seed dropping, and a pesticide tank. It is operated by a 45-hp tractor. As the equipment moves forward, the share point opens the furrow, the operator drops the setts through the chute, pesticide is sprinkled, and the fertilizer is applied. In the semi-automatic design, a rotating drum is used for each furrow with vertical compartments in it for feeding the setts. The sett is carried along by the rotating drum until it aligns with the opening provided in the stationary bottom plate underneath the drum. The work capacity of the machine is around 0.2 ha/h.

Rice Seeder

Direct seeding of rice in puddle soil offers the advantage of faster and easier planting, reduced labour and hence less drudgery, 7–10 days earlier crop maturity, more efficient water use and higher tolerance to water deficits, and often higher profits in areas with an assured water supply. Pre-germinated seeds (24 h soaking + 12–24 h incubation) are sown on to a puddle soil after 1–2 days of puddling. A shorter incubation time (12 h) is critical for easy flow of sprouted seeds from the perforated drums. The water should be drained before seeding and the puddle bed should be firm enough to support the seeder and to make shallow furrows for sowing. The seed rate depends on the rate of revolution of the drums and is normally 50–70 kg/ha. The rice seeds could also be drilled with a cup-type seed-metering mechanism in a well-prepared soil, provided weeds and water are properly managed.

Potato Planter

The potato planter facilitates furrow opening, tuber placement, and covering with the soil simultaneously. Belts with cup-type and picker-type tuber-metering systems are available for seeding. The picker has 12 notches spread uniformly on its periphery for holding an equal number of grabs for picker arms. The picker arms open out approaching the picking chamber and pick up a seed potato for subsequent release. The field capacity of the machine with two ridges varies from 0.2–0.4 ha/h with a tuber distance of 200–450 mm, row spacing of 600 mm, and depth of planting of 100–200 mm. The automatic unit saves 50–60% labour, 80–85% operation time, and 50–60% of the cost of the operation compared to the conventional method of placement of seeds in rows by making ridges manually.

Vegetable Transplanter

Vegetable transplanting is normally carried out manually all over the country. The labour requirements for transplanting vary between 10–30 man-days/ha for various vegetable crops, except onion. For onion, the labour requirements vary from 50–100 man-days/ha. A semi-automatic vegetable transplanter in which feeding of seedlings is done manually by the operator, and furrow opening and placing the seedlings in the furrows by the machine automatically, are carried out. Bare root nursery and cup-type nursery can be used. At present, the bare root nursery is adopted in India. Mechanical transplanters are generally tractor-operated multi-row models or self-propelled one-row models. A feasibility analysis showed that the adoption of these transplanters is economically feasible where the labour wage rate is more than Rs 60–80/day.

Rice Transplanter

Transplanting of rice is mostly practiced manually in India. It is time and labour consuming, besides being an arduous operation. Nearly 25–30 labour days are required for transplanting one ha of rice by hand using a root wash nursery. Mechanical rice transplanters have been introduced to mechanize the transplanting operation. The self-propelled rice transplanter (eight rows, single-wheel driven, 3-hp diesel engine) with three persons can transplant almost 1 ha a day using mat-type seedlings. This saves 75–80% of labour hours and 40–45% of the cost of transplanting compared to hand transplanting.

Tractor-Operated Bed Former-Cum-Seeder

A bed former-cum-seeder has been developed for sowing wheat on beds. The machine can make two beds in a single run and the width of each bed is adjustable (35–45 cm). The machine is capable of sowing one to three rows in each bed at a time along with the fertilizer application. It saves 15–20% water in comparison to flat sowing.

Tractor-Drawn Raised-Bed Pulse Planter

A tractor-operated raised-bed pulse planter was developed for making beds and planting pulse seeds in a single pass. Seeds are planted in one or two rows on beds with adjustable row spacings and plant spacings. Savings of seed, water, and energy are the hallmark of this precision planter. The planter can be used for other bold seeded crops with minor modifications (Kumar et al. 2020).

Manually Operated Seed-Drill

This has been designed to sow a single row of oilseed crops such as rapeseed and mustard and some other crops like wheat. In this, a seed hopper with a seed-metering mechanism is mounted on the wheel hand hoe. A machine is widely used for inter-row sowing of rapeseed and mustard in sugarcane and wheat crops. The machine can sow 0.8 ha/day with the help of two persons. A tractor-operated oilseed drill has been developed, which can sow crops like rapeseed and mustard with very low seed rates in the range of 3–4 kg/ha.

Interculture and Plant Protection Equipment

Power Weeder

The equipment consists of a 5-hp light-weight diesel engine mounted on the frame. The engine power is transmitted to the ground wheels and rotary through a gear reduction unit. The wheel setting is done as per crop row spacing. The rotary unit can be engaged/disengaged through the actuating clutch. The rotary weeder consists of discs mounted with curved blades in opposite directions alternatively in each disc. The rotating blades enable cutting of weeds and integrating into the soil. The width of coverage of the rotary weeder is 350 mm and the depth of operation can be adjusted.

Electro-Static Spraying

Agricultural spray particles are charged by applying an induction potential of up to 10 kV. These charged droplets experience an electrostatic force in addition to the normal gravitation and air drag forces which tend to attract the drops to the target plant and result in greater deposition of sprays even underneath the surface of the leaves where the insect population is higher. Application through this technique makes it possible to achieve pest control at a lower overall application rate.

Orchard Sprayer

Spraying of chemicals at heights in orchard crops poses problems in respect to the uniformity of application. The tractor-mounted orchard sprayer with air-assisted rotary atomizer has been found to be suitable for spraying of fungicides on small berries and mango trees up to 6 m height. The sprayer consists of a fluid tank, atomizer with hydraulic motor, and a flow control valve. The atomizer blows the chemicals up to the maximum swath providing a uniform and efficient spraying pattern. The flow control valve adjusts from low- to high-volume applications. The equipment operated by a 35-hp tractor can cover 0.20–0.50 ha/hr with 3 m penetration of droplets inside the plant canopy.

Self-Propelled High-Clearance Sprayer

This is a self-propelled unit suitable for spraying on tall crops like cotton. The machine consists of two rear steered wheels and two front lugged wheels which is powered by a 20-hp diesel engine through a gear box, tank, hydraulic pump, and boom fitted with 15 nozzles. The effective field capacity and field efficiency are 1.6–2.0 ha/hr and 70–80%, respectively.

Harvesting and Threshing Equipment

Grain Combine Harvester

Combine harvesting of field crops offers the advantage of timely harvesting with reduced grain losses and drudgery, and makes available the field for immediate sowing of subsequent crops. The machine being a self-propelled unit has the provision of cutting/reaping, conveying, threshing, cleaning, and delivering the grain to tanks/bags. It is generally suitable for harvesting rice, wheat, soybean, chickpea, and similar crops, and covers a wide area depending on the size of the cutter bar and field conditions. Combine harvesters with pneumatic wheels are suitable for wheat and other upland crops, whereas for rice in wet/water-logged field conditions, rubberized undercarriage systems are available, enabling minimum ground pressure. Provision for harvesting of a completely lodged crop with a crop divider attachment and provision for chopped/stripped straw as per requirements are available in many combines. A medium-sized combine including a 35-hp engine and with a reaping width of 1.4 m can give a field coverage of 0.25–0.30 ha/hr with good-quality grains and chopped/stripped straw. These combines are generally attached to a high-power engine and automatic hydrostatic transmission/controls to obtain the desired threshing speed, concave clearance, and grain movement.

High-Capacity Multi-Crop Thresher

The high-capacity multi-crop thresher consists of a spike tooth cylinder, three sets of aspirator blowers, cleaning sieves, and an automatic feeding and bagging system. The thresher is provided with accessories for threshing of different crops. It is operated by a 20-hp electric motor or 35-hp tractor. It is suitable for threshing of wheat, maize, sorghum, chickpea, pigeonpea, soybean, and sunflower. It saves 50% of the labour and operating time, and 54% of the operational cost as compared to the conventional spike tooth thresher.

Vertical Conveyor Reaper

Extensive work has been done on cereal harvesting equipment, and both animal-operated as well as tractor-operated reapers have been developed. However, the tractor-operated front-mounted vertical conveying reaper has been taken up by local industries because of its simplicity, windrowing quality, and low grain losses. This machine can cover about 0.4 ha/hr. The crop placed in the neat windrows has to be picked up manually for threshing by threshers.

Forage Harvesters

A flail-type forage harvester-cum-chopper has been developed for tall fodders. It is operated by a 35-hp tractor and can cover about 1.2–1.5 ha/day.

Tractor-Drawn Stubble Shaver

This is used for harvesting paddy straw after combining and is operated by a 35-hp tractor. The machine consists of two blades mounted on the ends of a rotating arm hinged with pivot pins. The power to the blades is supplied from the tractor PTO and a belt pulley drive giving a speed ratio of 1.8:1.0. The average field capacity of the machine varies between 0.48–0.80 ha/hr at a forward speed of 5–6 km/h.

Wheat Straw Combine

Wheat straw combines are used for the collection of wheat straw after harvesting the wheat crop with combines. This machine cuts and gathers the left-over straw from the field, chops it into fine straw called 'Bhusa' and then transfers it to a trolley. It has a 2 m wide cutter bar and is operated by a 45-hp tractor. The machine can cover about 0.4 ha/hr by working at a speed of 3–4 km/hr. Straw recovery in this case ranges between 50–55% in comparison to a manually harvested and mechanically threshed system. In addition to the above, straw combines also recover about 30–40 kg of clean grains/ha. The custom hiring rate of wheat straw combines is usually Rs 500/ha, whereas the straw recovery is worth Rs 1250/ha. Thus, there is a net saving of about Rs 750/ha with the use of wheat straw combines.

Energy Use in CA

Agricultural productivity is closely linked with energy inputs. The measure of energy flow in a crop production system provides a good indicator of the technological aspects of a crop production system in agriculture. For sustainability in agriculture, efforts should be made for efficient use of commercial energies, and harnessing renewable energy sources as supplementary and substituting commercial energy sources. Direct energy inputs to crop production systems are derived from power sources like humans, draft animals, engines, tractors, power tillers, and electric motors that are required to perform various operations as well as indirect energy inputs in the form of seeds, organic manures, fertilizers, pesticides, and growth regulators required for production.

Consumption of energy has been increasing at a steady rate to improve productivity in Indian agriculture. It is a critical input in intensive cropping. However, the energy-use efficiency is declining consistently. The adoption of high-yielding varieties, expansion of irrigation facilities, mechanization, fertilizer, diesel, and electricity consumption have pushed the demand for commercial energy to a new height. Among the field crops, legumes involve much less energy expenditure than cereals.

Tillage is the process of employing human, animal, and machine energy to alter the soil environment for weed control, crop establishment, and root proliferation to achieve higher yields and input-use efficiency. Water, fertilizers, fuel, and pesticides are other major energy inputs in intensive cropping. A highly energy-intensive rice--wheat system in the Indo-Gangetic plains uses more than 52,000 MJ/ha of energy to produce 8.0 t/ha grains. In this cropping system, energy expended in tillage constitutes only 7% of the total, compared to 42% in irrigation and 36% in fertilizers and manures (Prihar et al. 2000). Although energy expended in tillage is relatively low, it influences the efficiency of energy use in water and nutrients. A three-year field study on tillage, irrigation regimes, and N effects on wheat following rice on a sandy loam soil showed that crop responses to applied N and irrigation were governed by the tillage system. These relations showed that for comparable wheat yields, higher amounts of water and N are required with ZT than with conventional tillage. Although the energy required for tillage was less in ZT than in conventional tillage, total energy requirements in ZT for a given yield exceeded those in conventional tillage because of the higher energy input in water and N. This study suggested that within a given crop production system, energy expended in tillage should not be considered in isolation. Also, one form of energy can be substituted, within certain limits, by another, depending on the availability and price of the input.

More energy was consumed in fertilizer treatments for a soybean–chickpea crop sequence compared to the control, and increasing the levels of nutrients decreased the energy-use efficiency and productivity (Mandal et al. 2002). Seedbed preparation by power tiller (rotavator 2 passes + levelling) and tractor with improved implements (mouldboard plough 1 pass + disc harrow 2 passes + levelling) gave higher profit and energy-use efficiency than seedbed preparation using other methods. Sowing seeds by drilling in rows gave markedly greater returns and energy output than broadcast sowing (Sharma and Thakur 1989). Srivastava (2003) reported that tillage before planting, which required about one-third (936 MJ/ha) of the total operational energy (2795 MJ/ha), could be saved with ZT without adversely affecting the sugarcane yield. The largest

Table 10.2 Energy requirement (MJ/ha) in a greengram–mustard–cowpea cropping system

Crop\tillage	Tillage and crop establishment				
	ZT	CT	FIRB	BBF	Mean
Greengram	4,666	6,075	6,193	6,219	5,788
Mustard	10,919	12,327	11,806	12,149	11,800
Cowpea	9,553	10,961	9,744	10,437	10,174
Total	25,138	29,363	27,743	28,805	

Source: Behera and Sharma (2008).

contribution of CA to reducing the CO_2 emissions associated with farming activities is made by a reduction of the tillage operations. Erenstein and Laxmi (2008) compared studies in rice–wheat systems in the Indo-Gangetic plains and found seasonal savings in diesel for land preparation with ZT to be in the range of 15–60 l/ha, with an average of 36 l/ha or 81% saving across the studies, equivalent to a reduction in CO_2 emissions of 93 kg/ha in a year.

The adoption of ZT and conservation tillage has been increasing in recent years due to its effects on energy economy in the system. A study revealed that under the recommended fertilizer application, mustard crop required the highest energy of 11,800 MJ, followed by cowpea (85% of mustard) and green gram (50% of mustard). The total energy requirement of the green gram–mustard–cowpea cropping system under conventional tillage was 29,363 MJ, while it was 25,138 MJ for ZT, thus an energy saving of around 15–20% was recorded (Table 10.2).

Energy Relations in Cropping Systems

The adoption of CA not only helps to improve the soil quality and results in higher nutrient- and rainwater-use efficiency, but also provides a range of benefits through lower demand for chemical fertilizers, fuel, labour, and probably pesticides in the long term (Sangar et al. 2004). The benefits of CA include a reduction in the amount and cost of labour and energy required for land preparation, and sowing due to the fact that the soil becomes soft and easy to work. Sowing directly into the soil without any prior tillage operations implies less labour requirement. In fact, the reduction in cost and time required are usually the most compelling reasons for farmers to adopt conservation tillage. Farmers see ZT systems as a less laborious and less risky procedure, enabling fuel and machinery savings and cost reduction. Rotations can also spread labour needs more evenly over the year. Not tilling the soil and planting directly into a mulch of crop residues can reduce labour requirements at a critical time in the agricultural calendar, particularly in mechanized systems when a direct-seeding machine is used. Conservation tillage reduces the energy (for example, fuel for machines and calories for humans and animals) and time required.

Returns from crop cultivation are shrinking gradually due to the escalation of input costs. Under such a situation, CA systems have gained importance to make farming profitable by reducing the variable costs and enhancing resource-use efficiency. Karunakaran and Behera (2013) studied energy relations and irrigation water-use efficiency (IWUE) involving sequential tillage in a soybean–wheat system. The energy relations, namely input energy, output energy, net energy, and energy-use efficiency were statistically similar in most of the cases except for the energy-use efficiency for both soybean and wheat due to different sequential tillage and crop establishment techniques. Residue management significantly influenced the energy relations in both crops. Wheat + soybean residue recorded the highest gross energy, while it was the reverse with net energy and energy-use efficiency where there was no residue.

Maximum energy input was recorded with a flat system of crop establishment in a soybean–wheat cropping system (Table 10.3). The raised-bed system of crop establishment registered lower energy input to the system. The energy output among the sequential tillage and crop establishment techniques was non-significant. Among the residue management treatments a lower energy requirement was recorded with no residue while the maximum was recorded with the wheat + soybean residue treatment which had a 11.9–19.3% higher energy output than no residue.

Among the various CA practices in soybean, the lowest energy requirement was under ZT+NR and ZT+SR. This was due to the saving of energy in

Table 10.3 Energetics of a soybean (S)–wheat (W) cropping system as influenced by sequential tillage and crop establishment

Treatment	Input energy ($\times 10^3$ MJ/ha)		Net output energy ($\times 10^3$ MJ/ha)		EUE (output/input ratio)
	1st year	2nd year	1st year	2nd year	
Tillage and crop establishment					
CT-Flat (S)–ZT-Flat (W)	59.8	57.4	167.4	193.3	4.29
ZT-Flat (S)–CT-Flat (W)	59.8	57.5	172.3	200.4	4.45
CT-Bed (S)–ZT-Bed (W)	58.8	57.1	161.2	190.7	4.40
ZT-Bed (S)–CT-Bed (W)	58.8	57.1	154.6	182.2	4.21
SEm±			4.1	5.4	
CD (P=0.05)			NS	NS	
Residue management					
No residue	21.8	19.8	189.0	206.4	9.55
Wheat residue	59.3	57.3	164.3	188.2	3.03
Soybean residue	59.3	57.3	163.1	196.9	3.10
Wheat + soybean residue	96.8	94.8	139.1	175.1	1.65
SEm±			3.3	4.2	
CD (P=0.05)			9.8	12.3	

Source: Karunakaran and Behera (2016).

tillage operations. The highest energy requirement was recorded in ZT+SWR, followed by CT+NR. The maximum energy in ZT+SWR was due to the energy consumed in terms of residue at 3 t/ha and in CT+NR was due to the energy incurred in tillage operations needed for seed bed preparation. In residue-added plots, the high energy value of crop residue (12.5 MJ/g) was the reason for the maximum energy requirement (Karunakaran and Behera 2016). The maximum energy-use efficiency was recorded in ZT+SR and the minimum in ZT+SWR. Since crop residues have some energy value, their addition in large quantities makes treatments inefficient with respect to net energy balance and energy-use efficiency. Split application of fertilizers required more energy compared to basal applications. This was due to the energy involved in human labour for top dressing. The maximum output energy, energy balance, and energy-use efficiency were recorded with a basal application of 125% recommended dose of N (RDN), which was statistically on a par with 100% basal and 25% top dressing and significantly superior to the rest of the treatments.

Among the nitrogen management practices, the highest energy requirement was associated with 125% RDN compared to 100% RDN treatments. Among the N management practices, the highest energy was required under 125% RDN, more so when 25% N was applied through top dressing compared with 125% as basal. This was due to the energy involved in human labour for top dressing. The maximum output energy, energy balance, and energy-use efficiency were recorded with basal application of 125% RDN which was statistically on a par with 100% basal and 25% top dressing and significantly superior to the rest of the treatments. This was due to higher seed and stover yield in these treatments.

Maize and wheat crops require higher amounts of total energy for higher production levels. The high energy requirement in a maize–wheat system was attributed to the energy consumed in tillage operations and N fertilizer applications. A system of ZT–ZT in the sequence crops recorded the lowest energy requirement. This was due to the higher saving of energy in ploughing and for seed-bed preparation (Ram et al. 2010). ZT practices reduce the energy requirement due to saving of energy in tillage practices as well as in weeding operations (herbicides were used for weed management) than CT (Jain et al. 2007). The higher values of gross output energy and net output energy were recorded under CT-ZT treatment due to higher yield performance by crops as well as saving of input energy requirement in ZT practices in wheat. The maximum energy-use efficiency under the ZT–ZT sequence was due to having the lowest energy consumption due to continuous ZT accompanied with good yield performance of the wheat under ZT practices.

Conclusion

Efforts to develop, refine, and disseminate CA-based technologies have been underway for more than two decades in India. Significant progress has been made on ZT wheat under rice–wheat rotation in the Indo-Gangetic plains. There are more payoffs than trade-offs with the adoption of CA but the balance between the two has not been well understood not only by its adopters but also the promoters. CA technologies provide opportunities to reduce the cost of production and energy use, and enhance energy-use efficiency. In addition, CA saves water and nutrients, increases yields, promotes crop diversification, leads to efficient use of resources, and benefits the environment. Energy in agriculture is required for field operations which facilitate crop production and processing, besides indirect energy in terms of seeds, fertilizers, and chemicals. The efficacy of agricultural inputs and natural resources, i.e. seeds, fertilizers, chemicals, land, and water is increased through the adoption of appropriate agricultural machinery. ZT and CA practices have emerged as potential approaches for enabling energy economy and enhancing the energy-use efficiency in agriculture. The use of appropriate mechanization technologies not only ensures better productivity but also helps in conserving energy in crop production systems.

References

Behera, U.K. and Sharma, A.R. 2008. Effect of conservation tillage and crop establishment practices on performance, energy saving and carbon sequestration in greengram-mustard-cowpea cropping system. *International Conference on Conservation Farming and Watershed Management in Rainfed areas for Rural Employment and Poverty Eradication*. Soil Conservation Society of India, February 12–16, 2008, New Delhi.

Chamsing, A., Salokhe, V.M. and Singh, G. 2006. Energy consumption analysis for selected crops in different regions of Thailand. Agricultural Engineering International: the CIGR E-journal. Manuscript EE 06 013. Vol. VIII. November, 2006.

Erenstein, O. and Laxmi, V. 2008. Zero tillage impacts in India's rice–wheat systems: A review. *Soil and Tillage Research* 100: 1–14.

Jain, N., Mishra, J.S., Kewat M.L., et al. 2007. Effect of tillage and herbicides on grain yield and nutrient uptake by wheat (*Triticum aestivum*) and weeds. *Indian Journal of Agronomy* 52(2): 131–134.

Karunakaran,V. and Behera, U.K. 2013. Effect of tillage residue management and crop establishment techniques on energetic, water use efficiency and economics in soybean–wheat cropping system. *Indian Journal of Agronomy* 58(1): 42–47.

Karunakaran, V. and Behera, U.K. 2016. Tillage and residue management for improving productivity and resource-use efficiency in soybean (*Glycine max*)–wheat (*Triticum aestivum*) cropping system. *Experimental Agriculture* 52(4): 1–18.

Kumar, R., Sahoo, P.K., Kumar, A., et al. 2020. Design, development and performance evaluation of tractor-drawn raised-bed pulse planter for precision sowing of pigeonpea. *Indian Journal of Agricultural Sciences* 90(9): 1800–1809.

Mandal, K.G., Saha, K.P., Ghosh, P.K., et al. 2002. Bioenergy and economic analysis of soybean–based crop production systems in central India. *Biomass and Bioenergy* 23: 337–345.

Panesar, B.S. and Bhatngar, A.P. 1994. Energy norms for inputs and outputs of agricultural sector. In: *Energy Management and Conservation in Agricultural Production and Food Processing*, pp. 5–16. USG Publishers and Distributors, Ludhiana.

Prihar, S.S., Gajri, P.R., Benbi, D.K., et al. 2000. *Intensive Cropping - Efficient Use of Water, Nutrients and Tillage*. Food Products Press, New York.

Ram, H., Kler, D.S., Singh, Y., et al. 2010. Productivity of maize (*Zea mays*)–wheat (*Triticum aestivum*) system under different tillage and crop establishment practices. *Indian Journal of Agronomy* 55(3): 185–190.

Sahoo, P.K. and Behera, U.K. 2011. Energy use in production agriculture, 397–414. *In:* Resource Conserving Techniques in Crop Production. (Eds) Sharma, A.R. and Behera, U.K. Scientific Publishers, Jodhpur, India.

Sahoo, P.K. and Rajan, P. 2016. Development of Pneumatic Precision Planter for Vegetables. Technical Manual, The Institution of Engineers (India) 57: 1–6.

Sangar, S., Abrol, I.P., and Gupta, R.K. 2004. *Conservation Agriculture: Conserving Resources-Enhancing Productivity*. Centre for Advancement of Sustainable Agriculture, National Agriculture Science Centre (NASC) Complex Pusa Campus, New Delhi.

Sharma, R.S. and Thakur, C.L. 1989. Economic and energy factors in soybean cultivation. *Indian Journal of Agronomy* 34(3): 337–340.

Sidhu, H.S., Singh, S., Singh, T. et al. 2004. Optimization of energy usage in different crop production systems. *Journal of Institution of Engineers* 85: 1–4.

Srivastava, A.C. 2003. Energy savings through reduced tillage and trash mulching in sugarcane production. *Applied Engineering in Agriculture* 19(1): 40–45.

Part III

Soil Health and Greenhouse Gas Emissions

11 Greenhouse Gas Emission and Carbon Sequestration in Conservation Agriculture

Arti Bhatia, Avijit Ghosh, Amit Kumar, and Ranjan Bhattacharyya

Introduction

Agriculture is a fundamental sector that provides food for both people and animals, produces fibres for the textile sector, and supplies many other products and services essential for the existence of humanity. Like any other economic activity, agriculture is linked to the natural and social environment in which it is developed, and interacts with it. If there is any productive activity that depends directly on the climate and its variability, this is undoubtedly agriculture. A change of temperature and precipitation, or an increase in the concentration of atmospheric greenhouse gases (GHGs), especially CO_2, will significantly affect crop development and performance. With the application of conservation agriculture (CA) principles, it is possible to sequester large amounts of CO_2 per ha, compared to tillage-based systems (Lal 2004). Conservation agriculture is an approach to farming that seeks to increase food security, alleviate poverty, conserve biodiversity, and safeguard ecosystem services. Conservation agriculture practices can also contribute to making agricultural systems more resilient to climate change. In many cases, CA has been proven to reduce GHG emissions and enhance their role as carbon sinks.

Conservation agriculture was originally developed to combat wind and soil erosion in the United States. It is a farming approach that fosters natural ecological processes to increase agricultural yields and sustainability by minimizing soil disturbance, maintaining permanent soil cover, and diversifying crop rotations. The intent of CA practices is to optimize crop production while promoting soil health and providing ecosystem services, i.e. improved soil, water, and air quality. Conservation agriculture offers considerable environmental improvement of agricultural ecosystems without reducing yields. It has a double effect on the reduction of GHG concentration in the atmosphere and on increased SOC concentration through higher C inputs. Additionally, the drastic reduction in tillage operations, along with the minimal mechanical soil disturbances, lead to a reduction in CO_2 emissions through less fuel consumption and reduced soil organic matter (SOM) mineralization. In annual crops, CA governs the principles of zero tillage (ZT), permanent organic soil cover, and crop rotations,

DOI: 10.4324/9781003292487-14

while in permanent crops, the CA approach is based on ground cover between the tree and crop rows.

Despite the fact that CA has been successfully developed for many regions of the world, its adoption rate is much lower in India. Research during the past decade all over the world has demonstrated that CA can reduce GHG emissions as well as sequestering carbon in soils. However, studies on this aspect of CA in India are evolving with the necessity that more concentrated work be performed in the future, especially on quantification of the GHG balance under CA systems, taking into account all components, such as fuel consumption, herbicide use, C sequestration, and GHG emission. This chapter aims to discuss some basic principles and factors affecting GHG emission and C sequestration under conventional and conservation agriculture and to present some case studies in India on CA impacts on GHG emission and C sequestration.

Greenhouse Gas Emission

Agriculture has been considered to be one of the main contributors to GHG emissions and this continues to increase with increased crop production (Bhatia et al. 2010). Agricultural production practices contribute to C and N dynamics through the flux of carbon dioxide (CO_2), methane (CH_4), and nitrous oxide (N_2O). The major practices that contribute to GHG emissions are increased irrigation, increased use of fertilizers, intensive tillage, and burning of crop residues. However, agriculture is not only a contributor of GHG emissions but can also help in mitigation of climate change. The greatest challenge for agriculture in the 21st century is the development of sustainable alternative agricultural practices which increase productivity, and help in mitigating climate change and conserve natural resources. Gupta et al. (2016) reported that CA systems led to lowering of the global warming potential of cropped soil besides an increase in the benefit:cost ratio under zero-tilled wheat with rice residue retention followed by direct-seeded rice in the rice–wheat system of the Indo-Gangetic plains.

Nitrous Oxide

N_2O is a potent and long-lived GHG, with a global warming potential 310 times that of CO_2, and remaining in the atmosphere for up to 114 years. Fluxes of N_2O are produced in soils through processes of nitrification or denitrification, with higher emissions from fertilized soils, livestock manure, and biomass burning. Emissions of N_2O increase with the application of N fertilizers by increasing N availability in soil. In addition to direct emissions, N losses from volatilization and leaching contribute to 'indirect' N_2O emissions downstream. Nitrous oxide is emitted from soils when there is ample nitrate, soluble organic C, and wet soil conditions. Nitrification – the oxidation of ammonium to nitrate – occurs in aerobic conditions, while denitrification – the reduction of nitrate (NO_3^-) to N_2O and N_2 – takes place in anaerobic conditions. The relative contribution of these two N pathways to N_2O formation depends on episodic changes in soil aeration and the water-filled pore space (WFPS) (Bhatia et al. 2010).

Methane

Methane has a lifetime of 12 years and a global warming potential 21 times that of CO_2 over a 100-year horizon. Agricultural soils contribute to CH_4 emissions as a result of methanogenic processes in waterlogged conditions that are usually associated with rice production. Flooded rice production contributes 20% of total global CH_4 emissions. The C emissions in the form of CO_2 or CH_4 from rice soils are dictated by factors such as crop biomass, the amendments use, growing conditions, cultural practices, types of cultivars, fertilizer, soil types, and water management. The magnitude of CH_4 emissions is primarily a function of water management, with the addition of both mineral and organic fertilizers having a significant influence. The addition of organic fertilizers has the potential to increase emissions by over 50% relative to mineral fertilizers. CH_4 can be consumed (oxidized) by soil microorganisms and the total CH_4 flux from soils is therefore the difference between the production of CH_4 under anaerobic conditions and CH_4 consumption (Malyan et al. 2021). Agricultural soils, particularly those that have been fertilized, have a significantly lower CH_4 oxidation rate compared to natural soils.

Carbon Dioxide

The global atmospheric concentration of CO_2 has increased from a pre-industrial value of about 280 to 405 ppm in 2017. Naturally CO_2 is produced during the process of heterotrophic and autotrophic respiration. Heterotrophic respiration includes root, anaerobic, and aerobic microbial respiration, whereas autotrophic respiration includes above-ground plant respiration. An estimated 60 Pg C/year (1 Pg = 1×10^{15} g) is emitted to the atmosphere by autotrophic respiration and a similar amount is emitted as a result of heterotrophic respiration. CO_2 makes up 20% of the total GHGs emitted from agricultural soils, although agricultural soils are considered CO_2 neutral because in a conventional crop-growing condition, the emission and consumption of CO_2 due to plant respiration and photosynthesis, respectively, are balanced. However, rice–wheat cropped soils have been reported to be a major sink of atmospheric CO_2 during the crop growth period sequestering -944 g C/m^2 (Kumar et al. 2021). Emissions of CO_2 from soils appear to be highly variable in heterogeneous soil micro-sites, and they are influenced by the microbial processes, activity of roots, residue and litter content, and the microclimate. Soil conservation practices involve the retention of soil surface with crop residues, and may decrease CO_2 emissions with reduced tillage. The ZT system increases the C content in the soil, which is an effective strategy for CO_2 sequestration. The emission of CO_2 from of agricultural soils depends on the oxidative state of the soils. The incorporation of crop residues in the soil produces CO_2 if soil is aerobic, while anaerobic conditions produce CH_4.

CA and Emission of GHGs

Conservation practices mainly bring about two modifications: (i) minimal soil disturbance and (ii) retention of plant residues in soil. These two factors cause changes in soil properties and processes, physico-chemically as well as biologically. With the absence of intense soil disturbances, moisture retention improves, aggregates are stabilized, organic matter is protected, and microbial communities are less disturbed. With the absence of tillage, however, soil compactness increases, which may in the long term offer some problems in specific agro-ecosystems. Increased resistance to gaseous diffusivity under ZT causes low mobility of gases along the soil profile, thus affecting gaseous transport. Zero tillage is often accompanied by harvesting of ripened ears, while leaving the rest of the crop in the field. Residues added are colonized by decomposers, thus producing simpler compounds which act as substrates for many enzyme-mediated reactions. These processes ultimately guide the production or consumption of GHGs, relative proportions of the two leading to the observed net fluxes. Under reducing conditions, as is common in flooded rice fields, residue incorporation tends to make soil more anaerobic since organic compounds usually serve as electron donors. All these factors interact in complex ways, thus influencing GHG emissions, transport, and consumption. CH_4 emission rates are generally higher under continuous no-tilled than in continuously tilled soils. However, N_2O emission rates are higher in alternately tilled soils due to residue incorporation coupled with intermediate soil disturbances. The ZT permanent broad-bed systems with surface retention of crop residues lead to improvements in soil microbial activity and improved aggregation, and it is a feasible mitigation practice for reducing soil emissions per unit of grain produced (Bhattacharya et al. 2008). Alterations to drainage regimes and residue incorporation in rice production systems can also reduce CH_4 emissions. Crop management practices that increase SOC in semi-arid regions generally reduce CH_4 emissions. However, wetting and drying of soils may also enhance N_2O emissions and soil C mineralization, thereby reducing net GHG mitigation potential.

CA and Climate Change Mitigation

The net potential of CA to contribute to climate regulation and serve as a global warming mitigation strategy depends on the direction and magnitude of changes in soil C, N_2O, and CH_4 emissions associated with its implementation compared to conventional farmers' practices. Collectively, this is assessed in terms of the global warming potential (GWP) of the farming practices, which are soil, climate, and management dependent. For example, if there is an increase in soil C that is greater than the combined increase in N_2O or CH_4 emissions (expressed as CO_2 equivalents), the net GWP decreases. It is also important to note that there can be considerable impacts of CA compared to conventional agriculture, with changes in the intensity of mechanical tillage, less irrigation, and possibly less N fertilization and the associated reduced use of fossil fuels under CA.

Mitigating N_2O Emissions

Management practices that increase N-use efficiency result in less reactive N available for potential conversion to N_2O. Reducing the frequency of high N-demanding crops and including non-leguminous cover crops in rotation can reduce reactive N and thereby N_2O emissions. Moreover, inclusion of cover crops in rotation can contribute to increased C sequestration, representing an important GHG mitigation co-benefit. Zero tillage can, in some soils and climates, increase N_2O emissions compared to conventional tillage (CT) by increasing the soil water content and denitrification rates. Zero tillage has been found to increase SOC, and the improved physical properties have contributed to reduced N_2O emissions compared to both conventional and reduced tillage. Conservation agriculture has many attributes that can use the applied N more efficiently and reduce the risk of high N_2O emissions from soils.

With ZT, residues are returned to the soil resulting in surface mulches, which may lower evaporation rates, and hence increase soil moisture and increase labile organic C and consequently increase N_2O emissions compared to CT. Increased bulk density with NT/RT compared to CT may also increase emissions. However, continuous CA (residue retention along with NT/RT) maintained soil bulk density in many regions (Bhattacharyya et al. 2015). On the other hand, lower soil temperatures and better soil structure under ZT may reduce the incidence of soil saturation and reduce N_2O emissions. Some studies concluded that ZT only increased N_2O emissions in poorly aerated soils, and that N_2O emissions from ZT declined with time. Zero tillage in some soils increased N_2O emissions compared with CT by increasing soil water content and denitrification rates.

The frequency and magnitude of N_2O emissions are linked to the soil structure, which is a function of bulk density, soil C, and aggregation, all influenced by tillage practices and residue inputs. Nitrification is the main source of N_2O at low WFPS below 40%, while the contribution from denitrification increases above 65–75% WFPS. The N_2:N_2O ratio increases with little N_2O produced at WFPS above 80–90%. Soil bulk density is generally higher with ZT compared to conventional practices; therefore, the WFPS is higher. Hence, anaerobic conditions and denitrification are potentially induced sooner at the same water content with ZT. Residue management and crop rotations can affect N_2O emissions by altering the availability of NO_3^- in the soil and the decomposability of C substrates. The retention of crop residues and higher soil C in surface soils with CA play major roles in these processes. Under anaerobic conditions associated with soil water saturation, higher contents of soluble C or readily decomposable organic matter can significantly boost denitrification with the production of N_2O.

The quantity and quality of crop residues or cover crops of CA systems affect N_2O emissions. Legume residues can result in higher N_2O-N losses than those from non-legume, low-N residues. The N_2O emissions with legume N-rich residues compared to N mineral fertilizers, however, are lower per unit N added compared to the inorganic source. Low-quality cereal crop residues (C: N ratio, generally >25) combined with surface application of residues in CA systems could result in immobilization of N and ultimately decrease N_2O production compared to conventional systems. The net result of CA on N_2O emissions will thus depend on the crop rotation practices and the types and amounts of crop residue retention under CA systems compared to conventional ones. The inconsistent results of N_2O emissions with CA practices are potentially due to the lack of long-term observations at any site, and the high temporal and spatial variability in emissions.

Mitigating CH₄ Emissions

Transplanted flooded puddled rice is a large contributor of CH_4 emissions from agriculture. Reduced tillage or ZT is currently being promoted in the Indo-Gangetic plains (IGP) in the rice–wheat system. With this system, direct-drill seeded rice does not require continuous soil submergence, thereby reducing CH_4 emissions for lowland rice when it is grown as an aerobic crop (Pathak et al. 2009; Bhatia et al. 2013). The overall impact of RT in this environment, however, appears to be relatively minor. Grace et al. (2012) estimated an average of 29.3 Mg/ha GHGs emitted over 20 years in conventional rice–wheat systems across the IGP; this decreased by only 3% with the widespread implementation of CA.

The effect of tillage practices on the rate of CH_4 consumption, in general, depends on the changes in gas diffusion characteristics in soils. A decrease in CH_4 consumption and a potential net emission of CH_4 could be expected with RT or NT due to increased bulk density and WFPS. However, no significant tillage effects on CH_4 oxidation rates have been reported in the literature. Decreased CH_4 oxidations or increased CH_4 emissions with crop residue retention under CA have been observed. Residue retention provides a source of readily available C, which enhances CH_4 emissions from rice paddies that are generally under anaerobic conditions. Crop residues may affect CH_4 oxidation in upland soils and emission patterns in flooded soils differently depending on their C:N ratios. Residues with high C:N ratios have little effect on oxidation, while residues with narrow C:N ratios seem to inhibit oxidation. However, ZT has a greater ability to store SOC, to enhance water infiltration, to reduce water/nitrate runoff, and to stabilize aggregates in soils. Furthermore, it reduces CO_2 emissions by regulating the microbial decomposition of SOM which is otherwise stimulated by soil tillage.

Management strategies that can be aligned with ZT to keep soil in the oxidative state and promote aerobic organic matter decomposition are potential mitigation strategies for reducing CH_4 emissions from soils. Generally, reducing the duration of flooding is also being promoted as a practical solution to reduce CH_4

emissions in CA-based rice production systems, but these may be offset partially by increased N_2O emissions.

Mitigating CO_2 Emissions

Management practices that increase C inputs while reducing C losses can serve to enhance soil C sequestration. Enhancement of soil C sequestration can be achieved by maintaining plant residues on the soil surface, minimizing soil disturbance and erosion, improving the irrigation efficiency, adopting complex cropping systems that provide increased root biomass and/or continuous ground cover, reducing fallow periods, breeding crops for traits with larger root systems, and applying C-rich substrates to soils. Detailed processes involved in C sequestration under CA are discussed in the subsequent sections.

CA and C Sequestration

Soil organic C is the main component of organic matter (OM) and it is widely accepted as an indicator of soil quality, as it is of prime importance in all soil processes, improving its structure, fertility, and water-holding capacity. The reasons for an SOC decrease are:

i   The lower input of OM in the form of crop stubble.
ii The higher humus mineralization rate caused by intensive and inappropriate tillage. Tillage facilitates the penetration of air into the soil, and therefore, the mineralization of humus, a process that includes a series of oxidation reactions, generating CO_2 as the main by-product. One part of CO_2 gets trapped in the porous space of the soil, while the other part gets released into the atmosphere through diffusion mechanisms between zones of the soil with different concentrations.
iii The higher rate of erosion, which causes significant losses of OM and minerals. In intensive tillage-based systems, the preparation of soil for sowing leaves the soil exposed to erosive agents for a long period of time causing aggravated soil loss. Management practices that increase C inputs while reducing carbon losses can serve to enhance soil C sequestration.

Carbon sequestration is, in simplistic terms, the net difference between carbon inputs (photosynthesis, organic amendments such as manure) and outputs (decomposition, soil erosion) in agro ecosystems. Soil C sequestration can be enhanced by extending the effective growing season by growing perennial crops or by including intercrops or cover crops in the crop rotation, which would enhance the capture of CO_2 through photosynthesis and thereby increase the C inputs into the soils. Increasing the time periods where crops are actively growing would enhance C sequestration as long as there are no limiting factors (e.g. nutrient deficiencies, salinity, excess or insufficient water, soil pH). This can be achieved by using perennial crops, intercrops, or cover crops.

Cover Crops

This is the most representative agronomic CA practice in permanent crops, whereby the soil surface between the rows of trees remains protected from the water erosion generated by the direct impact of raindrops. At least 30% of the soil surface is protected by groundcover.

a   Reducing or eliminating tillage can reduce the mixing of the crop residues in the soils, and thereby result in lower decomposition rates and CO_2 emissions. Soil organic C outputs are a function of residue characteristics (e.g. C:N ratio) and their decomposition rates, which are often enhanced when the soil is cultivated.

b   Maximizing the return of C, by minimizing residue removal or burning can bolster soil carbon. Soil organic C can be maximized by growing crops with high biomass production (e.g. maize vs. soybean) or by using deep-rooted crops (e.g. wheat and barley) and by reducing the fallow period. In annual cropping systems, the main biomass inputs to the soil are decaying roots and crop residues.

Crop residues, including wheat straw and maize stover, are a feedstock for the production of cellulosic biofuel that could contribute to meeting advanced biofuel targets and decarbonization goals for the transport sector. Crop residues provide a number of environmental services when left in the field, including contributing to the formation of SOC, preventing erosion, reducing evaporation from the soil surface, improving soil structure, supporting living organisms, contributing nutrients to the soil, and providing water filtration and retention capacity. Whether and how much crop residue can be harvested without significant negative impacts on these ecosystem services is a critical question in understanding the potential for producing sustainable, low-carbon biofuel from this type of feedstock. Hence, a new quantitative assessment of the impacts of residue retention compared to removal on SOC sequestration is needed. It is generally accepted that the retention of crop residues in the field contributes to greater SOC formation and higher SOC levels compared to complete residue removal, but the degree of SOC benefit as well as the amount of residue necessary to provide SOC benefits remain unclear.

Crop Residues

Most of the crop residues produced in the tropics have huge potential to act as a C sink. Unfortunately, 25–40% of rice residues are burnt in the open air in order to clear the field for the next crop in intensive tropical rice-based cropping systems. It is a paradox that on the one hand we have a shortage of animal feed, biofuel, and manure, while on the other hand a considerable amount of crop residue is either wasted or burnt. This is not only a big loss of natural renewable resources, but also a source of GHG emissions and environmental pollution.

However, these residues can effectively be used as mulch and for the production of manure, ethanol, biodiesel, biochar, etc. as well as being employed in CA. In a rough estimate, if 20% of the world's rice straw was used for the production of ethanol annually, about 40 billion litres could be generated, which could replace 25 billion litres of fossil fuel-based gasoline. In this way, net GHG emissions would be reduced to the tune of 70 billion tons of CO_2 equivalent. Therefore, the effective and alternative management of crop residues would not only reduce GHG emissions, but simultaneously offset fossil fuel use. It could also sink atmospheric CO_2 for a larger period of time if it were judiciously retained and subsequently incorporated into topsoil, composted, or used for biochar production, thereby building a defence against climate change.

In general, CA moderates the impact of high temperature on and within the soil. It prevents the negative impact of heat on seed germination, gives high agronomic resilience by extending the seeding period, and improves seedling growth. It also suppresses weed growth and creates better conditions for root development and seedling growth. Improved infiltration and retention of soil moisture under CA result in less severe, less prolonged crop water stress and increased availability of nutrients. CA also adapts to climate change due to a reduction in the risks of pest and weed infestations and total crop failure. Wider diversity in plant production also contributes to minimizing these risks. By reducing organic matter decomposition and increasing C inputs coupled with increased humus formation and C accumulation, CA leads to enhanced C sequestration that mitigates GHG emissions, as mentioned earlier. Conservation agriculture increases micro-aggregates inside macro-aggregates and also augments labile C pools. Thus, CA improves system productivity and soil C sequestration. It also reduces energy needs on account of reduced fuel and mechanization usage, further contributing to climate change mitigation. Better physical soil quality ensures that the cropping system is optimized to cope with both heavy rainfall events and prolonged drought, events that are likely to increase in frequency in the future due to climate change. This diversity allows an ecosystem to remain stable when facing changes in environmental conditions. Through these processes, among many, CA adapts to and mitigates against global climate change.

Effect of CA on C Sequestration

Collectively, CA practices lead to an increase in water stable aggregates, greater SOC concentrations, and protection from wind and water erosion. Conservation agriculture-based crop management technologies include zero tillage (ZT) with residue recycling, laser-assisted precision land levelling, direct drilling into residues, and direct seeding. In the Himalayan region, year-round ZT under an irrigated rice–wheat system with two irrigations at critical growth stages (Bhattacharyya et al. 2008), year-round ZT with integrated nutrient management under an irrigated rice–wheat system (Bhattacharyya et al. 2012b), and 10-cm stubble retention (under CA) of rice and wheat crops for maximum yield and fodder production (VPKAS 2011) are novel technologies. Zero tillage enhanced macro-aggregate associated SOC and intra-aggregate particulate organic C under a rainfed finger millet–lentil system, but only in the topsoil. Plots with minimum tillage (MT; a 50% tillage reduction) improved SOC stock in the 0–15 cm layer, as well as soybean yield. Under direct-seeded rice–wheat systems, adoption of ZT with two irrigations in each crop improved topsoil physical properties and SOC content after four years with similar mean crop yields as with CT using four irrigations (Bhattacharyya et al. 2008). Conservation tillage improved soil aggregate stability and labile C pools in the surface layer, across different cropping systems both under rainfed and irrigated conditions in the Himalayas (Bhattacharyya et al. 2012a). Introduction of a legume crop improved C retention in surface soils under conservation tillage, even with only short-term adoption.

Unlike conventional farming methods, CA minimizes soil disturbance and recycles crop residues. Soil bulk density may be decreased, soil aggregation may be improved, and SOC may increase to reverse land degradation with CA. Specific results from four years of a wheat-based cropping system in the western IGP indicate that ZT had higher C retention potential than CT in the 0–30 cm soil layer with 8.6% and 10.2% of the gross C input retained under CT and ZT, respectively (Das et al.2013). In another study in the IGP, topsoil under ZT with bed planting had a greater concentration of macro-aggregates (0.25–8.0 mm) and mean weight diameter with a concomitant lower silt + clay-sized particles than under CT with bed planting and CT with flat planting after four years. Soil with both cotton/ maize and wheat residue retention had greater macro-aggregate concentration and mean weight diameter and similar bulk density than with residue removal (Bhattacharyya et al. 2013b). Soil aggregation is improved with larger aggregates and greater mean weight diameter.

The adoption of CA, as a complete package, is one of the major strategies for increasing SOC stock. Although crop residue incorporation initially leads to immobilization of inorganic N, the addition of 15–20 kg N/ha with straw incorporation eventually increases the yields of rice and wheat. Incorporation/ retention of rice residue in the soil returns essential organic C and N back to the field to favourably impact the soil structural status. Surface residue placement resulted in greater C retention than residue incorporation in a maize–wheat– greengram cropping system (Das et al. 2013). Zero tillage in particular can complicate manure application and may also contribute to nutrient stratification within the soil profile from repeated surface applications without mechanical incorporation. The annual rates of increase in SOC content over the initial condition under bed-planted NT and flat-bedded ZT plots were ~0.64 and 0.62 Mg C/ha, respectively, in the IGP. Similarly, crop residue-treated plots had ~0.54 Mg C/ha/year higher C retention in the 0–30 cm layer in a cotton/maize–wheat rotation.

In the same cropping system at the IGP, a critical biomass C input of ~2.7 Mg C/ha/year was needed to maintain SOC at equilibrium (with no change) under CT plots (with or without residue addition). Contrarily, ZT and residue-treated plots not only maintained initial SOC, but also improved it. Total C input by the maize–wheat rotation under residue removal plots was ~1.6 Mg/ha/

year and that under the cotton–wheat system was ~1.9 Mg/ha/year. Thus, residue removal treated plots under both cotton–wheat or maize–wheat systems require some amount of C addition to maintain the initial SOC, and ZT with partial or full residue retention is a system that can sequester/retain SOC, and thus possibly reduce CO_2 emissions.

The retention of crop residues before rice transplanting is an efficient management approach for increased productivity, nutrient-use efficiency, and sustainability without affecting soil quality. In northern India, some farmers, after wheat harvesting in April, grow short-duration green-manuring crops to improve soil health. Green manure *Sesbania* and green gram have been used to improve the N supply in rice–wheat production systems. Retention of green manures in rice helps to improve the soil nutrient status and promotes soil microbial activity (Singh and Dolly 2011). These alternative practices are known to exert positive effects on chemical, physical, and biological soil properties (Bhattacharyya et al. 2006, 2013a, 2016).

Farmers in this region mainly adopt transplanted rice because puddling has many advantages, including less weed density, and better soil chemical environment and nutrient availability in these alkaline soils due to the creation of anaerobic conditions. However, due to deterioration of soil physical health, direct-seeded rice (DSR)–zero-tilled wheat (ZTW) along with residue incorporation/retention could be an alternative technology for better soil health and system productivity. Then again, due to weed management and wheat residue management problems, along with reduced crop productivity in the initial years, farmers do not prefer zero-tilled DSR. Hence, tilled DSR-ZTW could be a viable technology in the region that avoids wet cultivation, manages weeds, and uses less water compared with wet cultivation and advocates tillage reduction in the cropping system by >50% (ZT in wheat and less intensive tillage for rice compared with conventionally tilled wheat and intensive tillage operations for wet cultivation in rice).

Relative performance of different CA practices like tilled DSR-ZTW with or without rice residues, tilled DSR-ZTW with or without brown manuring (BM), and tilled DSR-ZTW- zero-tilled mungbean (ZTMB) with or without rice residue retention on the net primary productivity and soil C pools compared with TPR-CTW were evaluated by Bhattacharyya et al. (2015). The results revealed that MBR+ DSR-ZTW + RR-ZTMB plots had significantly higher system productivity than other promising and novel CA practices like DSR + BM-ZTW+ RR and DSR + BM-ZTW plots, mainly due to the green gram productivity addition in the former treatment. The DSR-ZTW + RR plots had similar total SOC stock to TPR-CTW plots. However, the gain (over initial soil) in total SOC content of DSR-ZTW + RR plots was significantly higher (by 150 kg C/ha/year) than DSR-ZTW plots in the 0–15 cm layer. Most of the retained C under all treatments (including the MBR+ DSR-ZTW + RR-ZTMB) was observed in the topsoil. Both the C accumulation rate and mean rice productivity under DSR + BM-ZTW+ RR-treated plots were less than for MBR+DSR-ZTW+RR-ZTMB, but the former treatment also performed better than farmers' practice. Thus, the MBR+ DSR-ZTW + RR-ZTMB management practice has wide scope for adoption in north-western India.

C Sequestration Potential of ZT

A regional assessment of the impact of a change to zero-till was made for wheat-based production systems in the Indian states within the IGP. IPCC methodology was used to estimate the potential for climate change mitigation through soil C sequestration, applying the IPCC factors to the different soils and climatic conditions in the region. This modelling study led to calculated annual rates of SOC accumulation under zero-till in the range of 0.2–0.4 Mg C/ha, which is broadly consistent with annual rates measured in other regions of the world. The calculated annual rate of SOC accumulation in the entire region was <0.01 G t CO_2e per year, which is less than 1% of India's total annual GHG emissions. In the IGP, the most common cropping system is a rotation of wheat and flooded rice, with two crops being grown each year. All of the usable data on SOC changes from this region referred to individual components of CA and not from combinations of treatments. The overall predicted annual rate of increase of SOC stock compared to conventional practice was 0.37±0.045 Mg C/ha/year (Table 11.1). The individual rates for reduced tillage, residue retention, and crop diversification were 0.49, 0.16, and 0.47 Mg C/ha/year, respectively. All of these values were significantly different from conventional practice but differences between them were not significant. The majority of individual rates of increase, for all treatments, were between zero and about 0.4 Mg C/ha/year. The largest rate of increase (0.49 Mg C/ha/year was from reduced tillage, but this is probably an over-estimate). It is therefore likely that a value of around 0.3 Mg C/ha/year for the rate of SOC increase under ZT would be more representative, similar to the range of values (0.2–0.4 Mg C/ha/year) derived for the region in a modelling study using IPCC methodology (Grace et al. 2012). The relatively large predicted rate of SOC increase from crop diversification (0.47 Mg C/ha/year was only based on a very few data comparisons (Singh et al., 2005; Venkatesh et al., 2013) showing diverse results. In one case crop diversification comprised a legume replacing rice in the rice–wheat rotation and in the other a maize–wheat rotation was modified to include legumes.

Effect of CA on GHG Emissions

In India, only a few studies have considered the fluxes of all three major GHGs that are impacted by tillage management. Increases in SOM can increase N cycling in soils, which leads to higher N_2O emissions as nitrification is stimulated. As a result, more NO^{-3} will be formed in the soil, but when anaerobic micro-sites are formed when O_2 diffusion is inhibited, the reduction of N_2O to N_2 will be reduced, hence increasing the $N_2O:N_2$ ratio. However, ZT with residue retention improves the soil structure compared to conventional tillage so that fewer anaerobic micro-sites are formed, reducing denitrification and N_2O emissions. It

Table 11.1 SOC sequestration potential of ZT and residue management in different agro-ecosystems of India

Location	Cropping system	No. of years	Soil sampling depth (m)	Treatment	SOC stock (t/ha)	Annual rate of increase in SOC stock compared to conventional (t/ha/year)	References
Varanasi	Upland rice–wheat	2	0.15	Conventional till	7.61		Bazaya et al. (2009)
				Zero-till	9.03	0.71	
Ludhiana	Rice–wheat	11	0.15	No straw	11.57	0.22	Benbi et al. (2012)
				Rice straw	14.02		
Ludhiana	Rice–wheat	14	0.15	No straw	11.25	0.10	Bhandari et al. (2002)
				Wheat straw to rice	9.75		
Cuttack	Rice–rice	4	0.45	Urea	18.69	0.19	Bhattacharya et al. (2012)
				Urea + rice straw	19.45		
Almora	Lentil–finger millet	6	0.15	Conventional till	22.59	0.37	Bhattacharyya et al. (2012)
				Zero till (both crops)	24.81		
Almora	Upland rice–wheat	9	0.3	Conventional till	32.89	0.38	Bhattacharya et al. (2013a)
				Zero till	36.23		
		9	ESM basis	Conventional till	31.69	0.29	
				Zero till	34.30		
Modipuram	Rice–wheat	3	0.15	Conventional till	12.64	0.27	Gangwar et al. (2006)
				Zero-till	13.45		
				Straw removed	12.47	0.24	
				Straw returned	13.20		
West Bengal	Rice–wheat	25	0.45	No straw	53.2	0.31	Ghosh et al. (2012)
				Rice straw	61.0		
Solapur	Rabi sorghum	22	0.15	Leucaena loppings	14.4	0.65	Srinivasarao et al. (2012)
Tripura	Rice–rice	03	0.20	ZT + Gliricidia loppings	21.36	0.34	Yadav et al. (2017)

can be speculated that better aeration of soil as a result of increased SOM content and the resultant increased aggregate stability will inhibit denitrification and stimulate oxidation of CH_4. There are few studies on the quantification of GHG emissions from different conservation agricultural practices in India for quantification of their mitigation potential.

The potential of GHG emission reduction by the resource conservation technologies in India is given in Table 11.2. Bhattacharya et al. (2012), in eastern India, quantified the emissions of CH_4, CO_2, and N_2O, from flooded rice field, under different organic and inorganic fertilizer management systems. Maximum global warming potential (GWP) (10,188 kg CO equivalent per ha) was observed on the combined application of rice straw and green manure. Total C content and C storage in the topsoil were significantly increased for the rice straw + inorganic N fertilizer treatment; however, the combined application of rice straw and green manure increased gaseous C emission. The combined application of rice straw and an inorganic fertilizer was most effective in sequestrating SOC (1.39 Mg/ha), resulting in a higher grain yield.

In another study, SOC stock change was significantly higher under ZT transplanted rice than conventionally transplanted rice (Dash et al. 2017). CH_4 emission was significantly lower in ZT than conventionally transplanted rice (11.7%). In general, soil labile C fractions and enzymatic activities were significantly higher under resource conservation technologies (RCTs) over conventionally transplanted rice. The resource conservation technique of NT in transplanted rice could offer a low-carbon technology in the long term by minimizing GHG emissions and increasing soil C stock and also sustaining yield in tropical lowland rice.

Of late, the adoption of bed planting under CA is thought to have aggrading impacts, as bed planting is a cost-effective production technique and is also helpful in resource conservation. In one study the effects of a medium-term (five years) CA were studied on total soil N (TSN) changes in bulk soils and aggregates, N_2O and CO_2 emissions, GWP and total C fixed in soils under a maize–wheat system on the IGP (Bhattacharyya et al. 2018). In the maize–wheat system, the highest N_2O emission was observed in permanent narrow beds with residue retention (PNB + R) plots and the lowest was in CT plots. Greenhouse gas intensities (GHGi) (carbon dioxide equivalent per ton of grain yield) in the CT, PBB + R, and ZT + R plots were similar. Under ZT, N_2O emission was generally higher due to a reduction in the rate of diffusion in the presence of compact soils (Bhatia et al., 2013) and high soil moisture content. This condition promoted an anaerobic state, which benefited N_2O emission.

Sapkota et al. (2014) conducted on-farm trials in seven districts of Haryana to evaluate three different approaches to site-specific nutrient management (SSNM) based on recommendations from the Nutrient Expert® (NE) decision support system in ZT and conventional tillage (CT)-based wheat production systems. The global warming potential of wheat production was also lower with the ZT system than the CT system. The results suggested that the ZT system along with site-specific approaches for nutrient management can increase yield, nutrient-use efficiency, and profitability while decreasing GHG emissions from wheat production in the IGP. However, Bhatia et al. (2010) observed higher N_2O emissions under ZT in wheat than CT.

Pratibha et al. (2016) studied the GWP in rainfed semi-arid regions of Hyderabad, India, with different tillage practices including conventional tillage (CT), reduced tillage (RT), zero tillage (ZT), and residue retention levels in pigeonpea–castor systems. ZT and RT recorded 26% and 11% lower indirect GHG emissions (emissions from farm operations and input use) over CT, respectively. The percent contribution of CO_2 eq. N_2O emissions was higher in respect to total GHG emissions in both crops. Castor grown on pigeonpea residue recorded 20% higher GHG emissions over pigeonpea grown on castor residues. The fuel consumption in ZT was reduced by 58% and 81% compared to CT in pigeonpea and castor, respectively. Lower GWP and GHGI were observed with an increase in crop residues and decrease in tillage intensity in both crops. Chaudhary et al. (2017) carried out a 16-year field study in the IGP with different methods of rice cultivation, namely zero tillage (ZT), happy turbo seeder (HTS), bed planting (BP), reduced tillage (RT), conventional sowing (CS), direct sowing (DS), and broadcast method of sowing (BS), to evaluate the GWP. Among direct-sown unpuddled methods of rice cultivation, BP, RT, and ZT had lower GWP than farmers' practice. This study indicated that the direct-seeded method of rice cultivation was energy efficient with lower GWP, and thus may be recommended.

Pandey et al. (2012) assessed the impacts of four tillage practices in a rice–wheat cultivation system on fluxes of GHGs (CH_4, N_2O, and CO_2) and yield of rice. A reduction in tillage frequency led to significant reductions in fluxes of CH_4 and N_2O, but increased CO_2 while permutations of tillage and no tillage influenced the grain yield. Although no single tillage permutations showed a consistent increment in yield with accompanied emission reductions, tillage before sowing of wheat but no tillage before sowing of rice (RNT-WCT), may be considered as better agricultural practice for this region.

Conclusion and Future Outlook

CA has been successfully developed for many regions of the world, but these systems have not been widely adopted by Indian farmers for many reasons. Through greater CA adoption in India, there is an enormous potential to sequester SOC, which would: (i) help mitigate GHG emissions contributing to global warming, and (ii) increase soil productivity and avoid further environmental damage from the unsustainable use of excessive tillage systems, which threaten water quality, reduce soil biodiversity, and erode soils. Research over the past decade has demonstrated that CA can reduce GHG emission and help in sequestering C in soils. CA practices are also beneficial over traditional practices in terms of household income, food security, and natural resource management. Increased SOC sequestrations and soil quality coupled with decreased GHG emissions were observed under ZT and CA over CT in India, especially under irrigated agriculture.

Table 11.2 Resource conservation practices and GHG emissions in India

Crop (years of study)	Soil type	Organic C (g/kg)	Climate	CO$_2$ (kg/ha)	CH$_4$ (kg/ha)	N$_2$O (g/ha)	GWP (kg CO$_2$ equiv./ha)	References
Pigeonpea (5)	Typic Haplustalf	3.1	Rainfed semi-arid					Pratibha et al. (2016)
CT + no residue				4814		384	5198	
CT + 30 cm residue				5067		360	5427	
RT + no residue				4244		390	4634	
RT + 30 cm residue				5250		432	5682	
ZT + no residue				3442		320	3762	
ZT + 30cm residue				4242		401	4643	
Maize (5)	Typic Ustocrept	5.2	Sub-tropical semi-arid					Bhattacharya et al. (2018)
CT				792		846		
ZT				732		901		
ZT+R				802		961		
Wheat								
CT				772		705		
ZT				719		751		
ZT+R				800		841		
Castor (5)	Typic Haplustalf	3.1	Rainfed semi-arid					Pratibha et al. (2016)
CT + no residue				6419		394	6813	
CT + 30 cm residue				6756		580	7336	
RT + no residue				5658		483	6141	
RT + 30 cm residue				7001		500	7501	
ZT + no residue				4590		345	4935	
ZT + 30 cm residue				5656		340	5996	

Cropping system (treatment)	Soil type	pH	Climate					Reference
Transplanted puddled rice (4)	Aeric Endoaquept	4.9	Lowland, tropical					Bhattacharyya et al. (2012)
No N				1100.3	69.7	23 0	5862	
Rice straw + urea				1668.6	115.4	840	9418	
Rice straw + green manure				18.58.5	122.7	720	10188	
Rice (2)	Typic Ustochrept	5.9	Sub-tropical, semi-arid					Bhatia et al. (2013)
100% NPK (25% N by FYM)				1575	49.6	503	2772	
100% NPK (25% N by GM)				1853	46.0	579	2999	
100% NPK + previous crop residue (2 t/ha)				1642	42.7	549	2920	
100% N by organic source (50% FYM + 25% biofertilizer + 25% previous crop residue)				2081	57.1	537	3447	
Wheat (2)	Typic Ustochrept	5.9	Sub-tropical, semi-arid					Bhatia et al. (2010)
100% NPK (25% N by FYM)				2458		860	2725	
100% NPK (25% N by GM)				2493		922	2779	
100% NPK + previous crop residue (2 t/ha)				2167		820	2421	
100% N by organic source (50%FYM + 25% biofertilizer+ 25% previous crop residue)				2385		564	2560	
Wheat (2)	Typic Ustochrept		Sub-tropical, semi-arid					Bhatia et al. (2010)
CT						313		
ZT						355		

With the incorporation and retention of crop residues, there may be an increase in the global warming potential of cropped soils due to higher CO_2 emission; however, CA may lower the total on-farm GWP by reducing the fuel consumption (reduced tillage) and with higher N-use and water-use efficiencies. Few studies in India have quantified GHG balance under CA taking into account all components (fuel consumption, herbicide use, C sequestration and GHG emission, etc.). Research on these aspects is evolving with the necessity that more concentrated work may be performed in the future. Realizing GHG mitigation benefits requires tailoring CA principles within unique constraints (and opportunities) of working farms in varying climatic conditions. Research is needed to develop nuanced management approaches for mitigating GHG emissions from CA systems. Suitable economic incentives should be given to procure appropriate machinery so that farmers will be motivated to follow CA.

References

Bazaya, B.R., Sen, Avijit and Srivastava, V.K. 2009. Planting methods and nitrogen effects on crop yield and soil quality under direct seeded rice in the Indo-Gangetic plains of eastern India. *Soil and Tillage Research* 105: 27–32.

Benbi, D.K., Toor, A.S. and Kumar, S. 2012. Management of organic amendments in rice-wheat cropping system determines the pool where carbon is sequestered. *Plant and Soil* 360: 145–162.

Bhandari, A.L., Ladha, J.K., Pathak, H., et al. 2002. Yield and soil nutrient changes in a long-term rice-wheat rotation in India. *Soil Science Society of America Journal* 66: 162–170.

Bhatia, A., Kumar, A., Das, T.K., et al. 2013. Methane and nitrous oxide emissions from soils under direct seeded rice. *International Journal of Agricultural Statistical Science* 9(2): 729–736.

Bhatia, A., Sasmal, S., Jain, N., et al. 2010. Mitigating nitrous oxide emission from soil under conventional and no-tillage in wheat using nitrification inhibitors. *Agriculture, Ecosystems and Environment* 136: 247–253.

Bhattacharyya, R., Bhatia, A., Das, T.K., et al. 2018. Aggregate-associated N and global warming potential of conservation agriculture-based cropping of maize-wheat system in the north-western Indo-Gangetic Plains. *Soil and Tillage Research* 182: 66–77.

Bhattacharyya, R., Das, T.K., Pramanik, P., et al. 2013a. Impacts of conservation agriculture on soil aggregation and aggregate-associated N under an irrigated agroecosystem of the Indo-Gangetic Plains. *Nutrient Cycling in Agroecosystems* 96: 185–202.

Bhattacharyya, R., Das, T.K., Sudhishri, S., et al. 2015. Conservation agriculture effects on soil organic carbon accumulation and crop productivity under a rice–wheat cropping system in the western Indo-Gangetic Plains. *European Journal of Agronomy* 70: 11–21.

Bhattacharyya, R., Kundu, S., Pandey, S., et al. 2008. Tillage and irrigation effects on crop yields and soil properties under rice–wheat system of the Indian Himalayas. *Agricultural Water Management* 95: 993–1002.

Bhattacharyya, R., Pandey, S.C., Bisht, J.K., et al. 2013b. Tillage and irrigation effects on soil aggregation and carbon pools in the Indian sub-Himalayas. *Agronomy Journal* 105: 101–112.

Bhattacharyya, R., Pandey, A.K., Gopinath, K.A., et al. 2016. Fertilization and crop residue addition impacts on yield sustainability under a rainfed maize–wheat system in the Himalayas. *Proceedings of the National Academy of Sciences, India Section B: Biological Sciences* 86: 21–32.

Bhattacharyya, R., Prakash, Ved, Kundu, S., et al. 2006. Effect of tillage and crop rotations on pore size distribution and soil hydraulic conductivity in sandy clay loam soil of the Indian Himalayas. *Soil and Tillage Research* 82: 129–140.

Bhattacharya, P., Roy, K.S., Neogi, S., et al. 2012. Effects of rice straw and nitrogen fertilization on greenhouse gas emissions and carbon storage in tropical flooded soil planted with rice. *Soil and Tillage Research* 124: 119–130.

Bhattacharyya, R., Tuti, M.D., Bisht, J.K., et al. 2012b. Conservation tillage and fertilization impacts on soil aggregation and carbon pools in the Indian Himalayas under an irrigated rice–wheat rotation. *Soil Science* 177: 218–228.

Bhattacharyya, R., Tuti, M.D., Kundu, S., et al. 2012a. Conservation tillage impacts on soil aggregation and carbon pools in a sandy clay loam soil of the Indian Himalayas. *Soil Science Society of America Journal* 76: 617–627.

Chaudhary, V.P., Singh, K.K., Pratibha, G., et al. 2017. Energy conservation and greenhouse gas mitigation under different production systems in rice cultivation. *Energy* 130: 307–317.

Das, T.K., Bhattacharyya, R., Sharma, A.R., et al. 2013. Impacts of conservation agriculture on total soil organic carbon retention potential under an irrigated agro-ecosystem of the western Indo-Gangetic Plains. *European Journal of Agronomy* 51: 34–42.

Dash P.K., Bhattacharyya, P., Shahid, M., et al. 2017. Low carbon resource conservation techniques for energy savings, carbon gain and lowering GHGs emission in lowland transplanted rice. *Soil and Tillage Research* 174: 45–57.

Gangwar, K.S., Singh, K.K., Sharma, S.K., et al. 2006. Alternative tillage and crop residue management in wheat after rice in sandy loam soils of Indo-Gangetic plains. *Soil and Tillage Research* 88: 242–252.

Ghosh, S., Wilson, B., Ghoshal, S., et al. 2012. Organic amendments influence soil quality and carbon sequestration in the Indo-Gangetic plains of India. *Agriculture, Ecosystems and Environment* 156: 134–141.

Grace, P.R., Antle, J., Ogle, S., et al. 2012.Soil carbon sequestration rates and associated economic costs for farming systems of the Indo-Gangetic Plain. *Agriculture, Ecosystems and Environment* 146: 137–146.

Gupta, D.K., Bhatia, A., Das, T.K., et al. 2016. Economic analysis of different greenhouse gas mitigation technologies in rice–wheat cropping system of the Indo-Gangetic Plains. *Current Science* 110: 867–874.

Kumar, A., Bhatia, A., Sehgal, V.K., et al. 2021. Net ecosystem exchange of carbon dioxide in rice-spring wheat system of northwestern Indo-Gangetic plains. *Land* 10: 701.

Lal, R. 2004. Soil carbon sequestration in India. *Climate Change* 65: 277–296.

Malyan, S.K., Bhatia, A., Tomer, R., et al. 2021. Mitigation of yield-scaled greenhouse gas emissions from irrigated rice through Azolla, Blue-green algae, and plant growth-promoting bacteria. *Environment Science and Pollution Research* 28: 51425–51439.

Pandey, D., Agrawal, M. and Bohrab, J.S. 2012. Greenhouse gas emissions from rice crop with different tillage permutations in rice–wheat system *Agriculture, Ecosystems and Environment* 159: 133–144.

Pathak, H., Saharawat, Y.S., Gathala, M.K., et al. 2009. Simulating environmental impact of resource-conserving technologies in the rice-wheat system of the Indo-Gangetic Plains, pp. 321–333. In: Integrated Crop and Resource Management in the Rice-Wheat

System of South Asia. Ladha J.K. et al. (Eds.). International Rice Research Institute, Los Banos, Philippines.

Pratibha, G., Srinivas, I., Rao, K.V., et al. 2016. Net global warming potential and greenhouse gas intensity of conventional and conservation agriculture system in rainfed semi-arid tropics of India. *Atmospheric Environment* 145: 239–250.

Sapkota, T.B., Majumdar K., Jat, M.L., et al. 2014. Precision nutrient management in conservation agriculture based wheat production of Northwest India: Profitability, nutrient use efficiency and environmental footprint. *Field Crops Research* 155: 233–244.

Singh, Y.V. and Dolly, W.D. 2011. Influence of organic farming on soil microbial diversity and grain yield under rice-wheat-green gram cropping sequence. *Oryza* 48(1): 40–46.

Singh, V.K., Dwivedi, B.S., Shukla, A.K., et al. 2005. Diversification of rice with pigeonpea in a rice-wheat cropping system on a Typic Ustochrept: effect on soil fertility, yield and nutrient use efficiency. *Field Crops Research* 92: 85–105.

Srinivasarao, C., Deshpande, A.N., Venkateswarlu, B., et al. 2012. Grain yield and carbon sequestration potential of post monsoon sorghum cultivation in Vertisols in the semi-arid tropics of central India. *Geoderma* 175: 90–97.

Venkatesh, M.S., Hazra, K.K., Ghosh, P.K., et al. 2013. Long-term effect of pulses and nutrient management on soil carbon sequestration in Indo-Gangetic plains of India. *Canadian Journal of Soil Science* 93(1): 127–136.

VPKAS. 2011. Annual Report, 2011–12. ICAR-Vivekananda Parvatiya Krishi Anusandhan Sansthan Almora, Uttarakhand.

Yadav, G.S., Lal, R., Meena, R.S., et al. 2017. Energy budgeting for designing sustainable and environmentally clean/safer cropping systems for rainfed rice fallow lands in India. *Journal of Cleaner Production* 158: 29–37.

12 Soil Health Management and Conservation Agriculture

K.K. Bandyopadhyay, K.M. Hati, J. Somasundaram, and U.K. Mandal

Introduction

Sustaining crop productivity at a higher level is the key issue for Indian agriculture to meet the increasing demands for food and fibre for the growing population under the changing climatic scenario. Maintaining soil health is indispensable for sustaining agricultural productivity at a higher level. Though both soil health and soil quality are synonymous, soil health is a qualitative term and often used by producers, whereas soil quality is a quantitative term and is mainly used by scientists. Basically, soil quality is related to soil function, whereas soil health presents soil as a finite non-renewable and dynamic living resource. Soil health can be defined as the continued capacity of soil to function as a vital living system, within ecosystem and land-use boundaries, to sustain biological productivity, maintain the quality of air and water environments, and promote plant, animal, and human health. The main functions of soil include: (i) water flow and retention, (ii) solute transport and retention, (iii) physical stability and support, (iv) retention and recycling of nutrients, (v) buffering and filtering of potentially toxic materials, and (vi) maintenance of biodiversity and habitat. Soil health needs to be maintained and improved by following appropriate management practices to sustain productivity continuously at higher levels over the long term.

Conservation agriculture (CA), involving reduced tillage, residue retention, and crop rotation, has emerged as a paradigm shift in agricultural practices having a favourable effect on soil health, carbon sequestration, and sustainable agricultural production and mitigation of climate change (Naresh et al. 2016). The aims of CA are to: (i) conserve, improve, and make more efficient use of natural resources, and (ii) integrate the management of available soil, water, and biological resources combined with external inputs. CA contributes to environmental conservation as well as to enhancing and sustaining agricultural production. It can also be referred to as resource-efficient or resource-effective agriculture.

CA practices have been reported to improve soil health. However, soil health cannot be measured directly, although soil properties that are sensitive to changes in management can be used as indicators. Soil health includes three groups of mutually interactive attributes, i.e. soil physical, chemical, and biological

DOI: 10.4324/9781003292487-15

properties, which must be restored to their optimum levels for improving crop growth. The soil health approach is better applied when specific goals are defined for the desired outcome from a set of decisions. Therefore, the soil health evaluation process consists of the following activities: (i) selection of soil health indicators, (ii) determination of a minimum data set (MDS), (iii) development of an interpretation scheme of indices, and (iv) on-farm assessment and validation.

CA and Soil Physical Health

Soil physical health is the ability of a given soil to meet plant and ecosystem requirements for water, aeration, and strength over time and to resist and recover from processes that might diminish that ability. Concepts of soil physical health can be applied to individual soil horizons, profiles, or areas classified to a common soil type. Unless the soil physical health is maintained at its optimum level, the genetic realizable yield potential of a crop cannot be achieved even when all other requirements are fulfilled. Soil physical health also influences and is influenced by the chemical and biological health of the soil.

Soil Structure and Aggregation

Soil structure is a key factor in soil functioning, and it is an important factor in the evaluation of the sustainability of crop production systems. Soil structure is often expressed as the degree of stability of aggregates. Soil structural stability is the ability of aggregates to remain intact when exposed to different stresses, and measures of aggregate stability are useful as a means of assessing soil structural stability. Zero tillage (ZT) with residue retention improves dry aggregate size distribution compared to conventional tillage (CT). Even when CT results in good structural distribution, the structural components are weaker to resist water slaking than in ZT situations with crop residue retention, where the soil becomes more stable and less susceptible to structural deterioration. The reduced aggregation in CT is a result of direct and indirect effects of tillage on aggregation. Physical disturbance of soil structure through tillage results in direct breakdown of soil aggregates and increased turnover of aggregates and fragments of roots and mycorrhizal hyphae, which are major binding agents for macro-aggregates. The aggregate formation process in CT is interrupted each time the soil is tilled with the corresponding destruction of aggregates. The residues lying on the soil surface in CA protect the soil from raindrop impact. On the other hand, no such protection occurs in CT, which increases susceptibility to further disruption. Moreover, during tillage a redistribution of the soil organic matter takes place. Small changes in soil organic carbon can influence the stability of macro-aggregates.

Fresh residue forms the nucleation centre for the formation of new aggregates by creating hot spots of microbial activity where new soil aggregates are developed. The return of crop residue to the soil surface not only increases the aggregate formation, but also decreases the breakdown of aggregates by reducing erosion and protecting the aggregates against raindrop impact. Crops can affect soil aggregation by their rooting system because plant roots are important binding agents at the scale of macro-aggregates.

Soil Bulk Density and Penetration Resistance

The effect of tillage and residue management on soil bulk density is mainly confined to the top soil (plough layer). In deeper soil layers, soil bulk density is generally similar in ZT and CT. On a silt loam with a maize–soybean rotation, soil bulk densities were higher in the surface layer of ZT than CT after 23 years, but lower below 30 cm, reflecting the rupture action of tillage near the surface and the compacting and shearing action of tillage implements below tillage depths. Jat et al. (2013) found that under a maize–wheat cropping system in sandy-loam soil, the bulk density and penetration resistance were lower in a permanent raised bed treatment and no-till flat as compared to conventional tillage at 10–25 cm soil depth due to the formation of a plough layer in conventional tilled treatment. Bag et al. (2020) reported that although there was no significant difference in BD due to CT and ZT in the surface layers (0–15 cm), BD under ZT was decreased by 1.7%, 2.0%, and 3.7% compared with that of CT at 15–30 (1.72 Mg/ha), 30–45 (1.76 Mg/ha), and 45–60 cm (1.77 Mg/ha) soil depth, respectively, suggesting that NT can reduce the subsurface compaction in a 4-year tillage experiment. Due to the addition of crop residue mulch (CRM), BD was decreased by 4.3%, 5.1%, and 2.7% at 0–15, 15–30, and 30–45 cm soil depth compared to no mulch treatments. This may be attributed to higher earthworm activity, better porosity, and the presence of comparatively higher organic matter in mulch treatments (Acharya et al. 2005).

Lal (1997) reported that penetration resistance at 0–5 cm depth was the least for ZT + mulch treatment (116 kPa), the highest for ridge till treatment (348 kPa), and had a mean value of 243 kPa (Table 12.1). In comparison, penetration

Table 12.1 Tillage effects on penetration resistance (kPa) at different soil depths

Treatment	0–5 cm depth	5–10 cm depth
No till + mulch	116	336
No till + chisel	195	313
Ploughing	187	272
Discing	306	319
No till – mulch	312	316
Summer ploughing	200	249
Ploughing + mulch	281	338
Ridge till	348	421
Mean	243	321
LSD (0.05)		
Tillage	63	
Soil depth	31	
Tillage × soil depth	NS	

Source: Lal (1997).

resistance for the 5–10 cm depth was the least for summer ploughing treatment (249 kPa) and highest for ridge till treatment (421 kPa), with a mean value of 321 kPa. The altogether low penetration resistance of the surface soil was probably due to the sandy texture and low cohesion. Bag et al. (2020) reported that, after 4 years, there was a decrease in soil penetration resistance under NT compared with that of CT at 10–27 cm soil depth, indicating that NT could reduce soil compaction at these depths. With the addition of crop residue mulch, there was a decrease in soil penetration resistance up to 210 mm compared with no-mulch treatments. Mishra et al. (2015) revealed that during the third year of CA in puddled transplanted rice in a rice–wheat cropping system, the penetration resistance exceeded 2 MPa, whereas, in direct-seeded rice with brown manuring with zero-tilled wheat rotation, the penetration resistance remained below 1.5 MPa up to 0–60 cm soil depth, indicating ideal conditions for root growth. Therefore, CA practices provide a conducive soil physical environment for root growth and penetration.

Soil Porosity

Pores are of different size, shape, and continuity. These characteristics influence the infiltration, storage, and drainage of water, the movement and distribution of gases, and the ease of penetration of soil by growing roots. Pores of different size, shape, and continuity are created by abiotic factors (e.g. tillage and traffic, freezing and thawing, drying and wetting) and by biotic factors (e.g. root growth, burrowing fauna). Pore characteristics can change in both space and time following a change in tillage practices. These changes primarily reflect changes in the form, magnitude, and frequency of stresses imposed on the soil, the placement of crop residues, and the population of microorganisms and fauna in the soil. Total porosity is normally calculated from measurements of bulk density, so the terms bulk density and total porosity can be used interchangeably. A plough pan may be formed by tillage immediately underneath the tilled soil, causing higher bulk density in this horizon in tilled situations. A reduction in tillage would be expected to result in a progressive change in total porosity with time, approaching a new 'steady state'. However, initial changes may be too small to be distinguished from natural variation.

Pores with diameters >30 μm are referred to as macropores. Water flows primarily through these pores during infiltration and drainage and, consequently, these pores exert a major control on soil aeration. In addition, much of root growth is initiated in these pores. Pores with an equivalent diameter of 0.2–30.0 μm are referred to as mesopores, and are particularly important for the storage of water for plant growth. Micropores have an effective diameter of <0.2 μm. In general, micro- and meso-porosity are reported to be higher in ZT compared to conventional tillage, but in some cases, no effect of tillage is observed. Yoo et al. (2006) did not find consistent results at three locations (two with a silty clay loam and one with a silty loam soil). At one of the three locations (the silty loam soil), the volume of small macropores (15–150 μm) as well as large macropores (>150 μm) was smaller under ZT than under conventional tillage. At the other two locations, either small macroporosity (in the silty loam) or large macroporosity was smaller under ZT (in the silty clay loam). In the 0–5 cm layer of a 24-year experiment, the volume of pores >60 μm was significantly greater (>11%) under ZT with residue retention than under conventional tillage with the residue burnt (Zhang et al. 2007).

Hydraulic Conductivity

Although hydraulic conductivity is expected to be higher in ZT with residue retention compared to CT due to the larger macroporosity conductivity as a result of the increased number of biopores, the reported results are not consistent. This might be partly due to difficulty in measuring hydraulic conductivity when residue cover is present in ZT. Bag et al. (2020) reported that the addition of crop residue mulch significantly increased the saturated hydraulic conductivity (Ksat) of soil by 100.7% compared with no mulch treatment at 0–15 cm soil depth. However, the effect of crop residue mulching on Ksat in the sub-soil layers was not consistent. The presence of residue complicates the installation of measurement instruments or the removal of undisturbed samples and cores, and may cause high variation in conductivity values at small scales due to macropores and other structural attributes that are left intact due to the absence of tillage. Also, differences in soil sampling depth, amount of straw mulch, and site-specific characteristics (e.g., soil texture, slope, tillage) explain inconsistencies in the observed effects of tillage on hydraulic conductivity and water-holding capacity.

Infiltration and Runoff

Despite the inconsistent results on the effect of tillage and residue management on soil hydraulic conductivity, infiltration is generally higher in ZT with residue retention compared to CT and ZT with residue removal. This is probably due to the direct and indirect effects of residue cover on water infiltration. Soil macro-aggregate breakdown has been identified as the major factor leading to surface pore clogging by primary particles and micro-aggregates, and thus to the formation of surface seals or crusts (Lal and Shukla 2004). Jat et al. (2013) found a higher infiltration rate under conservation tillage as compared to conventional tillage due to minimum disturbance of soil pore continuity. The presence of crop residues over the soil surface prevents aggregate breakdown by direct raindrop impact as well as by rapid wetting and drying of soils.

Crop residues left on the topsoil with ZT and crop retention act as a succession of barriers, reducing the runoff velocity and giving the water more time to infiltrate. The residue intercepts rainfall and releases it more slowly afterwards. The 'barrier' effect is continuous, while the prevention of crust formation probably increases with time. Ball et al. (1997) found greater infiltration rates in ZT with residue retention after 26 years than after 9 years. It was observed that there was a significantly higher initial infiltration rate and equilibrium infiltration rate under ZT + mulch than conventional ploughing treatment.

Least Limiting Water Range

The concept of least limiting water range (LLWR) characterizes a single range of soil water content beyond which available water, soil aeration, and mechanical resistance impose significant limitations to root growth. This concept integrates three factors, i.e. soil water, soil aeration, and mechanical impedance, into a single variable, LLWR, and was found to be more sensitive to soil structural changes than available water (da Silva et al. 1994). The upper limit of LLWR is either soil water content at 10% aeration porosity (MC_{ap}) or soil water content at field capacity (MC_{fc}), whichever is lower, and the lower limit is either soil water content corresponding to 2 Mpa soil strength (MC_{2Mpa}) or soil water content at wilting point (MC_{wp}), whichever is higher. The structural quality could be considered as 'very good' for LLWR >0.20 m^3/m^3, 'good' for LLWR between 0.15–0.20 m^3/m^3, 'moderate' for LLWR between 0.10–0.15 m^3/m^3, and 'poor' for LLWR <0.10 m^3/m^3 (Kay and Anger 2002).

Aggarwal et al. (2013) reported that in a sandy-loam soil, both under bed planting and conventional tillage systems, θap decreased with an increase in BD, whereas θ2MPa increased appreciably with increase in BD (where θ volumetric moisture content of soil). On the other hand, θfc, and θpwp did not change much with an increase in BD. It was further observed that θfc was the upper limit and θ2MPa was the lower limit of LLWR, except in conventional planting where at higher BD (1.72 Mg/m^3), θap was the upper limit.

It was observed that the decline in LLWR was sharper in the conventional system than in the bed planting system, indicating that LLWR remained wider in the bed than the conventional system all throughout the crop growth. Wider LLWR in the bed system indicated better structural quality, more water availability, and less mechanical impedance to growing roots than in the conventional system. On the other hand, AWRC did not show such a variation with an increase in BD. The reduction in the range of available water due to deterioration of soil structure with time was best reflected in a decline in LLWR, whereas the available water retention capacity (AWRC) did not show significant temporal changes. These results indicated that LLWR is a better indicator of soil structure quality and water availability than AWRC.

The plant water stress period was computed as the number of days that the water content of soil was outside LLWR or AWRC during various growth periods. It was observed that when the stress period was calculated using LLWR as an index of water availability, a lower water stress period was obtained under bed planting as compared to CT. No significant variation in stress was observed among the treatments when the stress period was calculated using AWRC as the index of water availability. Hence, it could be suggested that in order to avoid stress, irrigation should be given as soon as the SWC reaches the lower limit of LLWR, i.e. θ2MPa and not the lower limit of AWRC, i.e. θpwp.

Mishra et al. (2015) reported that in a cotton–wheat system at the 0–15 cm soil layer, the plots under permanent broad-bed with residue (PBB + R) had nearly 14%, 17%, and 39% higher LLWR than CT (LLWR=12.3%), permanent narrow-bed with residue (PNB + R) (LLWR=12%) and ZT (LLWR=10.1%) plots, confirming that crop residue retention improved LLWR. The impact of PBB + R on the improvement in LLWR over CT plots in the sub-surface layer was much higher than in the surface layer. Residue addition invariably improved LLWR values in both soil layers (0-15cm and 15-30cm) under the cotton–wheat system. There was a more drastic reduction in LLWR in the ZT plots in the sub-surface layer than the CT plots.

CA and Soil Chemical Health

Conservation agriculture practices due to retention of residues and changes in the soil micro-environment influence soil chemical properties, namely soil pH, cation exchange capacity, total C and N distribution, and available macro- and micro-nutrient status of soil, which influence crop growth and productivity.

Soil pH

There are contrasting views about changes in soil pH due to tillage and residue application. Malhi et al. (2011) reported that tillage and straw management usually had little or no effect on soil pH in any soil layer. Kettler et al. (2000) found that the main effect of ploughing on soil pH was more significant at a soil depth of 0–7.5, and both ZT and sub-till treatments, which leave plant residues at or near soil surface, were of lower pH than mould board ploughing at all depths. The lower pH in ZT was attributed to accumulation of organic matter in the upper few centimetres under ZT soil, causing an increase in the concentration of electrolytes and a reduction in pH. Govaerts et al. (2007) found a higher pH in permanent bed with all the residues retained than with part or all of the residues removed. One possible way of protecting soil from acidification is by returning the crop residues to the soil as pH increases significantly with crop residue application.

Total Organic C, N, and C:N Ratio

Soil organic C is an important index of soil quality because of its relationship to crop productivity. Decomposition rates of soil organic matter are lower with minimal tillage and residue retention; consequently organic carbon content increases with time. Tillage can also influence the distribution of SOC in the profile with higher soil organic matter (SOM) content in surface layers with ZT than with CT, but a higher content of SOC in the deeper layers where residue is incorporated through tillage (Jantalia et al. 2000). Soil C storage is affected more by quantity than by the type or quality of organic inputs. The quality of the residues is determined primarily by the C:N ratio and can be modified by the amounts of lignin and polyphenolics in the material. Quality may affect short-term soil C storage and dynamics but does not seem to influence the longer-term C stabilization and storage in the soil. The quality of the residues may, however, affect soil fertility and thus the amount of residues produced for C inputs. For example, materials with a high C:N ratio, characteristic of cereal crop residues, reduce the available N in the soil due to N immobilization and could result in lower crop production, while residues with high N contents and low C:N ratios, as is the case with many legume residues and legume cover crops, increase soil N availability and possibly crop production.

It is generally recognized that differential effects of rotations on soil C are simply related to the amounts of above- and below-ground biomass (residues and roots) produced and retained in the system. Boddey et al. (2010) attributed higher soil C storage in ZT than CT to the inclusion of legume intercrops or cover crops in the rotations, and not due simply to higher production and residue inputs. They indicated that slower decomposition of residues and lower mineral N in ZT compared to CT result in higher root:shoot ratios and below-ground C input with ZT. Das et al. (2018) reported that a CA-based system with residue retention had higher SOC and C sequestration potential than a conventional system under a maize–wheat cropping system in the north-western Indo-Gangetic plains. Parihar et al. (2018) found that CA practices signifi-cantly increased the SOC stock and mineral N fractions at 0–30 cm soil depth in sandy-loam soils of inceptisols. In a long-term (15 years) study conducted in ZT in wheat under a rice–wheat cropping system in semi-arid regions of the Indo-Gangetic plains, it was found that ZT increased SOC significantly and the depth of its build-up increased with an increase in the fineness of the soil tex-ture (Singh et al. 2014). This is due to deeper penetration of wheat roots and less oxidation of *in situ* organic matter in the ZT system. A greater clay content retains more organic carbon due to formation of clay-humus complex; therefore, fine texture retained organic carbon at a deeper depth. This indicates that the adoption of CA in fine-textured soils has greater potential for C sequestration.

Crop residues provide a source of organic matter. When returned to soil, the residues increase the storage of organic C and N in the soil, whereas their removal results in a substantial loss of organic C and N from the soil system. Therefore, one would expect a dramatic increase in organic C in soil from a combination of ZT, straw retention, and proper/balanced fertilization. Naresh et al. (2016) found a significantly higher particulate organic carbon (POC) content under ZT probably due to higher biomass C. Results on particulate organic N (PON) con-tent after 3 years showed that in the 0–5 cm soil layer of a CT system, PON content increased from 35.8 mg/kg in CT to 47.3 and 67.7 mg/kg without CR, and to 78.3, 92.4, and 103.8 mg/kg with CR at 2, 4, and 6 t/ha, respectively. The corresponding increase of the PON content under a CA system was from 35.9

mg/kg in a CT system to 49 and 69.6 mg/kg without CR and 79.3, 93.0, and 104.3 mg/kg with CR at 2, 4, and 6 t/ha, respectively. A small improvement in PON content was observed after 4 years of the experiment. Fine-textured soils have more potential for storing C, and ZT enhances the C sequestration rate in soils by providing better conditions in terms of moisture and temperature for higher biomass production and reduced oxidation.

Intensification of cropping systems with high above- and below-ground biomass (i.e. deep-rooted plant species) input may enhance CA systems for storing soil C relative to CT. Gupta-Chaudhary et al. (2014) reported that conservation tillage (both RT and ZT) caused 21.2%, 9.5%, 28.4%, 13.6%, 15.3%, 2.9%, and 24.7% higher accumulations of SOC in >2 mm, 2.1–1.0 mm, 1.0–0.5 mm, 0.5–0.25 mm, 0.25–0.1 mm, 0.1–0.05 mm, and <0.05 mm sized particles than conventional tillage treatments. Direct-seeded rice (DSR) combined with ZT and residue retention had the highest capability to hold the organic carbon in the surface soil (11.57 g/kg soil aggregates) and retained the least amount of SOC in sub-surface (9.05 g/kg soil aggregates) soil. In comparison with transplanted rice (TPR), DSR enhanced SOC by 16.8%, 7.8%, 17.9%, 12.9%, 14.6%, 7.9%, and 17.5% in >2 mm, 2.1–1.0 mm, 1.0–0.5 mm, 0.5–0.25 mm, 0.25–0.1 mm, 0.1–0.05 mm, and <0.05 mm sized particles, respectively. The lower C:N ratio and polyphenol content of green manure are susceptible to rapid decomposition and yield lower values of the mean weight diameter (MWD) as compared to farmyard manure (FYM) and paddy straw with a greater C:N ratio and lingopolyphenol contents. Aulakh et al. (2013) found that in the 0–5 cm layer of CT system, there was an increase in TOC content from 3.84 g/kg in the control to 4.19–4.45 g/kg without CR, and to 4.40–5.79 g/kg with CR after 2 years. The corresponding TOC content values under a CA system were 4.55 g/kg in the control to 4.73–5.02 g/kg without CR and to 4.95–5.30 g/kg with CR. After 4 years, there was a further improvement in the TOC content from 1–26% in CT and none to 19% in CA treatments. Naresh et al. (2015) reported that the SOC concentration of the control was 0.54%, which increased to 0.65% in the RDF and 0.82% in the RDF+FYM treatments.

Available Nutrient Status

Jat et al. (2018) revealed that the CA system enhanced the content of available N, P, K, Zn, Fe, and Mn in surface soil, along with savings of 30% and 50% of N and K fertilizers, respectively. Thus, CA was reported to have beneficial effect on nutrient-supplying capacity of soils. Kushwa et al. (2016) conducted a long-term experiment on CA for 12 years in vertisols of central India under soybean–wheat cropping system. They found that the available P concentration was higher in ZT (11.9 mg/kg) and reduced tillage (RT) (10.8 mg/kg), followed by mouldboard tillage (MB) (9.6 mg/kg) and CT (7.9 mg/kg). The higher concentration of available P in CA practices was due to the higher organic matter and conversion of organic P present into available P.

CA and Soil Biological Health

CA improves the soil biological properties by enhancing carbon pools, enzyme activities, and growth of microflora and fauna for surface residues available to them as substrate.

Potentially Mineralizable N

Potentially mineralizable nitrogen (PMN), a measure of the capacity of soil to supply mineral N, constitutes an important measure of the soil health. Aulakh et al. (2013) showed that the PMN content after 2 years in the 0–5 cm soil layer of a CT system increased from 2.7 mg/kg/7 d in the control to 2.9–5.1 mg/kg/7 d without CR and to 6.9–9.7 mg/kg/7 d with CR. The corresponding increase of PMN content under the CA system was from 3.6 mg/kg/7 d in the control to 3.9–6.5 mg/kg/7 d without CR and to 8.9–12.1 mg/kg/7 d with CR. Doran and Zeiss (2000) reported that microbial biomass and PMN in the 0–7.5 cm surface layer of ZT soils were 34% higher than those of ploughed soils, although the opposite was true at the 7.5–15.0 cm depth. There was an increase of microbial biomass carbon (MBC) and mineralizable N in the surface soil with maize and cotton cropping sequences for 20 years under ZT and MT systems but little change in MBC concentration at 2.5–20.0 cm depths (Wright et al. 2005). There was a consistent increase in biological activity and N mineralization with ZT management. Interestingly, the CT soil mineralized as much N as the ZT systems but had less total soluble N (TSN) than ZT (Purakayastha et al. 2008). Tillage can greatly modify edaphic factors and thereby influences the rate of C mineralization.

Soil Microbial Biomass C and N

Microbial biomass C is an active component of SOM and constitutes an important soil health parameter as carbon contained within microbial biomass is stored energy for the microbial processes. The rapid build-up of microbial biomass in sub-tropical conditions implies that MBN could serve as a potential source of mineralizable N for plant nutrition in such soils. Thus, MBC and MBN, the measures of potential microbial activity, are strongly related to soil aggregate stability. Conversion to CA can improve soil biological quality with respect to microbial communities, microbial growth and decomposition processes, soil food web, and C dynamics. Residue management has a greater influence than the tillage system on microbial characteristics, and higher SMB-C and N levels are found in plots with residue retention than with residue removal, although the differences were significant only in the 0–10 cm layer (Spedding et al. 2004). Soil microbial biomass C and microbial biomass N are the most sensitive biological indicators, and are closely related to the cycling of C and N in the soil. Meanwhile, the conversion rate of soil microbial biomass C and N can directly or indirectly reflect changes in soil fertility.

The practice of crop residue retention and MT in association with basal fertilizer application, increases the supply of C and N, which is reflected in terms of increased microbial biomass, N-mineralization rates, and available-N concentrations in the soil (Kushwaha and Singh 2005). Silva et al. (2010) revealed that MBC and MBN values were consistently higher up to >100% under ZT in comparison to CT, and were associated with higher grain yields. The population and diversity of genomic patterns of the N_2 fixing *Bradyhizobium* increased with ZT compared to CT. Nunez et al. (2012) reported a bacterial diversity increase in ZT systems as compared to CT. Zero tillage proved to be more efficient than the other tillage systems (reduced and conventional tillage) in the conservation of organic carbon and MBC at the soil surface depth (Costantini et al. 1996). Higher contents of organic matter, MBC, MBN, and enzyme activities are observed in the surface layers of soils under CA than in soils under CT. The increase in microbial populations, diversity, and other biological indicators of soil health under CA practices can be attributed to a number of factors that favour microbial proliferation and activities.

Long-term ZT soils have significantly higher levels of microbes, more active C, SOM, and stored C than CT soils (Ghosh et al. 2012). A majority of the microbes in the soil exist under starvation conditions, and thus they tend to be in a dormant state, especially in tilled soils. Pankhurst et al. (2002) found that ZT with direct seeding into crop residue increased the build-up of organic C and SMB in the surface soil. This is attributed to higher levels of C substrates available for microorganism growth, better soil physical conditions, and higher water retention under ZT.

Soil Enzymatic Activities

Microbial activity-based indicators of soil quality respond to disturbances in a shorter period of time than those based on physical or chemical properties. As a consequence, microbiological properties, such as soil enzyme activities have been suggested as potential indicators of soil quality because of their essential role in soil biology, ease of measurement, and rapid response to changes in soil management. Soil enzyme activity can be used as an indicator of soil quality for assessing the sustainability of agricultural ecosystems. Roldan et al. (2005) reported that ZT soil had higher values of water-soluble C, dehydrogenase, urease, protease, phosphatase, and h-glucosidase activities and aggregate stability than tilled soils under sorghum, but had lower values than soil under native vegetation. This was mainly attributed to higher SOC and better microbial proliferation under conservation agriculture practices because of the addition of crop residues and minimum soil disturbance. With few exceptions, tillage had negative effects on the hydrolase activities considered in this study (urease, protease-BAA, phosphatase, and glucosidase), at all soil depths, mainly with the adoption of mouldboard. By decreasing the tillage frequency in the rice–wheat cropping system in the eastern Indo-Gangetic plains, the microbial biomass carbon and nitrogen as well as the activities of extracellular enzymes responsible for mineralization of carbon like β-D-glucosidase, cellobiohydrolase, alkaline phosphatase, and urease increased, especially under a no-tillage system, possibly due to enhanced SOM content (Pandey et al. 2014). The soil MBC and soil dehydrogenase enzyme activities (DHA) were found to be higher in ZT as compared to CT due to enrichment of soil organic matter from the crop residues (Yadav et al. 2019).

Labile Organic C and N Fractions in Total Organic C and N

Particulate organic matter (POM), dominated by undecomposed plant residues that retain recognizable cell structures including fungal hyphae, seeds, spores, and fungal skeletons, is an active fraction of SOM, which supplies nutrients to growing plants. POM-C and POM-N provide estimates of the intermediate pool of SOM between the active and passive pools, and provide substrate for microorganisms and influence soil aggregation. Light fraction organic matter (LFOM), composed primarily of plant-derived remains, and microbial and micro-faunal debris and other incompletely decomposed organic residues, is more sensitive to management practices than POM. Aulakh et al. (2013) found that an application of organic and inorganic fertilizers in a soybean–wheat cropping system under CA enhanced total organic C (TOC) from 3.8 g/kg in no–NP–FYM–CR control to 5.8 g/kg in surface layer and from 2.7 to 3.6 g/kg in the sub-surface layer after 2 years, leading to 41% and 39% higher TOC stocks over CT-control in the 0–15 cm soil layers of CT and CA, respectively. The changes in TOC stocks after 4 years were 52% and 59%. Likewise, the labile C and N fractions such as water-soluble C, particulate and light fraction organic matter, potentially mineralizable N, and microbial biomass were also highest under this integrated inorganic and organic treatment. Bhattacharyya et al. (2012) found higher 'labile' and 'less labile' SOC pools in ZT–ZT and ZT–CT plots than CT–CT plots under a rainfed lentil–fingermillet cropping system cultivated in a sandy clay loam soil of the Indian Himalayas.

Emissions of Gaseous and Aerosol Species

The Ministry of New and Renewable Energy (MNRE 2009), Govt. of India, has estimated that about 500 M t of crop residues are generated every year and of this 91–141 M t of residues are burnt in this country. The mechanized harvesting and threshing of rice using combine harvesters is a common practice in NW India. In the process, residues are left behind the combine harvesters in a narrow strip (windrow) in the field. Disposal or utilization of the left-over residue in the short window of 10–20 days for timely planting of wheat crop is a difficult task. Therefore, farmers commonly opt for burning of rice residue in the combine-harvested fields due to a lack of access to user-friendly, cost- and time-effective options. It is estimated that in the NW states of India about 23 M t of rice residues are burnt annually (NAAS 2017). This leads to deterioration of air quality in adjacent areas, resulting in various health hazards besides a deterioration of soil health.

Venkataraman et al. (2006) inventoried the emissions from open biomass burning including crop residues in India using a moderate resolution imaging spectroradiometry (MODIS) active fire and land cover data approach. Sahai et al. (2011) estimated that burning of 63 M t of crop residue emitted 4.86 M t of CO_2 equivalents of GHGs, 3.4 M t of CO, and 0.14 M t of NOx. However, by practicing CA, crop residue burning and the associated pollution can be minimized. ZT reduces the C emissions of farm operations compared to CT. While the amount of C that can be sequestered in soil is finite, the reduction in net CO_2 flux to the atmosphere by reducing fossil fuel use can continue indefinitely. The net global warming potential (GWP) taking into account soil C sequestration, emissions of GHG from soil, and fuel used for farm operations and the production of fertilizer and seeds was near neutral for ZT with crop residue retention (40 kg CO_2/ha/year), whereas in the other management practices, it was approximately 2000 kg CO_2/ha/year.

The incorporation of cereal residues into paddy fields at the optimum time before rice transplanting can help to minimize the adverse effects on rice growth and CH_4 emissions. The incorporation of wheat straw before transplanting of rice showed no significant effect on N_2O emissions due to immobilization of mineral N by the high C:N ratio of the straw incorporated (Ma et al. 2007). However, an increase in N_2O emissions from fields with mulch compared to those with incorporated residue has been observed in sub-tropical Asian rice-based cropping systems. Baggs et al. (2003) speculated that the timing of residue return such that the N becomes available when needed by the upland crop should minimize N_2O emission as compared with residue return at the beginning of the pre-season fallow. Crop residue management (CRM) is unlikely to have significant overall effects on CH_4 emission in upland crops like wheat. For CH_4 to be produced, there must be at least a small number of anaerobic microsites for methanogenic bacteria to grow, so any treatment that makes the soil more anaerobic is likely to increase the risk of CH_4 emission, including a rainfall event or mulch application. As in flooded systems, any action that causes residue to decompose before becoming anaerobic will lessen the risk of CH_4 emission. From the perspective of mitigating GHG emissions from wheat crop in the RW cropping system, residues are not the primary crop management concern. When soil is at or near field capacity, there is little CH_4 formation and N_2O emission, and the effect of CRM is negligible.

Cultivation of the direct-seeded upland rice–mustard cropping system in the north-eastern region under no tillage with 100% residue retention had 13% less CO_2 equivalent emissions than CT with 100% residue incorporation treatment (Yadav et al. 2019). Chaudhary et al. (2017) revealed that cultivation of rice in the Indo-Gangetic plains through a turbo happy seeder, bed planting, reduced tillage, and zero tillage had lower GWP than transplanted rice. Flux of GHGs, i.e. CO_2, CH_4, and nitrous oxide from rice–wheat cropping system soils in the Indo-Gangetic plains was 10–15% less and GWP per unit of wheat yield was 10 times lower in the CA-based system than the CT-based system (Sapkota et al. 2015). This may be because tillage induced disturbances in the CT-based system resulted in greater decomposition, causing higher CO_2 emissions. Parihar et al. (2018) reported that in inceptisols, the CA-based practices in maize-based cropping systems have lower nitrous oxide emissions than CT due to an increased oxygen diffusion rate which decreased the denitrification and, thus, the production of nitrous oxide.

Issues Related to the Adoption of CA

Evidently, there are several positive effects of CA practices on soil health and sustainable crop production. However, the following constraints inhibit its wider adoption among the farming community:

- Converting to CA needs higher management skills. There is a lack of information on how to implement CA (knowhow). The necessary technologies are often unavailable to farmers.
- The initial years might be very difficult for farmers. Therefore, they might need support from other farmers or from extension services, and perhaps even financial support to invest in new machinery such as zero-till planters.
- Poor management of residues and alternative competing demands for residues.
- Lack of adequate seeding equipment. Few farmers take the risk of buying new machinery. Machinery dealers might not wish to promote CA.
- Lack of knowledge among farmers for weed management under CA practices.
- Lack of crop genotypes adaptable for CA.
- Lack of knowledge on management practices, i.e. water, nutrient, and pesticides for CA.
- Cultural background (tradition, prejudice) and mindset of the farmers to till the soil. Inadequate policy support for promoting CA practices among farmers.

Conclusion and Future Outlook

Soil health degradation due to unscientific management of land resources is a significant threat to the sustainable intensification of agricultural production. CA practices result in improvements to the main indicators of soil physical, chemical, and biological properties, enhance C sequestration, and minimize GHG emissions. However, the benefits of soil health improvement may not be immediately translated to crop yield, but have a significant role in improving input use efficiency and long-term sustainability of crop yields. There are certain issues limiting the adoption of CA on a large scale such as the availability of technical knowledge, machinery, weed and pest management, residue management, competing demands of residues, and economic and policy support. Specialized equipment and knowhow for CA should be made available to farmers. Legislation should be enacted to stop burning of crop residues by farmers and some incentives to farmers may be given to adopt CA rather than burning of crop residues. CA practices hold promise for improving soil health and sustainable intensification of crop yields and hence need to be validated, and site-specific CA practices should be promoted for diverse soil, crop, and agro-climatic situations. These issues need to be addressed in a systematic manner with a proper timeline so that the benefits of this technology in terms of enhanced productivity, soil health, and mitigating adverse effects on climate are realized in Indian agriculture.

References

Acharya, C.L., Hati, K.M. and Bandyopadhyay, K.K. 2005. Mulches, pp. 521–532. In. Hillel, D. et al. (Eds.) *Encyclopedia of Soils in the Environment*. Elsevier Academic Press, Amsterdam.

Aggarwal, P., Kumar, R. and Yadav, B. 2013. Hydrophysical characteristics of a sandy loam soil under bed planted maize (*Zea mays*). *Indian Journal of Agricultural Sciences* 83(5): 491–496.

Aulakh, M., Garg A. and Shrvan, K. 2013. Impact of integrated nutrient, crop residue and tillage management on soil aggregates and organic matter fractions in semiarid subtropical soil under soybean-wheat rotation. *American Journal of Plant Science* 4: 2148–2164.

Bag, K., Bandyopadhyay, K.K., Sehgal, V.K., et al. 2020. Effect of tillage, residue and nitrogen management on soil physical properties, root growth and productivity of wheat. *Indian Journal of Agricultural Sciences* 90(9): 1753–1757.

Baggs, E.M., Stevenson, M., Pihlatie, M., et al. 2003. Nitrous oxide emissions following application of residues and fertiliser under zero and conventional tillage. *Plant and Soil* 254: 361–370.

Ball, B.C., Campbell, D.J., Douglas, J.T., et al. 1997. Soil structural quality, compaction and land management. *European Journal of Soil Science* 48: 593–601.

Bhattacharyya, R., Tuti, M.D., Kundu, S., et al. 2012. Conservation tillage impacts on soil aggregation and carbon pools in a sandy clay loam soil of the Indian Himalayas. *Soil Science Society of America Journal* 76(2): 617–627.

Boddey, R.M., Jantalia, A.C.P.J., Conceicao, P.C., et al. 2010. Carbon accumulation at depth in Ferralsols under zero-till subtropical agriculture. *Global Change Biology* 16: 784–795.

Chaudhary, V.P., Singh, K.K., Pratibha, G., et al. 2017. Energy conservation and greenhouse gas mitigation under different production systems in rice cultivation. *Energy* 130: 307–317.

Costantini, A., Cosentino, D. and Segat, A. 1996. Influence of tillage systems on biological properties of a Typic Argiudoll soil under continuous maize in central Argentina. *Soil and Tillage Research* 38: 265–271.

da Silva, A.P., Kay, B.D. and Perfect, E. 1994. Factors influencing least limiting water range of soils. *Soil Science Society of America Journal* 58: 1775–1781.

Das, T.K., Saharawat, Y.S., Bhattacharyya, R., et al. 2018. Conservation agriculture effects on crop and water productivity, profitability and soil organic carbon accumulation under a maize-wheat cropping system in the North-western Indo-Gangetic Plains. *Field Crops Research* 215: 222–231.

Doran, J.W. and Zeiss, M.R. 2000. Soil health and sustainability: managing the biotic component of soil quality. *Applied Soil Ecology* 15: 3–11.

Ghosh, S., Brian, W., Subrata, G., et al. 2012. Organic amendments influence soil quality and carbon sequestration in the Indo-Gangetic plains of India. *Agriculture, Ecosystems and Environment* 156: 134–141.

Govaerts, B., Sayre, K.D., Lichter, K., et al. 2007. Influence of permanent bed planting and residue management on physical and chemical soil quality in rain fed maize/wheat systems. *Plant and Soil* 291: 39–54.

Gupta-Choudhury, Shreyasi, Srivastava, Singh, S., et al. 2014. Tillage and residue management effects on soil aggregation, organic carbon dynamics and yield attribute in rice–wheat cropping system under reclaimed sodic soil. *Soil and Tillage Research* 136: 76–83.

Jantalia, C.P., Resck, D.V.S., Alves, B.J.R., et al. 2000. Tillage effect on C stocks of a clayey Oxisol under a soybean-based crop rotation in the Brazilian Cerrado region. *Soil and Tillage Research* 95: 97–109.

Jat, H.S., Datta, A., Sharma, P.C., et al. 2018. Assessing soil properties and nutrient availability under conservation agriculture practices in a reclaimed sodic soil in cereal-based systems of North-West India. *Archives of Agronomy and Soil Science* 64(4): 531–545.

Jat, M.L., Gathala, M.K., Saharawat, Y.S., et al. 2013. Double no-till and permanent raised beds in maize–wheat rotation of north-western Indo-Gangetic plains of India: Effects on crop yields, water productivity, profitability and soil physical properties. *Field Crops Research* 149: 291–299.

Kay, B.D. and Anger, D.A. 2002. Soil structure, pp. 249–296. In: *Soil Physics* Companion. A Warrick. (Ed.).

Kettler, T.A., Lyon, D.J., Doran, J.W., et al. 2000. Soil quality assessment after weed-control tillage in a no-till wheat-fallow cropping system. *Soil Science Society of America Journal* 64: 339–346.

Kushwa, V., Hati, K.M., Sinha, N.K., et al., 2016. Long-term conservation tillage effect on soil organic carbon and available phosphorous content in vertisols of central India. *Agricultural Research* 5(4): 353–361.

Kushwaha, C.P. and Singh, K.P. 2005. Crop productivity and soil fertility in a tropical dryland agro-ecosystem: impact of residue and tillage management. *Experimental Agriculture* 41: 39–50.

Lal, R. 1997. Long-term tillage and maize monoculture effects on a tropical Alfisol in western Nigeria. I. Crop yield and soil physical properties. *Soil and Tillage Research* 42(3): 145–160.

Lal, R. and Shukla, M.J. 2004. *Principles of Soil Physics*. New York, Marcel Dekker.

Ma, J., Li, X.L., Xu, H. et al. 2007. Effects of nitrogen fertiliser and wheat straw application on CH_4 and N_2O emissions from a paddy rice field. *Australian Journal of Soil Research* 45: 359–367.

Malhi, S.S., Nyborg, M., Goddard, T., et al. 2011. Long-term tillage, straw and N rate effects on some chemical properties in two contrasting soil types in Western Canada. *Nutrient Cycling in Agroecosystems* 90: 133–146.

Mishra, A.K., Aggarwal, P., Bhattacharyya, R. et al. 2015. Least limiting water range for two conservation agriculture cropping systems in India. *Soil and Tillage Research* 150: 43–56.

MNRE. 2009. Ministry of New and Renewable Energy Resources, Government of India, New Delhi. www.mnre.gov.in/biomassrsources.

NAAS. 2017. Innovative viable solution to rice residue burning in rice-wheat cropping system through concurrent use of super straw management system fitted Combines and Turbo Happy Seeder. *Policy Brief No.*, 216 p. National Academy of Agricultural Sciences, New Delhi.

Naresh, R.K., Gupta, Raj K., Singh, S.P., et al. 2016. Tillage, irrigation levels and rice straw mulches effects on wheat productivity, soil aggregates and soil organic carbon dynamics after rice in sandy loam soils of subtropical climatic conditions. *Journal of Pure Applied Microbiology* 10: 1061–1080.

Naresh, R.K., Gupta, Raj, K., Pal, G., et al. 2015. Tillage crop establishment strategies and soil fertility management: resource use efficiencies and soil carbon sequestration in a rice-wheat cropping system. *Ecology, Environment & Conservation* 21: 127–134.

Nunez, E.V., Bens, F., Valenzuela-Encinas, C. et al. 2012. Bacterial diversity as affected by tillage practice in raised bed-planting system. *African Journal of Microbiology Research* 6(43): 7048-7058.

Pandey, D., Agrawal, M. and Bohra, J.S. 2014. Effects of conventional tillage and no tillage permutations on extracellular soil enzyme activities and microbial biomass under rice cultivation. *Soil and Tillage Research* 136: 51–60.

Pankhurst, C.E., McDonald, H.J., Hawke, B.G., et al. 2002. Effect of tillage and stubble management on chemical and microbiological properties and the development of suppression towards cereal root disease in soils from two sites in NSW, Australia. *Soil Biology and Biochemistry* 34: 833–840.

Parihar, C.M., Parihar, M.D., Sapkota, T.B., et al. 2018. Long-term impact of conservation agriculture and diversified maize rotations on carbon pools and stocks, mineral nitrogen fractions and nitrous oxide fluxes in inceptisol of India. *Science of Total Environment* 640: 1382–1392.

Purakayastha, T.J., Huggins, D.R. and Smith, J.L. 2008. Carbon sequestration in native prairie. *Soil Science Society of America Journal* 72: 534–540.

Roldan, A., Salinas-Garcia, J.R., Alguacil, M.M., et al. 2005. Soil enzyme activities suggest advantages of conservation tillage practices in sorghum cultivation under subtropical conditions. *Geoderma* 129: 178–185.

Sahai, S., Sharma, C., Singh, S.K., et al. 2011. Assessment of trace gases, carbon and nitrogen emissions from field burning of agricultural residues in India. *Nutrient Cycling in Agroecosystems* 89: 143–157.

Sapkota, T.B., Jat, M.L., Aryal, J.P., et al. 2015. Climate change adaptation, greenhouse gas mitigation and economic profitability of conservation agriculture: Some examples from cereal systems of Indo-Gangetic Plains. *Journal of Integrative Agriculture* 14(8): 1524–1533.

Silva, A.D., Babujia, L.C., Franchini, J.C., et al. 2010. Microbial biomass under various soil- and crop- management systems in short- and long- term experiments in Brazil. *Field Crops Research* 119: 20-26.

Singh, A., Phogat, V.K., Dahiya, R., et al. 2014. Impact of long-term zero till wheat on soil physical properties and wheat productivity under rice–wheat cropping system. *Soil and Tillage Research* 140: 98–105.

Spedding, T.A., Hamel, C., Mehuys, G.R., et al. 2004. Soil microbial dynamics in maize-growing soil under different tillage and residue management systems. *Soil Biology and Biochemistry* 36: 499–512.

Venkataraman, C., Habib, G., Kadamba, D., et al. 2006. Emissions from open biomass burning in india: integrating the inventory approach with higher solution Moderate Resolution Imaging Spectroradiometer (MODIS) active fire and land count data. *Global Biogeochemical Cycles* 20: 13–20.

Wright, A.L., Hons, F.M. and Matocha, J.E. 2005. Tillage impacts on microbial biomass and soil carbon and nitrogen dynamics of corn and cotton rotations. *Applied Soil Ecology* 29: 85–92.

Yadav, G.S., Lal, R., Meena, R.S., et al. 2019. Conservation tillage and nutrient management effects on productivity and soil carbon sequestration under double cropping of
rice in north eastern region of India. *Ecological Indicators* 105: 303–315.

Yoo, G., Nissen, T.M. and Wander, M. 2006. Use of physical properties to predict the
effects of tillage practices on organic matter dynamics in three Illinois soils. *Journal of
Environmental Quality* 35(4): 1576–1583.

Zhang, S.L., Simelton, E., Lovdahl, L., et al. 2007. Simulated long-term effects of different
soil management regimes on the water balance in the Loess Plateau, China. *Field Crops
Research* 100: 311–319.

13 Soil Microbes and Conservation Agriculture

Geeta Singh

Introduction

The major soil factors affecting Indian agriculture are nutrient depletion, loss of soil organic matter, and the overall decline in soil quality. The loss of soil structure and fertility can be retarded by adoption of a suitable agri-management that favours soil microbial activities/properties and thereby prevents soil degradation. As cultivation and tillage are important determinants of agriculture sustainability of crop production, there is a need to promote and adopt alternative agri-technologies that can help overcome these limitations. To achieve this objective, conservation agriculture (CA), defined as minimal soil disturbance and permanent soil cover (mulch) combined with rotations, has been identified as a more sustainable cultivation system. Alterations in tillage, crop residue, and crop rotation have a qualitative and quantitative impact on soil microbiota. In a meta-analysis utilizing 139 observations from 62 studies across the globe, it was concluded that zero tillage (ZT) promotes larger microbial communities, microbial biomass and greater enzymatic activity (Zuber et al. 2016) with a few exceptions. These practices induce significant changes in the quantity and quality of plant residue entering the soil, their spatio-temporal distribution, the ratio between above- and below-ground inputs, and nutrient dynamics, all of which collectively influence soil microorganisms and their metabolic processes (Parihar et al. 2016). This chapter focuses primarily on the status of soil microbiological parameters related to soil quality and health due to CA practices in different agro-ecosystems of India.

Soil Quality

Soil quality is the capacity of a soil to function within ecosystem boundaries to sustain biological productivity, maintain environmental quality, and promote plant and animal health. It is a complex interactive effect of physical, chemical, and biological parameters, expressed as the soil quality index (SQI). It is an indicator of the sustainability of the cropping system (Mandal et al. 2005). A high SQI value is desirable as it denotes high and sustainable levels of production.

DOI: 10.4324/9781003292487-16

Microorganisms as Soil Quality Indicators

Soil microorganisms are the primary agents responsible for soil quality and soil health. Their complex interactions mediate the transformation of organic matter, nutrient cycling, and regulating soil fertility, and thereby influencing crop performance for sustainable crop production.

The contribution of soil microorganisms to crop production is the net outcome of a multitude of complex interactions. The subset of soil microbiomes associated with plant roots are described as rhizosphere microorganisms. These may be free-living or they may exhibit a symbiotic association. Collectively, these assist in nutrient acquisition, phytohormone production, and phytotoxic compound degradation, and impart tolerance to biotic and abiotic stresses (Choudhary et al. 2010; Meena et al. 2010; Singh et al. 2013). Additionally, they also produce numerous biomolecules, e.g. organic acids, and siderophores that aid the mobilization of nutrients and facilitate their uptake by crop plants (Singh et al. 2010). In return, the plant provides a favourable environment for microbial growth and a continuous supply of carbon-rich rhizodeposition (Singh and Mukerji 2006). There is growing awareness about the significance of soil microorganisms in maintaining soil quality and soil health. Their high sensitivity to environmental factors and agri-management practices qualify them as early and potential indicators of soil quality. Conventionally, the quantification of soil microbial activities is undertaken by estimating the microbial biomass carbon (MBC) and microbial biomass nitrogen (MBN), and enzymatic activities. The major microbial enzymes involved in the nutrient cycling in soil systems are presented in Table 13.1.

Some microbes (algae, photosynthetic bacteria) also enhance organic matter input in soil that augments the availability of macro- and micro-nutrients. The contribution of soil microorganisms in nutrient acquisition is best exemplified by N_2-fixing nodules formed on leguminous roots by bacteria of the genus *Rhizobium* and the existence of a biotrophic association of arbuscular mycorrhizal fungi (AMF) with the majority of crop plants. Microbial activity in soil also accounts for losses of nutrients (e.g. C, N) to the atmosphere through bioprocesses like respiration, methanogenesis, nitrification, and denitrification. thereby contributing to GHG emissions. The adoption of a suitable agri-management practice can modulate microbial processes that aid in nutrient transformation and carbon sequestration, and assist in a reduction of GHG emissions.

Soil Enzymes

Quantification of the functional role microbes play in nutrient cycling is mainly achieved through assessment of the rates of soil enzymatic activity. Soil enzymes catalyse biochemical reactions involved in organic matter mineralization; and thus, serve as sensitive indicators of SOM decomposition (Sinsabaugh et al. 2008; Singh et al. 2015a). Two bio-indicators, namely the hydrolases [fluorescein diacetate (FDA)] and oxidoreductases [dehydrogenase (DHA)] are typically

Table 13.1 Different soil microbial parameters linked with nutrient transformation

Nutrient transformation	Specific enzyme assay	Additional useful technique for soil	Agricultural significance/indicator
Carbon	Degrading cellulose (β-1,4-glucosidase), chitin (β-1,4-N-acetylglucosaminidase [NAG]), and lignin (phenol oxidase and peroxidase)	• Soil respiration • Microbial biomass carbon • Metabolic quotient (qCO$_2$): respiration/MBC	• SOC mineralization potential • MBC is a useful indicator of soil quality • Metabolic quotient: differentiate response of soil microorganisms to soil management practices
Nitrogen	N mineralization: Amidohydrolases (asparaginase, l-glutaminase, amidase, and urease) and arylamidase N-acetyl glucosaminidase (NAG), arylamidase, l-glutaminase, and l-asparaginase, arginine deaminase, urease, ammonia monooxygenase, nitrate reductase	Soil MBN	i. N mineralization ii. Loss of the applied N and availability of N for crop plants
Phosphorus	Acid phosphatases, alkaline phosphatases	Available P in soil	Indication of availability of inorganic P in soil. It reflects the mineralization of organic P
Secondary and micronutrient	Abundance of arbuscular mycorrhizal fungi in soil as spores, plant root infection index (%)	Glycoprotein: glomalin	Availability of micronutrients (Fe, Zn)

used to reflect the total microbial activity in soil. DHA represents the complete range of oxidative activity of soil microflora, while FDA is the cumulative expression of microbial enzymes including protease, lipase, and esterases. Thus, FDA activity represents the hydrolytic potential of soil microorganisms. Soil enzymes, arylsulfatase, and ß-glycosidase are involved in the degradation of organic matter and are highly sensitive to perturbations due to soil management. Acid and alkaline phosphatases are produced by a wide group of soil microbiota that release inorganic phosphate from organic P sources, thereby improving crop phosphorus availability. Alkaline phosphatases are mainly produced by bacteria, fungi, and earthworms (Tarafdar and Marschner 1994). Acid phosphatases originate either from plant roots (and associated mycorrhiza or other fungi) or from bacteria (Meena et al. 2010). Abiotic factors prevailing in the soil environment such as tillage, moisture, nutrient availability, and compaction significantly affect soil microbial enzymatic activity (Sharma et al. 2011). The three cardinal components of CA, namely ZT, crop rotations, and residue inputs, have been shown to exert a positive influence on microbial enzymatic activity, implying a favourable microclimate. Minimum soil disturbances, e.g. in ZT, cause fungal hyphae to remain intact. As fungi play a key role in the cycling of C and N, and contribute to the soil enzymes, the assessment of the impact of CA practices on soil microbiology is often undertaken in terms of quantification of microbial enzymes. The results of some of these studies are described below.

Alkaline Phosphatase

ZT practices substantially increase phytase activity in soil due to the build-up of microbial biomass at the expense of decomposing surface plant biomass (Yadav and Tarafdar 2004). Application of farmyard manure (FYM) and wheat straw in conjunction with inorganic fertilizer significantly increased the activity of alkaline phosphatase in the 0–10 cm soil layer as compared with the application of inorganic fertilizer alone in a double ZT rice–wheat system (Singh and Ghosal 2013).

Soil N Transformation

Soil N transformations, e.g. N mineralization rate in the soil, are the determinant for the quantity of plant-available N, and this process is mediated by soil microbial enzymes. The major enzymes include urease and amidohydrolases, N-acetyl glucosaminidase (NAG), arylamidase, glutaminase, and asparaginase. Tabatabai (1994) observed higher activity of enzymes arylamidase, L-asparaginase, L-glutaminase, amidase, urease, and L-aspartase activity under ZT than ridge-till and chisel ploughing.

Transformation of C in Soil

Cellulase, glucosidase, and xylanase are involved in C cycling and yield energy for the soil microorganisms. Cellulases are a group of enzymes that catalyse the breakdown of β-1,4-linked glucose units present in cellulose. Hence, cellulases perform a crucial role during the initial phase of decomposition of crop residues (organic matter) in soil. Xylanase (endo-1,4-β-xylanase) enzymes are mainly responsible for the decomposition of polysaccharide hemicelluloses present in crop residues (Sinsabaugh et al. 2008). The enzyme β-glucosidase acts at the final stage of decomposition of cellulosic crop residues, and sugars are the final product formed. β-Glucosidase enzyme is sensitive to any changes in the management practices in soil and is directly related to the amount of organic matter, and so is considered to be a promising soil quality indicator for assessing the changes induced by tillage practices (Sharma et al. 2012). The differential accumulation of organic C, N, and inorganic nutrients in top soil under different tillage practices as well as accumulation of inorganic nutrients tend to increase soil enzymatic activities. The biomass C input from crop residues serves as a readily available substrate for the various microbial enzymes, and thus accounts for the observed increase in their activities.

Soil Microbial Biomass

Nutrient availability and the productivity of agroecosystems are regulated by the size and activity of the soil microbial biomass (SMB). This is achieved by mineralization of soil organic matter (SOM), of which SMB is a small but dynamic fraction. Microbial biomass C and N describe the microbial catalytic potential and repository for C and N. Soil microbial biomass C (Cmic) constitutes only 1–3% of total soil C and the biomass N (Nmic) up to 5% of total soil N – they are the most labile C and N pools in soil. The turnover time for N immobilized in the microbial biomass is approximately 10 times faster than that derived from plant material (Moore et al. 2000). The determination of Nmic is, therefore, important for the quantification of N dynamics in agricultural ecosystems because it controls soil inorganic N availability and loss. Due to their high sensitivity, these serve as a tool to measure changes much earlier than they are seen in terms of SOC and N. In CA, the rate of biomass-C input as plant residues (surface cover) strongly influences the quantity of SMBC in soil.

Unlike in temperate conditions, the tropical Indian agroecosystems experience much faster degradation of SOM, resulting in low soil carbon pools. Against this background, the CA practices were found to favour microbial biomass growth, thereby maintaining higher SOM and SMB levels. The magnitude of increase in microbial biomass under ZT compared to CT differs according to the prevailing climatic conditions. In tropics, soil samples from ZT plots were found to have significantly higher MBC than CT. The greatest difference between ZT and CT is often reported in the surface layer (0–5 cm) (Meena et al. 2008; Pandey et al. 2014). For instance, the MBC in ZT was on average 83% higher than in CT plots (Das et al. 2014). The observed increases in SOM and SMB in the CA systems are a net outcome of residue retention, a reduction of SOM degradation, and the impact of crop rotation in terms of the rhizo-depositions contributed by crop plants. The rate of organic C input from crop biomass is generally considered as the dominant factor controlling the amount of SMB. On the other hand, soil disturbance induced by conventional tillage results in a decline in SMB at a higher rate than that observed in SOM. Consequently, it results in a decline in the percentage of total C as MBC. In contrast, conventional tillage results in a decline in SMB at a higher rate than that observed in SOM. The threshold ratio of MBC:SOC is used as an indicator of C accumulation, and this value differs according to the climate. Under CA practice, crop residue retention on the soil surface results in enrichment of SOM and MBC that stimulates microbial activity in the surface layer. This ensures a higher magnitude of MBC and consequently enhances soil nutrient availability and biological health (Sharma et al. 2011; Singh et al. 2018). In addition, other factors such as lowering of soil temperature, moisture conservation, soil aggregation, and elevated C content in the ZT system due to accumulation of surface litter also account for the higher increase of MBC under ZT compared with CT systems in tropical soils.

Metabolic Quotient

The metabolic quotient (qCO_2) is the ratio of basal respiration to microbial biomass, and a low qCO_2 value is an indicator of a stable agricultural system. Prasad et al. (1994) observed an increase of 106% in MBC and a decrease in the qCO_2 of 39% in different ecosystems of India. In tropical conditions like India, the range of qCO_2 is 3.6–5.5 under CT and 2.16–3.60 mg CO_2-C/g MBC per day under ZT. The higher stability of ZT relative to CT is due to the abundance of large aggregates that protect the microorganisms from adverse conditions. This offers storage and protection to the SOC resulting in a reduction of oxidative losses of carbon. In contrast, top soils under CT with lower SOM, MBC contents, altered soil microbial community, and metabolic rates with reduced C substrates availability to microorganisms, register a higher magnitude of metabolic quotient than ZT systems. The cumulative effects of an increase in MBC and a decrease in qCO_2 are indicative of a stable soil ecosystem under ZT.

Tillage

Tillage-induced soil disturbance can cause a stratification of nutrients, and therefore alters microbial communities. Conventional tillage mixes crop residues and nutrients into the soil profile, with disruption of soil aggregates exposing organic matter, and enhanced aeration facilitating microbial growth at deeper depths than possible under ZT soils. These changes are accompanied by altered physical parameters such as soil water, temperature, aeration, equilibrium of reactions, and also an increase in soil erosion. In ZT soils with higher SOM, MBC offers a favourable microclimate for microbial proliferation relative to CT (Das et al. 2014; Bhattacharyya et al. 2015). The relatively low redox potential, larger proportions of microbial biomass, and carbohydrate-C per unit of organic C under ZT accounts for the higher dehydrogenase activity in CA-based systems (Singh et al. 2015a).

Crop Rotation

Crop rotations influence soil biological activities through their effects on the quantity, structure, and distribution of soil organic matter. There is a significant positive influence on soil quality due to differences in plant root morphologies (e.g. density and penetration depth), physiology (nature of root exudates), amount and quality of residue generated, and influence on soil aggregation/ microbial habitat that collectively accounts for differences in soil chemistry, structure, and rhizosphere microbial populations (Singh and Mukerji 2006). It is reported that microbial biomass C and N show significant responses to crop rotations because root-associated microorganisms, rhizodeposition of root mass, and exudates greatly influence C turnover in soils as well as net C accumulation. Monocultural systems, in general, contain significantly lower concentrations and qualities of SOM and MBC, and have less soil structural stability.

Soil amendment with crop residues having varied chemical compositions derived from crop rotations contribute to enriching labile SOC pools, and altering the composition and functioning of soil microbiota. A positive relation between organic matter inputs as crop residues labile SOM, MBC, and enzyme activities in the surface layers is reported. The nature of crops included in the CA system also influence the carbon sequestration potential of soils due to differences in chemical properties between the different cropping systems.

Surface Retention of Crop Residues

Crop residues are plant biomasses left in the field after crops have been harvested and economically important produce has been threshed out. The multiple benefits of crop residue retention on the soil surface include efficient recycling of nutrient sources, reduction in runoff and soil erosion, decreased soil evaporation, and energy conservation in terms of reduced land preparation costs. In addition, there is a positive effect on soil quality indices, including improved soil biological properties (i.e. species diversity of soil biota, SOC sequestration, and microbial biomass C), chemical properties (i.e. nutrient cycling, cation exchange capacity, soil reaction) (Beri et al. 1995), and physical properties (i.e. structure, infiltration rate, plant available water capacity) (Chakraborty et al. 2008). The combined effects of soil physical, chemical, and biological benefits accrued by the surface crop residue retention in CA systems favour soil microbial parameters. The abundance and nature of soil microflora are directly linked to the quality and quantity of crop residues. Soil treated with crop residues held 5–10 times more aerobic bacteria and 1.5–11.0 times more fungi than soil in which it was either burnt or removed. Similarly, it has been found that residue retention resulted in a higher number of plant growth-promoting bacteria (Bacillus, Pseudomonas) as well as stimulated enzymatic activities (Sharma et al. 2015). In a short-term study undertaken in the irrigated region of the NW IGP, complete use of crop residues as mulch for alternate season crops improved the dehydrogenase enzyme activity, with improvements of about 156% and 147% over a no-residue control (Singh et al. 2018). Under the rainfed ecosystem of the NW-IGP, a short-term (3-year) study involving a pearlmillet–chickpea cropping system revealed that the plots with residue retention had ~40% higher soil glycoprotein glomalin content compared with that with residue removal (Table 13.2a,b). It was also found that

Table 13.2(a) Effect of crop rotation and residue management on selected soil enzyme activities in the 0–15 cm soil layer after 3 years of cropping

Treatment	Dehydrogenase (μg TPF/g/ 24 h)	FDA (μg fluorescein g/hr)	Glucosidase (μmol p-nitrophenol/ g/h)	Acid phosphatase (μmol p-nitrophenol/ g/h)	Alkaline phosphatase (μmol p-nitrophenol/ g/h)
Crop rotation					
Pearlmillet–wheat	247.4[c]	18.0[c]	109.5[b]	126.8[a]	124.7[b]
Pearlmillet–chickpea	459.2[a]	21.0[b]	117.0[ab]	127.4[a]	149.3[a]
Pearlmillet–mustard	412.7[b]	23.3[a]	121.7[a]	122.4[a]	154.6[a]
Residue management					
No residue	185.4[b]	13.3[c]	106.2[c]	104.7[b]	116.6[c]
Crop residue	475.3[a]	27.9[a]	126.9[a]	146.8[a]	175.7[a]
Leucaena residue	458.7[a]	21.1[b]	115.2[b]	125.2[b]	136.3[b]

Notes: FDA = Fluorescein di-acetate.

Means followed by a similar letter within a column for a management practice are not significantly different at $P<0.05$ according to Tukey's HSD.

Table 13.2(b) Effect of crop rotation and residue management on selected soil enzyme activities in the 0–15 cm soil layer after 3 years of cropping

Treatment	pH	Soil respiration (mg CO_2- C/g soil/week)	Microbial quotient (10-2 per hr)	Microbial biomass C (mg/kg soil)	Total SOC (g/kg soil)
Crop rotation					
Pearlmillet–wheat	7.48[a]	54.35[a]	4.36[a]	64.87[b]	5.1[a] (4.0[a])
Pearlmillet–chickpea	7.46[a]	53.20[a]	3.32[b]	83.38[a]	5.3[a] (4.2[a])
Pearlmillet–mustard	7.42[a]	51.30[a]	2.97[b]	89.77[a]	5.2[a] (4.2[a])
Residue management					
No residue	7.47[a]	46.63[b]	3.28[a]	63.62[b]	5.0[b] (3.8[b])
Crop residue	7.45[a]	56.05[a]	3.39[a]	90.84[a]	5.5[a] (4.3[a])
Leucenea residue	7.44[a]	56.17[a]	3.71[a]	83.56[a]	5.5[a] (4.2[a])

Source: Singh et al. (2018).

Notes: Data in parentheses indicate Walkley-Black C (WBC).

Means followed by a similar letter within a column for a management practice are not significantly different at $P<0.05$ according to Tukey's HSD.

retention of crop residues promotes N_2 fixation in soil by free-living diazotrophes, e.g. *Azotobacter chrococcum* and *A. agilis*, as these organisms are favoured by a high SOC content. A higher soil microbial population also augments the synthesis and activities of soil enzymes responsible for the conversion of unavailable to available forms of nutrients.

Residue Management in CA

In India, the rice–wheat cropping system is one of the major cropping systems. Post-harvest residue left in the field is burnt *in situ*, leading to a deterioration of soil and environmental quality (Erenstein and Laxmi 2008). Attempts have been made to use the native lingocellulose-degrading fungi (isolated from rice–wheat field soils), namely *Aspergillus flavus*, *Aspergillus terreus*, *Alternaria alternata*, and *Penicillium janthinellum*, which were found to efficiently enhance the degradation of crop residues with no adverse effects on subsequent wheat growth. A simultaneous increase in enzymatic activities, build-up of SOM, and loss of dry mass was also found, which indicated that in CA-based RWCS the use of native soil microflora for the management of crop residues holds promise (Choudhary et al. 2016).

Indian experiences with respect to the impact of CA experiments on soil microbiology are mainly confined to the IGP, and only few studies in other agro-ecosystems have been documented. Selected examples of the impact of CA on soil microbiota are presented in Table 13.3.

Conclusion

Anthropogenic activities are leading to a serious decline in soil productivity and fertility. Conventional tillage involving mechanical disruption of soil leads to a loss of organic matter content, water, and nutrients, and degradation and erosion of soil. CA involving the absence of soil inversion, retaining >30% of crop residues on the soil surface, and adopting crop rotations helps to maintain soil moisture, increases soil organic matter content, reduces soil erosion, and promotes soil fertility and biological activity. Tillage practices have been shown to influence microbial biomass accumulation, microbial community structure, taxonomic composition, and microbial abundance and activity by changing the physico-chemical properties of soil. The impact of CA on soil biological health has been assessed by microbe-mediated processes that are directly or indirectly linked to the macro- and micro-nutrients transformation and turnover. A positive impact of CA management has been observed on soil health as well as environmental indices associated with climate change. A significant difference in the soil microbial community structure and predicted function in response to alternative tillage systems has been reported in the different agro-ecosystems of India. CA has been found to be a sustainable and environment-friendly management system for cultivating crops. Indian experiences show that CA is an effective component to improve soil quality and sustainability.

Table 13.3 Selected field studies on the impact of CA practices on soil microbial indices in different agroecosystems in India

S. no.	Climate/soil type	Experimental details	Results	References
1	Himalayan sub-temperate, characterized by moderate summer (May–June), extreme winter (December–January); sandy clay loam soil	In a 3-year study, the impact of different tillage practices on biological activity, major nutrient transformation, rotation of two grain crops per year (lentil–finger millet) with four different tillage practices: zero–zero (ZZ), conventional–conventional (CC), zero–conventional (ZC), and conventional–zero (ZC) tillage were evaluated	Biological parameters were influenced by tillage practices. Soil carbohydrate content increased from 3.1 to 4.9 mg/g soil under alternate ZT and conventional tillage practice behaved differently in terms of biological attributes as compared with continuous ZT or conventional tillage. ZT increased the SOC content in the top soil layer from 6.8 to 7.5 mg/g soil. Dehydrogenase activity increased significantly under continuous ZT practice. Alkaline phosphatase and protease activity was greater (by 9.3–48.1%) in the ZT system over conventional practice. In contrast, cellulase activity was higher (by 31.3–74.6%) in conventional practice over other practices	Mina et al. (2008)
2.	Sub-tropical semi-arid with hot and dry summers and cold winters. Annual rainfall (~750 mm), mean annual maximum and minimum temperature (~35 and 18°C, respectively). Alluvial soil Typic Haplustept. pH 8.0 and silty clay loam, organic C 0.62%, and available N, P and K 250, 56 and 406 kg/ha, respectively	Alternate tillage practices adopted in RWCS: rice transplanted after conventional puddling the field; non-puddled rice/direct-seeded rice was cultivated after ploughing the field. Tilled wheat was sown after conventional ploughing, no-tilled wheat with zero tillage. Integrated nutrient management was adopted	Fertilizer and tillage treatments significantly affected MBC content. Soil organic amendment was superior to 100% NPK and crop residue treatments by increasing MBC by 19% and 4%, respectively, compared to 100% NPK treatment in puddled soil. In non-puddled soil, the corresponding increases were 6% and 9%, respectively. Addition of C to soil through INM (FYM, green manure, crop residues and biofertilizer) stimulated microbial activity and increased the soil MBC content. In wheat crop the maximum value of MBC was recorded in the FYM treatment (185 mg/kg) in ZT plots	Banerjee et al. (2006)

(continued)

3	Sub-tropical semi-arid with hot and dry summers and cold winters. Annual rainfall (~750 mm), mean annual maximum and minimum temperature ~35 and 18°C respectively. Alluvial soil Typic Haplustept. pH 8.0 and silty clay loam, organic carbon 0.62%, and available N, P, and K 250, 56, and 406 kg/ha, respectively	Alternate tillage practices adopted in RWCS: rice transplanted after conventional puddling the field; non-puddled rice was direct-seeded after ploughing the field. Conventional tilled, ZT wheat. Integrated nutrient management was adopted	Resource-conserving practices exerted a positive effect on the population of beneficial soil microbes and crop yield. Increased populations of *Azotobacter* (5.01–7.74%), *Bacillus* (3.37–6.79%), and *Pseudomonas* (5.21–7.09%) were observed due to improved structure and increased organic matter in the soil	Sharma et al. (2015)
4	Semi-arid climate, with an average annual rainfall of 650 mm (70–80% of which received during July–September) with the mean annual evaporation of 850 mm. Rainfall along the period of the cropping cycle (July to June) ranged from 533 to 1507 mm. The mean daily minimum temperature of 0–4°C in January, mean daily maximum temperature of 40–46°C in May–June, and mean daily relative humidity of 67–83% during the experimentation years. Sandy-loam (Typic Haplustept) soil in north-western India	A long-term field study involving tillage permutations (permanent raised-bed, zero tillage and conventional tillage) and four diversified intensive maize-based crop rotations: MWMb: maize–wheat–mungbean MCS: maize–chickpea–sesbania MMuMb: maize–mustard–mungbean MMS: maize–maize–sesbania	MBC was higher by 48.9% and 44.9% in ZT and PB compared to CT, respectively. However, the ZT and PB were statistically at par ($P<0.05$) with respect to soil MBC. The highest soil MBC (436.1 mg/g soil) was recorded in MCS. The order of MBC was: MCS>MWMb>MMuMb>MMS. A significant improvement in soil enzymatic activities, i.e. fluorescein diacetate, dehydrogenase, b-glucosidase and alkaline phosphatase was also recorded in the CA-based treatments. DHA and FDA were significantly higher under ZT and permanent bed than conventional tillage. A higher DHA and FDA was found under MCS as compared with MWMb, MMuMb, and MMS	Parihar et al. (2016)

Table 13.3 Cont.

S. no.	Climate/soil type	Experimental details	Results	References
5	Sub-tropical semi-arid with hot and dry summers and cold winters. Annual rainfall (~750 mm), mean annual maximum and minimum temperature ~35 and 18°C, respectively	In an 8-year field trial, combined impact of tillage, water, and nutrient management on rice and wheat was studied in the Indo-Gangetic plains in India	Non-puddling significantly enhanced DHA (5%), MBC (3%), and PMN (5%) over puddling. Puddling benefited soil respiration (48%) and metabolic quotient (41%) in rice. DHA after harvesting of wheat crop was 11% higher under previously non-puddled soil as compared with puddled soil. ZT resulted in higher values of soil biological indicators under wheat cultivation	Bhaduri et al. (2017)
6	Inceptisol classified as Typic Haplustept	Maize–wheat cropping system with crop residue retention (four) in main plot, i.e. residue removal (no residue), 25%, 50%, 75% crop residue; and P fertilizer rate (five) in sub-plot namely S1: No-P, S2: 50%, S3: 100% (recommended dose of phosphorus (RDP), S4:150% RDP, S5: 50% RDP + PSB, and AM each with three replications	Crop residue retention at 50% and 75% significantly enhanced fluorescein diacetate activity of soil (0–5 cm), FDA had significant and positive relation with P fertilization. Crop residue retention had no significant effect on FDA in 5–15 cm soil depth. Highest FDA (378.3 μg fluorescein/g dry soil/h) was recorded with 75% CR, while lowest FDA (349.6 μg fluorescein/g dry soil/h) in 25% CR. Significant increase in FDA was recorded at 50% CR and 75% CR over control. Crop residue retention had no significant effect on FDA in 5–15 cm soil depth	Kumawat et al. (2017)
7	Inceptisol classified as Typic Haplustept	Zero tillage without residue (ZT–R), ZT with soybean residue (ZT+SR), ZT with soybean and wheat residue (ZT+SWR) (soybean residue is applied to the wheat crop and wheat residue to the	Application of 25% higher N over the RDN either as basal or as top dressing along with ZT plus crop residue gave the highest growth and improved soil microbial properties in wheat	Ronanki and Behera (2018)

	Sub-tropical semi-arid with hot and dry summers and cold winters. Annual rainfall (~750 mm), mean annual maximum and minimum temperature ~35 and 18°C, respectively.	preceding soybean crop) and conventional tillage without residue (CT–R), along with four N management practices in the subplot (100% RDN) as basal, 125% RDN as basal, 100% basal + 25% top dressing at crown root initiation (CRI) stage and 75 % basal + 25% top dressing at CRI stage were evaluated A 7-year field experiment to study the interactive effect of bed planting (BP) and conventional tillage (CT) with three water regimes and eight N management options on biological properties under a maize–wheat sequence	BP method significantly increased the soil DHA (52.5%) and respiration (72.0%) over CT. N increased soil respiration and DHA with corresponding increase in the SMBC was recorded in BP, whereas in CT, DHA declined with an increase in the SMBC	Singh et al. (2011)
8	Sub-tropical semi-arid with hot and dry summers and cold winters. Annual rainfall (~750 mm), mean annual maximum and minimum temperature ~35 and 18°C, respectively. sandy clay loam soil	Field experiments to investigate the effect of tillage, irrigation regimes, i.e. sub-optimal, optimal and supra-optimal water supply, and INM on soil enzymatic activities after cultivation of wheat	Soil glucosidase (54.5%), urease (88.8%), acid phosphatase (97.4%), and alkaline phosphatase (85.3%) activities increased significantly under conservation tillage compared with conventional tillage fields. A combination of inorganic and organic nutrient combinations with conservation tillage and optimal water supply significantly improved soil enzymatic activities	Sharma et al. (2012)

(continued)

Table 13.3 Cont.

S. no.	Climate/soil type	Experimental details	Results	References
9	Typic Haplustept, neutral and non-saline soil	CA-based practices permanent bed (PB) and ZT with 30% crop residues retained on the soil surface (each year) and a conventional tillage (CT) in main plots with four sub-plots consisting of cropping systems, i.e. maize–wheat–mungbean (MWMb), maize–chickpea–sesbania (MCS), maize–mustard–mungbean (MMuMb), and maize–maize–sesbania (MMS)	Dehydrogenase activity (DHA) measured during different seasons was invariably higher under CA (PB and ZT) compared with CT irrespective of soil depth. Long-term effect of CA practices, namely PB, ZT vis-à-vis CT was evaluated on WBC and DHA activity under different maize-based cropping systems. WBC contents under PB and ZT were greater than CT	Chaudhary et al. (2017)
10	Experimental site (located between 25.59° N, 85.13° E at an altitude of 50 m asl) was in Bihar, India. Old alluvium, texture of surface soil was silty clay. Middle IGP of India having subtropical humid climate. During the experimental period, lowest and highest rainfall was (626 mm) and (1172 mm) respectively with an average value of 902.2 mm. The temperature ranged from 6.5–39.8°C	A 7-year (2009–2016) study to assess tillage, cropping systems, and residue management practices in four cropping systems. TPR-CTW: conventional till puddled transplanted rice–conventional tilled wheat; TPR/MTNPR + R-ZTW + R-CTMB + R: conventional till puddled transplanted rice/machine transplanted non-puddle rice with residue–ZT wheat with residue–conventional till mungbean	Microbial parameters (MBC, FDA) increased with an increase in residue carbon addition. Fresh residues supplied readily mineralizable and hydrolysable C for better microbial growth. Legume crop taken along with rice and wheat with CA	Samal et al. (2017)

with residue; ZTDSR + R-ZTW + R-ZTC/ZTMB + R: ZT direct-seeded rice with residue-ZT wheat with residue-ZT cowpea/ZT mung bean with residue; NPTPR/ZTDSR + R-CT(P + M)/ZTM + R-ZTC/ZTM + R: non-puddle transplanted rice/ZT direct-seeded rice with residue-conventional till potato and maize intercrop/ZT mustard with residue–ZT cowpea/ZT maize with residue

| 11 | Inceptisol classified as Typic Haplustept | Field experiment on a maize–wheat cropping system. Crop residue retention in the main plot, T1: residue removal (no-residue), T2: 25% crop residue, T3: 50% crop residue, T4: 75% crop residue, and P fertilizer rate (five) in subplot treatments were S1: No-P, S2: 50% recommended dose of phosphorus (RDP), S3: 100% RDP, S4:150% RDP, S5: 50% RDP + PSB & AM | A significant increase of FDA in 0–5 cm soil layer in a maize–wheat system with 50% and 75% residue retention of each crop. FDA had significant and positive relation with CR and P fertilization. No significant effect on FDA in 5–15 cm soil depth was observed | Kumawat et al. (2017) |

(*continued*)

Table 13.3 Cont.

S. no.	Climate/soil type	Experimental details	Results	References
12	Rainfed dryland agroecosystems. Soil were inceptisol and suborder orchrepts subgroup. Udic Ustocrepts sandy loam texture	Two-year, double no-till rice-wheat system involving input of herbicide alone, and in combination with chemical fertilizer, crop residues (wheat straw, *Sesbania*), and FYM was evaluated for soil microbial activities in terms of soil enzymes (*β*-glucosidase, alkaline phosphatase, and urease)	Application of FYM and wheat straw in conjunction with inorganic fertilizer significantly increased the activities of *β*-glucosidase, alkaline phosphatase in 0–10 cm soil layer as compared with the application of inorganic fertilizer alone in a double ZT rice-wheat system. The enzyme activities were influenced by the C:N ratio of soil amendments	Singh and Ghosal (2013)
13	Sub-tropical, semi-arid and the annual rainfall is 750 mm; about 80 % of which occurs from June to September	Maize–wheat/chickpea/ mustard/linseed–greengram rotations under conventional or no-tillage with or without residue treatments	Residue incorporation with conventional tillage (CT+R) or retention on the surface under ZT+R were at par but had higher SOC than conventional tillage (CT). ZT also strongly affected the MBC, microbial quotient (ratio of MBC to SOC) and metabolic functionality of the microbial community (i.e., ratio of DHA to MBC). These parameters followed the order: ZT+R>CT+R>ZT>CT, and, as in case of SOC, ZT+R under linseed had the highest MBC, DHA and MQ. Soil organic C or potential N mineralization in CT+R or ZT+R were almost at par but higher than only CT or only ZT systems. Linseed showed the highest potential in C sequestration in the soils. Metabolic functionality of the microbial communities was found to be more strongly associated with the pool of microbial biomass or metabolic quotient than the total soil organic C.	Patra et al. (2011)

(continued)

| 14 | Sandy-loam soil. Semi-arid climate with average rainfall of 582 mm | In a farmer participatory research trial in Karnal, eight combinations of cropping systems, tillage, crop establishment method, and residue management effects on key soil physico-chemical and biological properties were evaluated. Maize–wheat (MW) system with ZT and residue (Rm) MW/ZT+Rm), rice–wheat with conventional tillage without residue (RW/CT–R). MW/ZT + residue | Higher proportion of bacteria to fungi in CT than ZT systems indicated the significant role of fungi in decomposing crop residues and nutrient mineralization

MW with ZT and residue (Rm) MW/ZT+Rm) registered 208%, 2635, 210%, and 48% improvements in soil MBC and N, DHA and APA. RW system in RW/ZT+Rm registered 83%, 81%, 44%, and 13%, respectively, as compared with (RW/CT–R).

MW/ZT + Residue also recorded the highest microbial population, namely bacteria, fungi, and actinomycetes.

Soil MBC, APA, BD, and micro-arthropod population were identified as the key indicators and contributed significantly towards soil quality index (SQI). MW system with ZT and Rm yielded the highest SQI (1.45) and the lowest score (0.29) being in (RW/CT–R). The SQI was higher by 90% in MW compared to RW, 22% in ZT compared to CT, and 100% in residue recycling compared with residue removal | Chaudhary et al. (2018) |

Table 13.3 Cont.

S. no.	Climate/soil type	Experimental details	Results	References
15	Typic Chromusters	A 2-year study, the influence of tillage-residue management, N-levels and weed management on the population of bacteria, fungi, and actinobacteria in soil under rice–wheat-greengram system. Continuous practice of ZT with preceding crop residue retention improved soil microbial biomass. N fertilization made no change in soil microbial activities in the first cropping cycle but there was improvement in the second cropping cycle when the 100% RDN was applied over 125% RDN. Direct seeding in rice–wheat–greengram system under ZT in the presence of preceding crop residue along with recommended dose of N and application of recommended practice of weed management proved to be a promising technology for improving soil biological properties	ZT with preceding crop residue retention increased population of bacteria by 65–83%, fungi by 28–32%, and actinobacteria by 22–37% compared with conventional tillage with or without preceding crop residue. Tillage-residue management had significant influence on microbial properties of soil. ZT with preceding crop residue retention followed in all three crops in a system markedly improved soil microorganism communities by stimulating the growth of bacteria, fungi, and actinobacteria	Rathore et al. (2016)

| 16 | Semi-arid climate conditions of South Asia. Typic Ustochrept sandy loam soil at Ludhiana (30°56´N and 75°52´E) in the IGP in the northwestern India | A 3-year field experiment on irrigated RWS was conducted on ZTW and rice residue retention | All soil enzymes were positively correlated with each other and with the soil chemical properties and grain yield. The beneficial effect of RCTs on soil quality was mainly confined to the soil surface layer. Improved soil enzyme activities and chemical properties in the surface 0–5 cm soil layer enhanced productivity of the rice–wheat system. DHA under ZTW+R was 6% and 14% higher as compared to ZTW-R and CTW-R, respectively. DHA was 9% higher under ZTW-R than CTW-R. FDA and APA at 0–5 cm depth were 9% and 13% higher in ZTW+R as compared to ZTW-R and CTW-R. PA was increased by 9% and 24% under ZTW+R in the 0–5 cm layer compared to ZTW-R and CTW-R, respectively. Three enzyme activities (FDA, APA, and PA) were significantly higher under ZTW-R than CTW-R. The increase in UA under ZTW+R was 8% and 13% higher as compared to ZTW-R and CTW-R in 0–5 cm soil | Kharia et al. (2017) |
| 17 | Vertisols Jabalpur | A 2-year field study on a soybean–wheat cropping system. Effect of conservation tillage and weed management practices on the total bacteria, fungi, actinomycetes, and dehydrogenase activity were studied | Tillage systems significantly influenced the microbial population. ZT+crop residue (soybean) fb crop residue (wheat) had higher bacterial and fungal population and dehydrogenase activity during both the seasons. However, the actinomycetes population was higher in ZT + crop residue (soybean) fb ZT (wheat). There was no adverse effect of herbicides use in the soybean–wheat cropping system on microbial population | Singh et al. (2015b) |

(continued)

Table 13.3 Cont.

S. no.	Climate/soil type	Experimental details	Results	References
18	Hyperthermic typic Haplustepts soils of NW-IGP. Sandy loam. Delhi	Under limited irrigation in rainfed conditions cropping systems: pearlmillet–wheat, pearlmillet–chickpea, pearlmillet–mustard, with crop residues at 5 t/ha dry biomass, *Leucaena* twigs at 10 t/ha green biomass and control (no mulch/ residue removal) under ZT were evaluated for their impact on soil microbial parameters. The crop residues as well as *Leucaena* twigs were retained on the soil surface and all crops were grown exclusively under ZT and ZT, limited irrigation rainfed conditions	Crop residue and *Leucaena* mulching improved DHA by ~about 156% and 147% over the no residue treated plots. Glomalin, dehydrogenase, FDA, glucosidase, and alkaline phosphatase were significantly (P <0.05) affected by crop rotations and residue management, and the interaction effects for glomalin, dehydrogenase, FDA, and alkaline phosphatase were significant. However, acid phosphatase, soil ergosterol content, and soil respiration were affected by residue management only, and microbial quotient and microbial biomass carbon by crop rotation only. The residue-retained plots had ~40% and 13% higher soil glomalin contents compared with residue removal (~290 µg/g/soil) and *Leucaena* added plots, respectively	Singh et al. (2018)
19	Eastern Indo-Gangetic plains	Four permutations of conventional tillage and ZT under rice–wheat system, namely tillage before sowing/ transplantation of each crop (RCT–WCT), tillage before transplantation of rice and ZT before sowing of wheat (RCT–WZT), tillage before sowing wheat and ZT before sowing of rice (RZT–WCT) and no tillage before sowing of each crop (RZT–WZT)	Microbial biomass carbon and nitrogen and activities of β-d-glucosidase (BG), cellobiohydrolase (CBH), polyphenol oxidase (PPO), nitrogen [urease (UR)], glycine-amino peptidase (GAP) BG, CBH, ALP, and UR increased with a reduction in tillage frequency, becoming the highest under RZT–WZT and the lowest under RCT–WCT	Pandey et al. (2014)

| 20 | Tripura India (latitude of 23°54′ 24.02″N and 91°18′58.35″ E and altitude of 162 m above mean sea level). The average annual rainfall 2200 mm. Sandy-clay-loam soil Typic Kandihumults, rice–rice system under rainfed hill ecosystems of eastern Indo-Gangetic plains | Rice (wet season)–rice (dry season) system (RRS). The replacement of farmers practice with conservation effective tillage (ZT) and integrated nutrient management (INM) practice along with 30% residue retention. ZT + 25% N through green leaf manure (GLM) + 60 kg N, 9 kg P, 17 kg potassium (K), 2 kg B and 5 kg Zn/ha (INM) + 30% RR + cellulose decomposing microorganism (CDM) | ZT combined with nutrient and residue management increased BMC (17.8%) and DHA (44%) over reduced tillage (RT) + 40 kg N and 9 kg P/ha +30% RR in a rice–rice cropping system | Yadav et al. (2017) |

References

Banerjee, B., Aggarwal, P.K., Pathak, H., et al. 2006. Dynamics of organic carbon and microbial biomass in alluvial soil with tillage and amendments in rice–wheat systems. *Environmental Monitoring and Assessment* 119: 173–189.

Beri, V., Sidhu, B.S., Bahl, G.S., et al. 1995. Nitrogen and phosphorus transformations as affected by crop residue management practices and their influence on crop yield. *Soil Use and Management* 11: 51–54.

Bhaduri, D., Purakayastha, T.J., Patra, A.K., et al. 2017. Biological indicators of soil quality in a long-term rice–wheat system on the Indo-Gangetic plain: combined effect of tillage–water–nutrient management. *Environmental Earth Science* 76: 202.

Bhattacharyya, R., Das, T.K., Sudhishri, S., et al. 2015. Conservation agriculture effects on soil organic carbon accumulation and crop productivity under a rice–wheat cropping system in the western Indo-Gangetic Plains. *European Journal of Agronomy* 70: 11–21.

Chakraborty, D., Nagarajan, S., Aggarwal, P., et al. 2008. Effect of mulching on soil and plant water status, and the growth and yield of wheat (*Triticum aestivum* L.) in a semi-arid environment. *Agricultural Water Management* 95: 1323–1334.

Chaudhary, A., Meena, M.C., Parihar, C.M., et al. 2017. Effect of long-term conservation agriculture on soil organic carbon and dehydrogenase activity under maize-based cropping systems International *Journal of Current Microbiology and Applied Sciences* 6(10): 437–444.

Chaudhary, M., Datta, A., Jat, H.S., et al. 2018. Changes in soil biology under conservation agriculture based sustainable intensification of cereal systems in Indo-Gangetic plains. *Geoderma* 313 (supplement C): 193–204.

Choudhary, M., Sharma, P.C., Jat, H.S., et al. 2016. Crop residue degradation by fungi isolated from conservation agriculture fields under rice–wheat system of North-West India. *International Journal of Recycling of Organic Waste in Agriculture* 5: 349–360.

Choudhary, R.L., Kumar, D., Shivay, Y.S., et al. 2010. Performance of rice (*Oryza sativa*) hybrids grown by the system of rice intensification with plant growth promoting rhizobacteria. *Indian Journal of Agricultural Sciences* 80(10): 853–857.

Das, T.K., Bhattacharyya, R., Sudhishri S., et al. 2014. Conservation agriculture in an irrigated cotton–wheat system of the western Indo-Gangetic Plains: Crop and water productivity and economic profitability. *Field Crops Research* 158: 24–33.

Erenstein, O. and Laxmi, V. 2008. Zero tillage impacts in India's rice-wheat systems: a review. *Soil and Tillage Research* 100: 1–14.

Kharia, S.K., Thind, H.S., Sharma, S., et al. 2017. Tillage and rice straw management affect soil enzyme activities and chemical properties after three years of conservation agriculture based rice-wheat system in North-Western India. *International Journal of Plant and Soil Science* 15(6): 1–13.

Kumawat, C., Sharma, V.K., Meena, M.C., et al. 2017. Fluorescein diacetate activity as affected by residue retention and P fertilization in maize under maize-wheat cropping system. International *Journal of Current Microbiology and Applied Sciences* 6(5): 2571–2577.

Mandal, B., Ghoshal, S.K., Ghosh, S., et al. 2005. Assessing soil quality for a few long-term experiments– an Indian initiative. In: Proceedings of International Conference on Soil, Water & Environmental Quality-Issues and Challenges, New Delhi.

Meena, B.L., Saha, S., Kumar, N., et al. 2008. Changes in soil nutrient content and enzymatic activity under conventional and zero-tillage practices in an Indian sandy clay loam soil. *Nutrient Cycling in Agroecosystems* 82: 273–281.

Meena, K.K., Mesapogu, S., Kumar, M., et al. 2010. Coinoculation of the endophyticfungus *Piriformospora indica* with dynamics and plant growth in chickpea. *Biology and Fertility of Soils* 46: 169–174.

Moore, J.M., Klose, S. and Tabatabai, M.A. 2000. Soil microbial biomass carbon and nitrogen as affected by cropping systems. *Biology and Fertility of Soils* 31(3): 200–210.

Pandey, D., Agrawal, M. and Bohra, J.S. 2014. Effects of conventional tillage and no tillage permutations on extracellular soil enzyme activities and microbial biomass under rice cultivation. *Soil and Tillage Research* 136: 51–60.

Parihar, C.M., Yadav, M.R., Jat, S.L., et al. 2016. Long-term effect of conservation agriculture in maize rotations on total organic carbon, physical and biological properties of a sandy loam soil in north-western Indo-Gangetic Plains. *Soil and Tillage Research* 161: 116–128.

Patra, A.K., Purakayastha, T.J., Kaushik, S.C., et al. 2011. Conservation tillage, residues management and cropping systems effects on carbon sequestration and soil biodiversity in semi-arid environment of India. In: Proceedings of World Congress of Conservation Agriculture, Brisbane, Australia.

Prasad, P., Basu, S. and Behera, N. 1994. Comparative account of the microbiological characteristics of soils under natural forest, grassland and crop field from Eastern India. *Plant and Soil* 175: 85–91.

Rathore, A.K., Sharma, A.R., Sarathambal, C., et al. 2016. Nitrogen and weed management effect on soil microbial properties in rice-based cropping system under conservation agriculture. *Indian Journal of Weed Science* 48(4): 360–363.

Ronanki, S. and Behera, U.K. 2018. Effect of tillage, crop residues and nitrogen management practices on growth performance and soil microbial parameters in wheat. *International Journal of Current Microbiology and Applied Sciences* 7(1): 845–858.

Samal, S.K., Rao, K.K., Poonia, S.P., et al. 2017. Evaluation of long-term conservation agriculture and crop intensification in rice-wheat rotation of Indo-Gangetic Plains of South Asia: Carbon dynamics and productivity. *European Journal of Agronomy* 90: 198–208.

Sharma, P., Singh, G. and Singh, R.P. 2011. Conservation tillage, optimal water and organic nutrient supply enhance soil microbial activities during wheat (*Triticuma aestivum*) cultivation. *Brazilian Journal of Microbiology* 42(2): 531–542.

Sharma, P., Singh, G. and Singh, R.P. 2012. Conservation tillage and optimal water supply enhance microbial enzymes (glucosidase, urease and phosphatases) activities in fields under wheat cultivation during various nitrogen management practices. *Archives of Agronomy and soil Science* 59(7): 911–928.

Sharma, P., Singh, G., Sarkar, S.K., et al. 2015. Improving soil microbiology under rice-wheat crop rotation in Indo-Gangetic Plains by optimized resource management. *Environmental Monitoring and Assessment* 187(3): 150.

Singh, A. and Ghoshal, N. 2013. Impact of herbicides and various soil amendments on soil enzymes activities in a tropical rainfed agroecosystem. *European Journal of Soil Biology* 54: 56–62.

Singh, D., Yadav, D.K., Sinha, S., et al. 2013. Genetic diversity of iturin producing strains of Bacillus species antagonistic to *Ralstoniasolanacerarum* causing bacterial wilt disease in tomato. *African Journal of Microbiology Research* 7(48): 5459–5470.

Singh, G. and Mukerji, K.G. 2006. Root exudates as determinants of rhizospheric microbial diversity, pp. 39–49. In: Microbial activity in the rhizosphere. Eds. Mukerji, K.G. et al. Springer-Verlag.

Singh, G., Bhattacharyya, R., Das, T.K., et al. 2018. Crop rotation and residue management effects on soil enzyme activities, glomalin and aggregate stability under zero tillage in the Indo-Gangetic Plains. *Soil and Tillage Research* 184: 291–300.

Singh, G., Biswas, D.R. and Marwaha, T.S. 2010. Mobilization of potassium from waste mica by plant growth promoting rhizobacteria and its assimilation by maize (*Zea mays*) and wheat (*Triticum aestivum* L.): a hydroponics study under phytotron growth chamber. *Journal of Plant Nutrition* 33: 1–16.

Singh, G., Kumar, D. and Sharma, P. 2015a. Effect of organics, biofertilizers and crop residue application on soil microbial activity in rice-wheat and rice-wheat mungbean cropping systems in the Indo-Gangetic plains. *Cogent Geoscience* 1(1): DOI: 10.1080/23312041.2015.1085296.

Singh, G., Kumar, D., Marwaha, T.S., et al. 2011. Conservation tillage and integrated nitrogen management stimulates soil microbial properties under varying water regimes in maize wheat cropping systems in northern India. *Archives of Agronomy and Soil Science* 57(5): 507–521.

Singh, P., Sarathambal, C., Kewat, M.L., et al. 2015b. Conservation tillage and weed management effect on soil microflora of soybean–wheat cropping system. *Indian Journal of Weed Science* 47(4): 366–370.

Sinsabaugh, R.L., Lauber, C.L., Weintraub, M.N., et al. 2008. Stoichiometry of soil enzyme activity at global scale. *Ecology Letters* 11: 1252–1264.

Tabatabai, M.A. 1994. Soil Enzymes. In: Methods of Soil Analysis, pp. 775–833. (Eds.) Weaver R.W. et al. Part 2. *Microbiological and Biochemical Properties*. Science Society of America Inc.

Tarafdar, J.C. and Marschner, H. 1994. Phosphatase activity in the rhizosphere and hyposphere of VA mycorrhizal wheat supplied with inorganic and organic phosphorus. *Soil Biology and Biochemistry* 26: 387–395.

Yadav, B.K. and Tarafdar, J.C. 2004. Phytase activity in the rhizosphere of crops, trees and grasses under arid environment. *Journal of Arid Environments* 58(3): 285–293.

Yadav, G.S., Datta, R., Pathan, S.I., et al. 2017. Effects of conservation tillage and nutrient management practices on soil fertility and productivity of rice (*Oryza sativa* L.)–rice system in north eastern region of India. *Sustainability* 9(10):1–17.

Zuber, S.M., María, B. and Villamil, M.B. 2016. Meta-analysis approach to assess effect of tillage on microbial biomass and enzyme activities. *Soil Biology and Biochemistry* 97: 176–187.

14 Conservation Agriculture in Agroforestry Systems

Inder Dev, Asha Ram, Naresh Kumar,
A.R. Uthappa, and A. Arunachalam

Introduction

Agricultural production intensification worldwide has impacted global C, water, and nutrient cycles. Land-use changes to agricultural production continue to contribute significantly to atmospheric CO_2, accounting for as much as 24% of global greenhouse gas (GHG) emissions. Carbon dioxide is the main driver for anthropogenic-induced climate change, along with other GHGs like methane and nitrous oxide that are more potent in terms of global warming potential (GWP). Several factors have contributed to increasing GHGs in the atmosphere that have resulted in unprecedented rains, droughts, and even changes in the monsoon pattern. As much of the world is concerned today with climate change, a sort of 'climate emergency' is evolving. Rising temperature due to global warming affects flowering and leads to the build-up of pests and disease. Floods and excess rain over a short duration of time cause extensive damage to crops. Extreme weather events have caught the attention of agrarian experts and scientists alike and they are now focusing on climate smart agriculture practices to arrest the impacts of climate change. Countries all over the world are looking for viable strategies to sequester atmospheric CO_2.

In the C cycle, a variety of processes take place over time scales ranging from a few hours to millions of years. The long-term cycle is distinguished by the exchange of C between rocks and the surficial system, which consists of the oceans, atmosphere, biosphere, and soil. This cycle is the real controller of the concentration of CO_2 in the atmosphere. The short-term C cycle (which includes photosynthesis, respiration, air–sea exchange of CO_2, and humus accumulation in soil) over decades and centuries is of greater importance than the long-term cycle in forest, agroforestry systems (AFS), and agricultural ecosystems (Dhyani et al. 2020). Fixation of atmospheric CO_2 in plants through photosynthesis and the return of part of that C to the atmosphere through plant, animal, and microbial respiration as CO_2 under aerobic and CH_4 under anaerobic conditions are important processes of the short-term C cycle. The responsible factors for CO_2 emission in the atmosphere are vegetation fire, burning of fossils and fuels, and burning and land cleaning for cultivation, but much of this emitted C is recaptured in subsequent regrowth of vegetation (Nair et al. 2009; Lorenz and Lal

DOI: 10.4324/9781003292487-17

2018). The Paris Climate Change Agreement (popularly known as the Global Conference on Climate Change) has given emphasis to C capture and storage to mitigate the adverse effects of climate change. Under this Agreement, India has agreed to develop an additional 'carbon sink' of 2.5–3.0 billion tonnes of CO_2 equivalent through additional forest and tree cover (Ghosh et al. 2021).

Terrestrial ecosystems are a significant C sink on Earth, accounting for about 20–30% of the total anthropogenic CO_2 emissions to the atmosphere. When compared with oceans, it can be readily managed to either increase or decrease C sequestration by restoring or degrading the vegetation available on lands. Any increase in the concentration of radioactively active trace gases in the atmosphere is now recognized as modifying global climate, affecting terrestrial ecosystems both functionally and structurally. A land-use and management option that optimizes sustainable production and enhanced C sequestration in the soil is the need of the hour, particularly with reference to tropical and sub-tropical terrestrial systems, where the soil is hungry for C. Conservation agriculture and agroforestry are gaining importance due to their climate change mitigation and adaptation potential.

Conservation Agriculture

According to the FAO website (www.fao.org/wsfs/forum2050 (FAO 2009)) agricultural production will need to be enhanced by 70% using scientifically sound, environment-friendly, and socially acceptable technologies to meet the global food demand by 2050. A degraded environment and inadequate options for coping with extreme weather events (frequent droughts, floods, hail storms, high fluctuation in the atmospheric temperature, etc.) will create a decline in productivity and lead to greater instability in agricultural production (crop, forestry, livestock, and fisheries) systems. The Indo-Gangetic plains have been the major contributor towards the self-sufficiency of India. However, nutrient depletion owing to the use of high-yielding varieties, faulty agricultural practices, and the absence of the sufficient addition of fertilizers back to the soil have led to soil degradation and micronutrient deficiencies. The region has been heavily exploited due to indiscriminate use of chemicals including fertilizers and pesticides, which has resulted in soil degradation and unsustainable use of natural resources. Declining productivity, soil quality, and water table, and increasing waterlogging and salinity have now become major problems in this region. Furthermore, in most cultivated soils of India, the soil organic carbon (SOC) concentration is less than 5 g/kg, whereas in uncultivated virgin soils, it is in the range of 15–20 g/kg (Bhattacharyya et al. 2000). This means that excessive soil tillage, burning of crop residues, and intensive monocropping systems over the decades have led to soil degradation. In India, the response ratio (kg grain per kg of nutrient) in foodgrain crops in irrigated areas has continuously declined over time. During the Green Revolution (1970s), the response ratio was ~14, and this came down to ~4 in 2010 (Biswas and Sharma 2008). Accordingly, conserving soil health has been the top priority of government policies to produce sufficient food for the growing population. Soil health and its productive potential for a long time have been governed by the interactions among the biological, chemical, and physical properties of the soil (Jat et al. 2021).

Although the productivity of cereals such as rice, wheat, and maize has been increasing over time, the rate of increment has reduced over time also. This is mainly because the natural resource base has deteriorated since the Green Revolution due to intensive tillage-based exploitative farming and unscientific management of resources (water, fertilizers, and pesticides). Promoting best agriculture management practices, particularly conservation agriculture-based, improved seeds and balanced fertilization, integrated soil and crop management, as well as supporting increased investment in agricultural research for development are some of the ways for obtaining a sustainable production system (Gathala et al. 2013). Conservation agriculture can capture interactions existing among management activities for sustainable agriculture that would contribute substantially to achieving the Millennium Development Goals (MDGs) of zero hunger and improved environmental resource management.

Currently, CA is increasingly being promoted in the region as climate smart agriculture practices (CSAPs) by integrating it with precise water and nutrient management (Sidhu et al. 2019) with the aim of increasing resilience to climate change and enhancing the mitigation potential of cropping systems. In the era of climate change, CA is an answer to the production of sustainable food with a minimum adverse impact on environment. CA, though advocated a long time ago, is now spreading as a response to the concern for sustainable agriculture at the global level.

Agroforestry Systems

Agroforestry is the integration of woody perennials into farmlands or rangelands and, in doing so, it diversifies and sustains production with increased benefits for farmers and the environment. It encompasses a wide range of trees that are grown on farms and in rural landscapes, and includes the generation of science-based tree enterprise opportunities. Agroforestry emerged as a well-established nature-based approach to sustainable land management, not only for conserving natural resources, but also for ecological and environmental considerations. Practicing agroforestry is a win–win opportunity to combine the twin objectives of climate change adaptation and mitigation and a sustainable production system (Dhyani et al. 2020). Zomer et al. (2009) assessed that 48% of all agricultural land had at least 10% tree cover, and about 1.2 billion rural people practice agroforestry on their farms.

Agroforestry in India meets almost half of the demand for fuel wood, two-thirds of the small timber, 70–80% of wood for plywood, 60% of the raw material for paper pulp, and 9–11% of the green fodder requirement of livestock, besides meeting the subsistence needs of households for food, fruit, fibre, and medicine. Through practicing agroforestry, the biomass productivity per unit area can safely be increased from <2.0 t/ha/year to 10 t/ha/year by carefully selecting tree–crop combinations. Although a primary objective of agroforestry has been conservation-based biomass enhancement and diversification, in recent times, several studies have further suggested that agroforestry/tree farming plays an important role in C sequestration both above and below ground (Nair et al. 2009; Dhyani et al. 2016). Eventually, agroforestry has the potential for enhancing agroecosystem resilience to extreme climatic conditions by also contributing to higher CO_2 sequestration in the degraded lands, which will help to maintain a global C balance (Basu 2014). India has announced the landmark National Agroforestry Policy (NAP 2014) that will mainstream the growing of trees on farms to meet a wide range of developmental and environmental goals. India submitted voluntary pledges for climate change action in the form of the Intended Nationally Determined Contributions (INDCs) to the United Nations Framework Convention on Climate Change (UNFCCC). India's INDCs target a 33–35% reduction of emission intensity of its GDP by 2030 from 2005 levels through the creation of an additional C sink of 2.5–3.0 billion tonnes of CO_2 equivalent through afforestation and an increase in the share of non-fossil fuel energy in power generation up to 40% of the total installed power capacity by 2030 (Dhyani et al. 2016).

Trees are integral to our traditional farming systems for their innumerable benefits. However, over time, with shrinking land holdings, annual crops have replaced trees for various reasons. Trees complement farming in terms of manure, fodder, and the fuel needs of the farmer. They form the backbone for practicing integrated farming systems, which is necessary for self-reliant and sustainable agriculture. Agroforestry systems, due to diverse options and products, provide opportunities for employment generation in rural areas. an increased supply of wood in the market triggered a substantial increase in the number of small-scale industries dealing with wood and wood-based products not long ago. Various improved agroforestry practices and systems such as home gardens, block plantation, energy plantation, shelterbelts, and improvements or alternatives to shifting cultivation are some of the specialized agroforestry systems that have been developed. The prominent agroforestry systems from different regions of the country are presented in Table 14.1.

Mitigation and Adaptation to Climate Change through Agroforestry

Sequestration of atmospheric CO_2 through agroforestry has long been considered as a means of climate change mitigation. It is true that compared to a monoculture or system where crop residues have been removed, agroforestry along with CA adds litter and residues to the system, which can enhance SOC. Due to its multiple plant species and soil types in which agroforestry system prevails, it has huge potential as a mitigation strategy to address climate change (Montagnini and Nair 2004; ICAR 2006). Dhyani et al (2020) and Newaj et al. (2014) reviewed the available literature on the C sequestration potential of various tree species in agroforestry systems (Table 14.2) in India. In such studies the most common tree density was in the range of 300–800 trees/ha, and the reported C

Table 14.1 Prominent agroforestry systems in different regions of India

Agro-climatic zone	Agroforestry systems	Tree component	Crop/grass
Western Himalayas	Silvipasture (RF)	*Grewia optiva*	*Setaria* spp.
		Morus alba	*Setaria* spp.
	Agrihorticulture	*Malus pumila*	Millets, wheat
	Agrihorticulture	*Prunus persica*	Maize, soybean
Eastern Himalayas	Agrisilviculture	*Anthocephalus cadamba*	Rice
	Agrihorticulture	*Alnus nepalensis*	Large cardamom/ coffee
	Silvipasture	Bamboos, *Parkia roxburghii*, *Morus alba*	–
	Silvipasture	*Bauhinia variegata, Ficus* spp., *Morus alba*	Napier
Lower Gangetic plains	Agrisilviculture (Irri)	*Eucalyptus* spp., *Albizia lebbeck*	Rice
	Agrihorticulture (Irri)	Mango/banana/litchi	Wheat, rice, maize
	Silvipasture	*Morus alba, Albizia lebbeck*	*Dicanthium, Pennisetum*
Middle Gangetic plains	Agrisilviculture (Irri)	*Populus deltoids*	Sugarcane–wheat
	Agrisilviculture (Irri)	*Eucalyptus* spp.	Rice–wheat
	Agrisilviculture	*Dalbergia sissoo*	Sesamum
	Agrihorticulture (Irri)	Mango/citrus	Rice–wheat
	Silvipasture	*Albizia lebbeck*	*Chrysopogon, Dicanthium*
Trans Gangetic plains	Agrihorticulture (Irri)	*Emblica officinalis*	Blackgram/ greengram
	Agrisilviculture	*Azadirachta indica*	Blackgram– wheat/mustard
	Silvipasture	*Bauhinia variegata, Albizia lebbeck*	*Cenchrus, Pennisetum*
Upper Gangetic Plains	Agrisilviculture (Irri)	*Populus deltoids*	Wheat, bajra fodder
	Agrisilviculture (Irri)	*Eucalyptus* spp.	Rice-wheat
	Silvipasture	*Bauhinia variegata, Albizia lebbeck*	*Chrysopogon, Poa*
Eastern plateau and hills	Agrisilviculture	*Gmelina arborea*	Rice, linseed
	Agrisilviculture	*Acacia nilotica*	Rice
	Silvipasture	*Acacia mangium, A. nilotica,* bamboos	–
	Silvipasture	*Leucaena leucocephala*	*Chrysopogon, Pennisetum, Dicanthium*
Central plateau and hills	Agrihorticulture (Irri)	*Psidium guajava*	Bengalbram/ groundnut
	Agrihorticulture (RF)	*Emblica officinalis*	Blackgram/ greengram

(*continued*)

Table 14.1 Cont.

Agro-climatic zone	Agroforestry systems	Tree component	Crop/grass
	Agrisilviculture	Acacia nilotica/Leucaena leucocephala/Azadirachta indica/ Albizia lebbeck	Soybean, blackgram– mustard/wheat
	Silvipasture (RF- and degraded lands)	Albizia amara, Leucaena leucocephala, Dichrostycus cinerea	Chrysopogon, Stylosanthes hamata, S. scabra
	TBOs (RF)	Jatropha curcas	–
Western plateau and hills	Agri-horti- silviculture (Irri)	Tectona grandis, Achrus zapot	Rice, maize
	Agrihorticulture	Areca catechu	Black pepper, cardamom
	Silviculture	Prosopis juliflora, Ailanthus excelsa	–
	Silvipasture	Acacia mangium, Albizia amara	Cenchrus
Southern plateau and hills	Agrisilviculture (RF)	Eucalyptus, Casuarina equisetifolia, Ailanthus excelsa	Cotton, groundnut
	Agrisilviculture (Irri)	Eucalyptus tereticornis, Melia dubia	Chilli
	Silviculture (RF)	Leucaena leucocephala, Acacia leucopholea	–
		Eucalyptus	–
	Agrihorticulture	Tamarindus indica	Chilli
	TBOs	Pongamia pinnata	–
East coast plains and hills	Agrisilviculture (RF)	Ailanthus excelsa, Acacia leucophloea	Cowpea
	Silviculture	Casuarina equisetifolia, Leucaena leucocephala	–
	TBOs	Pongamia pinnata	–
	Silvipasture	Artocarpus spp.	Chrysopogon, Napier, Cenchrus
West coast plains and hills	Agrisilviculture (RF)	Acacia auriculiformis	Black pepper
	Agrihorticulture (RF)	Artocarpus heterophyllus	Black pepper
	Agrisilviculture (RF)	Acacia auriculiformis	Rice
	Agrihorticulture	Cocos nucifera/Areca catechu	Rice
	Agrisilviculture	Casurina equisetiofolia	Rice
	Silvipasture	Hardwickia binnata, Albizia lebbeck	Cenchrus
Gujarat coast plains and hills	Agrisilviculture	Azadirachta indica, Ailanthus excelsa	Cowpea, greengram
	Silviculture	Prosopis juliflora, Acacia nilotica	–
	Silvipasture	Leucaena leucocephala	Cenchrus, Setaria

Table 14.1 Cont.

Agro-climatic zone	Agroforestry systems	Tree component	Crop/grass
Western dry region	Agrisilviculture	Prosopis cineraria, Tecomella indica, Acacia nilotica, Azadirachta indica	Pearlmillet
	TBOs	Jatropha curcas	–
	Silvipasture	Albizia lebbeck, Hardwickia binnata	Cenchrus
All islands	Agrihorticulture	Cocos nucifera	Rice
	Silvipasture	Bauhinia spp., Erythrina indica, Leucaena leucocephala	Cenchrus, Pennisetum

Source: Dhyani et al. (2009)

Note: Irri: irrigated; RF; rainfed, TBOs: tree borne oilseed

Table 14.2 Carbon sequestration potential (CSP) of different trees in India

Location	Agroforestry system	Tree species	No. of trees/ha	Age (year)	CSP (Mg C/ha/ year)
Uttarakhand	Agrisilviculture	D. hamiltonii	1000	7	15.91
Himachal Pradesh	Agrihorticulture	Fruit trees	69	–	12.15
Khammam, Andhra Pradesh	Agrisilviculture	L. leucocephala	4444	4	14.42
			10000	4	15.51
Uttarakhand	Agrisilviculture	P. deltoids	500	8	12.02
SBS Nagar, Punjab	Agrisilviculture	P. deltoids	740	7	9.40
Dehradun, Uttarakhand	Silviculture	E. tereticornis	2500	3.5	4.40
			2777	2.5	5.90
Kurukshetra, Haryana	Silvipasture	A. nilotica	1250	7	2.81
		D. sissoo	1250	7	5.37
		P. juliflora	1250	7	6.50
Chandigarh	Agrisilviculture	L. leucocephala	10666	6	10.48
Tripura	Silviculture	T. grandis	444	20	3.32
		G. arborea	452	20	3.95
Tarai region Uttarakhand	Silviculture	T. grandis	570	10	3.74
			500	20	2.25
			494	30	2.87
Jhansi, Uttar Pradesh	Agrisilviculture	A. procera	312	7	3.70
Jhansi, Uttar Pradesh	Agrisilviculture	A. pendula	1666	5.3	0.43
Jhansi, Uttar Pradesh	Silviculture	A. procera	312	10	1.79
		A. amara	312	10	1.00
		A. pendula	312	10	0.95
		D. sissoo	312	10	2.55
		D. cinerea	312	10	1.05
		E. officinalis	312	10	1.55
		H. binata	312	10	0.58
		M. azedarach	312	10	0.49

(continued)

Table 14.2 Cont.

Location	Agroforestry system	Tree species	No. of trees/ha	Age (year)	CSP (Mg C/ha/ year)
Hyderabad, Andhra Pradesh	Silviculture	L. leucocephala	2500	9	10.32
		E. camaldulensis	2500	9	8.01
		D. sissoo	2500	9	11.47
		A. lebbeck	625	9	0.62
		A. albida	1111	9	0.82
		A. tortilis	1111	9	0.39
		A. auriculiformis	2500	9	8.64
Hyderabad, Andhra Pradesh	Agrisilviculture	L. leucocephala	11111	4	2.77
			6666	4	1.90
Raipur Chhattisgarh	Agrisilviculture	G. arborea	592	5	3.23
Coimbatore Tamilnadu	Agrisilviculture	C. equisetifolia	833	4	1.57
Kerala	Home garden	Mixed tree spp.	667	71	1.60

Source: Dhyani et al. (2020)

sequestration potential varied from 0.39–11.47 Mg C/ha/year. Studies conducted in different parts of the world reported the C sequestration potential of different AFS in the range of 0.29–15.21 Mg C/ha/year above ground, and 30–300 Mg C/ha up to 1 m depth in soil (Nair et al. 2009). Thus, the existing trees on farmers' fields not only add some income to small and marginal farmers, but also help in mitigating global warming by enhancing the C sequestration potential (Ajit et al. 2013; Dhyani et al. 2016).

Soil is one of the major C sinks on Earth because it is the source of organic matter, which plays a vital role in the maintenance of soil fertility. A land use system is the action of modifying and managing the natural environment into a settled environment, such as wood, pastures, forests, and arable fields. Carbon sequestration through soil involves adding a maximum amount of C to the soil. Below-ground biomass of trees in the form of roots comprise about one-fifth to one-fourth of the total living biomass, and there is constant addition of organic matter to the soil through decaying dead roots (Dhyani and Tripathi 2000), which leads to improvements in the C status of the soil. Accumulation of 2.91% organic C was observed under areca nut + jackfruit + black pepper + cinnamon (tejpatra) followed by 1.85% under areca nut + betelvine + miscellaneous trees, as against 0.78% only in degraded land in the same period. MPTS like Alnus nepalensis, Parkia roxburghii, Michelia oblonga, Pinus kesiya, and Gmelina arboria with greater surface cover, constant leaf litter fall, and extensive root systems increased SOC by 96.2%, aggregate stability by 24.0%, available soil moisture by 33.2%, and in turn reduced soil erosion by 39.5%. Soils under Acacia auriculiformis, Leucaena leucocephala, and Gmelina arborea had a

Table 14.3 Soil organic carbon (SOC) stock in various agroforestry systems

Agroforestry system	Location	Age (year)	Soil depth (cm)	Soil C (Mg/ha)	References
Agrisilviculture (*Gmelina arborea* + eight field crops)	Chhattisgarh, Central India	5	0–60	27.4	Swamy and Puri (2005)
Home gardens	Kerala, India	35	0–100	101–126	Saha et al. (2009)

high humification rate, while soils under the canopy of *Acacia auriculiformis*, *Michelia champaca*, *Tectona grandis*, and *Dalbergia sissoo* showed low humification of organic matter. Such improvements in soil quality under tree-based AFS have a direct bearing on the long-term sustainability and productivity of soil (Subba Rao and Saha 2014) (Table 14.3).

CA in Agroforestry Systems

CA-based cropping systems benefit from the addition of a tree component, which supports the system's functioning, diversification, and resilience. This practice is aimed at improving the CA through the provision of fodder, fuel, timber, biomass, nutrients, fencing, and fruits, among other products and services. Residue retention, one of the key principles of CA, can be improved by trees and shrubs which provide additional biomass for surface retention. Here, the supporting elements are leaves, shade, and additional groundcover from tree prunings that can overcome the shortcomings of CA systems. In addition, trees help in improving the soil fertility, which supports the production of field crops (Baudron et al. 2017). This broadens the concept of crop rotations to incorporate the role of fertilizer and fodder trees to more effectively enhance soil fertility and provide needed biological and income diversity in the system. Recently, the CA and agroforestry research and development communities have mutually recognized the value of integrating fertilizer trees and shrubs into CA in AFS to enhance both fodder production and soil fertility (FAO 2010, 2011). Agroforestry provides more diverse and variable habitats, which fulfil the principle of crop diversification. A wide range of tree species have been shown to integrate well into CA farming systems (ICRAF 2009). CA with trees or conservation agroforestry have been promoted under the umbrella of evergreen agriculture (Ram et al. 2016). CA with agroforestry trees increases the resilience of the farm enterprise to climate change through greater drought resilience, and they sequester more C. CA systems tend to sequester a maximum of 0.2–0.4 t C/ha/year. CA practices with AF systems accumulate C both above and below ground in the range of 2–4 t C/ha/year, an order of magnitude higher than with CA alone. This is particularly true for systems incorporating fertilizer trees, such as *Faidherbia* or *Gliricidia* (Makumba et al. 2006).

Integration of *Faidherbia albida* is seen as a successful example of conservation agroforestry that shows how both crops and trees can coexist and give more benefits as compared to annual crops. *Faidherbia* trees have a unique characteristic by which they shed their leaves during the cropping season, while they have a full canopy during the dry period (winter season). Garrity et al. (2010) reported that with the integration of *Faidherbia* with CA, crop yields can be enhanced while external fertilizer use can be reduced. The deep rooting trees benefit the soil through recycling of nutrients from deeper layers to the soil surface (Ram et al. 2017). However, the reverse phenology (shedding of the leaves in the cropping season) only happens when trees have access to soil moisture and when they are not over-pruned. Another good example of integration of agroforestry species is the use of *Gliricidia* in CA systems (Lewis et al. 2011). *Gliricidia* was planted in rows 5 m apart and 1 m between rows, and the trees were pruned every year to make use of the nutrients in the leaves. This practice helped in providing fodder leaves to cattle, ground cover, and soil fertility. *Faidherbia* and *Gliricidia* are leguminous trees and their leaves are high in N content. Maize or groundnut can be planted in the 5 m inter-row space, and benefit from the pruned leaves to an extent that NPK fertilizer can be reduced. Due to high price of mineral fertilizer and/or lack of access to it, greener technologies such as this have been adopted in southern Africa and other parts of world. Ram et al. (2017) estimated that a leguminous tree in agroforestry produces on average 18.7 kg litter/ha/year. Considering an average density of 12.44 trees/ha in agroforestry, 0.055 Tg of N is added through litter fall of the agroforestry system in India. The average number of BNF trees was 1.5 trees/ha, fixing about 11.18 kg N/ha/year. Therefore, BNF trees in an area of 17.45 M ha in India can fix N of up to 0.195 Tg/year. Accordingly, CA systems can significantly benefit from the biomass contributions of tree-based components. However, agroforestry systems are more likely to be adopted in situations of high soil fertility decline, with erosion problems, and where fuelwood and fodder are scarce and in high demand.

Complementarity between CA and Agroforestry

Agroforestry is a multi-faceted subject, which along with CA sustains the yield with minimal environmental damage. Some of the exciting areas where CA-based agroforestry system works in tandem are described in the following.

Ecosystem Services

Reduced tillage (RT) or zero-till (ZT), permanent organic soil cover by retaining crop residues, and crop rotations, including cover crops, aim to increase crop yields by enhancing several regulating and supporting ecosystem services (ESS). CA in AFS can provide a wide spectrum of ecosystem services. It can result in soil conditions that lead to reduced erosion and runoff, and improved water quality compared to conventional practices. Likewise, water-holding capacity and storage are enhanced with CA providing some buffer to crop production during drought conditions. Different ecosystem services generated by CA AFS include: food and fodder, water, oxygen, energy (provisional services); C sequestration, climate and weather regulation, waste decomposition, water and air purification, pest and disease control, drought and flood risk, erosion control (regulating services); recreation, aesthetic (cultural services); and nutrient dispersal and cycling, primary production (supporting services).

Climate Regulation

Agroforestry-based CA practices of retentions of trees, crop rotations, and surface residue retention are intended to increase C inputs to decrease decomposition through increased soil aggregation and protection of soil C from decomposers. The balance of those two processes and the resulting soil C vary with soil, climate, and other management practices. Agroforestry plantations with CA are intensively managed, and produce far more biomass than conventional forests. Ajit et al. (2016) estimated an average C sequestration potential of 0.21 Mg C/ha/year of the agroforestry system, representing varying edapho-climatic conditions.

Soil and Water Conservation

Integration of trees in crop lands helps in improving soil bio-physico-chemical processes. Trees generally have their roots well below the crop zone, and use nutrients from the lower soil layers, resulting in increased nutrient- and water-use efficiencies with the least competition with crops. CA with agroforestry practices of residue retention, ZT, and certain crop rotations increase SOM in the top soil, which in turn impact soil physical properties and processes that reduce erosion and runoff. Accordingly, it leads to improved N-use efficiencies and fewer N losses to the environment. Reducing runoff and water erosion with CA in agroforestry result in lower transport of sediments, nutrients, and pesticides, and thereby better water quality.

CA with Agroforestry in Bundelkhand Region: A Case Study

The Bundelkhand region of India is prone to severe droughts due to undulating topography, shallow soil depth, low water-holding capacity, and high evapotranspiration owing to intense radiation. Here very little or no residue is available for surface application due to its competing uses as fodder. Agroforestry is the only option in which biomass generation can be integrated along with crop production. Keeping in view the technological gap, efforts to adopt and promote CA practices are increasing in intensively cropped areas; however location-specific CA practices in conjunction with the agroforestry intervention are essentially required. The development of agroforestry-based location-specific CA systems is more appropriate and holistic for efficient utilization of natural resources for sustainability. Therefore, a research project has been undertaken at ICAR-CAFRI involving agroforestry systems based on teak and bael since 2014.

Crop Productivity in Bael- and Teak-based CA Systems

In a bael-based CA system, the seed yield of mungbean varied from 715 kg/ha (MT) to 760 kg/ha (CT). Similarly, in a teak-based CA system, the seed yield of mungbean varied from 640 (MT) to 671 kg/ha (CT). Among the residue management treatments, the addition of crop residues resulted in maximum positive effects on mungbean seed yield, and it varied from 684 kg/ha (without crop residue addition) to 765 kg/ha (with crop residue). A similar trend was also observed in a teak-based CA system. In the case of urdbean, CT had a better effect on the seed yield than MT, and it varied from 344 kg/ha in MT to 354 kg/ha in CT in a bael-based CA system. The addition of crop and leucaena residue increased the seed yield of urdbean from 324 (without crop/leucaena residue) to 333 kg/ha (with leucaena residue), and to 360 kg/ha (with crop residue). The application of residues also increased the seed yield of urdbean in a teak-based CA system. Barley and mustard crops grown in a bael-based agroforestry system gave a higher yield in CT (barley 2940 kg/ha and mustard 1334 kg/ha) than MT (barley 2880 kg/ha and mustard 1298 kg/ha). A similar trend was observed for yield when grown in teak-based agroforestry system (CT – barley 3015 kg/ha; CT – mustard 1304 kg/ha and MT – barley 2956 kg/ha; MT – mustard 1273 kg/ha). The application of crop and leucaena residue in bael- as well as teak-based agroforestry systems resulted in higher yields.

Soil Fertility in a Teak-based CA System

The initial levels of SOC, N, P, and K were 0.21%, 130, 4.7, and 185 kg/ha, respectively, at a 0–30 cm soil depth, while in the sub-surface (30–60 cm), the corresponding values for these nutrients were 0.17%, 117, 4.15, and 175 kg/ha. After harvesting of mustard and barley during 2017, the soil samples in different tillage and residue management treatments were analysed for SOC, N, P, and K (Table 14.4). The available N content in 0–30 cm depth of soil ranged from 150 kg/ha in CT (Bg-M) to 168 kg/ha in MT (Bg-M). Available P and K ranged between 5.8–7.5 kg/ha and 215–233 kg/ha, respectively. In the sub-surface layer (30–60 cm), SOC, N, P, and K declined with depth and followed a similar trend as in the case of the 0–30 cm depth. Residue addition showed a remarkable change in the status of N, P, and K. The available N in leucaena and crop residue-added plots increased by 31.3% and 15.6%, respectively, over no residue. Further, the available P ranged between 7.2–8.4 kg/ha in the same treatment combination. No major change was observed in available K with residue addition. Thus, the residue addition brought about a visible change in the soil fertility status of major nutrients.

MT improved the SOC in both surface and sub-surface layers compared with CT. It varied from 0.35% in CT (G-B) to 0.44% in MT (G-B) in the 0–30 cm soil layer. The addition of crop and leucaena residues increased the SOC by 20.0% and 42.5%, respectively, over the no-residue plots. In the 30–60 cm depth, there was a decrease in SOC concentration due to a lower amount of organic matter and residue addition (Table 14.3).

Table 14.4 Effect of tillage and residue management on depth-wise distribution of nutrients in teak-based agroforestry system

Soil depth (cm)	Parameters	Initial	Tillage practice				Residue management		
			CT		MT		WCR	CR	LR
			BgM	GB	BgM	GB	BgM	BgM	BgM
0–30	Organic C (%)	0.21	0.38	0.35	0.42	0.44	0.40	0.48	0.57
	Available N (kg/ha)	130	150	156	168	165	160	185	210
	Available P (kg/ha)	4.7	5.9	5.8	7.5	7.3	7.2	7.9	8.4
	Available K (kg/ha)	185	215	222	227	233	225	230	234
30–60	Organic C (%)	0.17	0.29	0.26	0.36	0.34	0.32	0.41	0.42
	Available N (kg/ha)	117	145	149	158	152	155	175	178
	Available P (kg/ha)	4.15	5.5	5.3	7.1	7.2	6.9	7.3	8.0
	Available K (kg/ha)	175	200	214	224	226	217	235	240

*CT, conventional tillage; MT, minimum tillage; G, greengram; B, barley; Bg, blackgram; M, mustard; WCR, without crop residue; CR, with crop residue; LR, with leucaena residue.

Table 14.5 Effect of tillage and residue management on depth-wise distribution of nutrients in bael-based agroforestry system

Soil depth (cm)	Parameters	Initial	Tillage practice				Residue management		
			CT		MT		WCR	CR	LR
			BgM	GB	BgM	GB	BgM	BgM	BgM
0–30	Organic C (%)	0.15	0.26	0.21	0.29	0.25	0.26	0.33	0.36
	Available N (kg/ha)	107	119	127	139	131	136	145	156
	Available P (kg/ha)	2.8	3.6	3.9	4.9	4.2	4.0	4.2	4.9
	Available K (kg/ha)	122	146	154	169	172	147	158	169
30–60	Organic C (%)	0.09	0.17	0.15	0.22	0.2	0.22	0.25	0.29
	Available N (kg/ha)	94	116	123	126	119	119	125	132
	Available P (kg/ha)	2.4	3.2	3.8	4.1	3.9	3.1	3.5	3.9
	Available K (kg/ha)	111	129	124	142	138	129	134	143

*CT, conventional tillage; MT, minimum tillage; G, greengram; B, barley; Bg, blackgram; M, mustard, WCR; without crop residue; CR, with crop residue; LR, with leucaena residue.

Soil Fertility in a Bael-based CA System

The initial SOC and available N, P, and K in a bael-based CA system were 0.15%, 107, 2.8, and 122 kg/ha in 0–30 cm depth and 0.09%, 94, 2.4, and 111 kg/ha, in 30–60 cm depth, respectively. After 3 years, the SOC and available N, P, and K increased in both tillage treatments; however they were higher in MT as compared to CT (Table 14.5). The application of residue brought about remarkable changes in soil fertility status. SOC and available N, P, and K were found to be higher in leucaena residue plots, followed by crop residue applied plots.

Role of Soil Fauna in a Teak-based CA System

Soil organisms are essential components of agro-ecosystems, making vital contributions to soil functions and soil processes. Without soil organisms, the soil would be a sterile medium that may not be able to sustain crop production. Soil biota provide essential benefits for the functioning of agro-ecosystems, which are important for long-term sustainability. They support essential soil processes and play a key role in improving the soil fertility and thereby enhance crop productivity. Soil organisms help in the decomposition of crop residues, increase the availability of nutrients for plant growth, and contribute to soil carbon storage. In an agroforestry system, tree species significantly add leaf litter to the soil which improves the soil fertility through the actions of soil fauna. Maintaining soil biodiversity for sustainable agriculture is connected with maintaining available organic matter and essential nutrient sources in the soil.

A study was conducted at Jhansi to estimate the soil fauna in different residue application treatments in a CA-based AF system during pre-sowing and pre-harvesting of the crops. It was observed that before sowing of the crop, the population of micro-, meso-, and macro-fauna was at its maximum in the soil with the application of crop residue as compared to the application of leucaena residue and without any residue application (Figure 14.1) and a similar trend was noticed during pre-harvesting of the crops (Figure 14.2).

The micro-fauna population was higher in MT compared to CT, however, the meso- and macro-fauna populations did not differ significantly in MT and CT (Figure 14.3). Interestingly, the actinomycetes population was low in leucaena

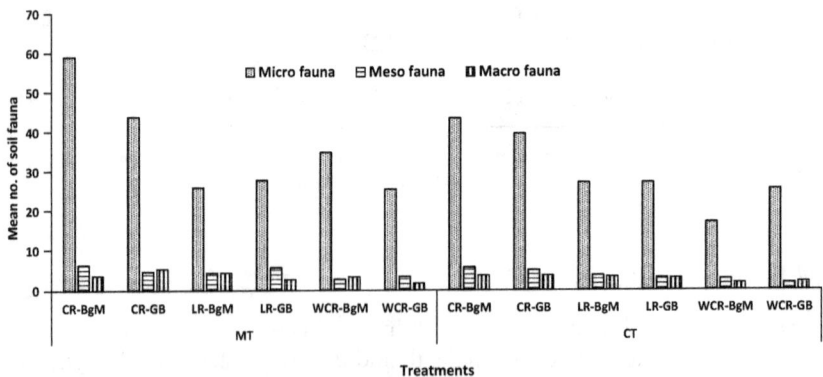

Figure 14.1 Abundance of soil fauna in different tillage and residue management practices during pre-sowing of crops.

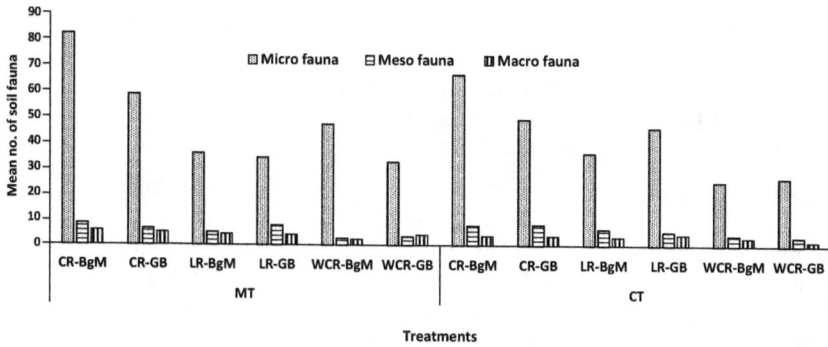

Figure 14.2 Abundance of soil fauna in different tillage and residue management practices during pre-harvesting of the crops*CT, conventional tillage; MT, minimum tillage; G, greengram; B, barley; Bg, blackgram; M, mustard; WCR, without crop residue; CR, with crop residue; LR, with leucaena residue.

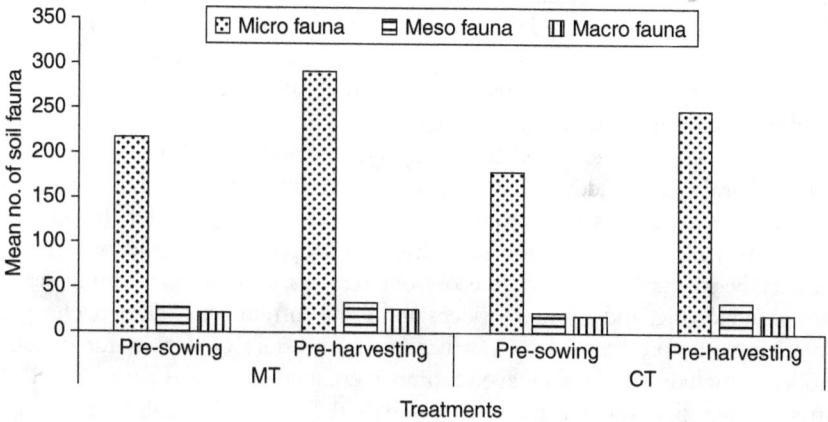

Figure 14.3 Abundance of soil fauna in minimum tillage (MT) and conventional tillage (CT) systems during pre-sowing and pre-harvesting periods.

residue-added treatments in pre-sowing as well as pre-harvesting treatments. Long-term studies are necessary to understand the effect of leucaena residues on actinomycetes populations.

Srinivasa Reddy et al. (1999) reported that higher soil fauna was recorded in soil conservation practices (enriched with organic manures) compared to without manure. Collembola were found to be relatively more abundant (44.9%) in the conservation method, whereas insect and non-insect fauna were predominant in recommended practices. Soil samples from fields with cereals, vegetable crops, ornamental plants, fruit trees, and forest plantations showed that crypto-stigmatid mites were the most abundant, followed by mesostigmatids, prostigmatids, and astigmatids. The relative densities of cryptostigmatids and mesostigmatids were very similar in annual and perennial crops. In terms of abundance, prostigmata were more abundant in the upper layer and mesostigmata in the deeper layer (Walia and Mathur 1994).

Conclusion

Conservation agriculture (CA) has emerged as a climate smart practice, which could enable sustainability in food production in vulnerable climatic conditions. CA-based agroforestry is also playing an effective role in lowering the vulnerability and enhancing the resilience of agriculture against climate extremes. The practice of residue application in CA improves the soil fertility and soil faunal abundance. In agroforestry-based CA, crop yields were at a par in conventional and minimum tillage. However, crop and leucaena residue applications increased the nutrient status and crop yields substantially. The micro-, meso-, and macrofauna populations also increased in minimum tillage and residue application. Although the yield enhancement under different CA practices was substantial in the initial years, it is expected that this will further increase considerably as these practices have positive impacts on soil bio-physico-chemical properties.

Agroforestry is a multi-faceted subject, which along with CA sustains the yield with minimal environmental damage. Combining CA with agroforestry increases C sequestration, and improves the availability of nutrients and soil faunal abundance. In addition, it helps in biodiversity conservation and reduces the pressure on forests for various products, such as fruit, fodder, fuel-wood, and timber. Accordingly, it increases livelihood opportunities and nutritional security, besides providing various ecosystem services, such as provisioning, regulating, supporting, and cultural services under the current era of climate change. Practicing agroforestry with CA will ease the pressure on forests for various products, including fodder, fuelwood, timber, etc. and also act as a source of leaf litter residue. Simultaneously, it will diversify the ecosystem, maintain sustainability, and increase nutritional security as well as the livelihood opportunities in the era of climate change.

References

Ajit, Dhyani, S.K., Handa, A.K., et al. 2016. Estimating carbon sequestration potential of existing agroforestry systems in India. *Agroforestry Systems* 91: 1101–1118.

Ajit, Dhyani, S.K., Newaj, R., et al. 2013. Modeling analysis of potential carbon sequestration under existing agroforestry systems in three districts of Indo-Gangetic plains in India. *Agroforestry Systems* 87(5): 1129–1146.

Basu, J.P. 2014. Agroforestry, climate change mitigation and livelihood security in India. *New Zealand Journal of Forestry Science* 44: 1–10.

Baudron, F., Duriaux Chavarría J-Y, Remans, R., et al. 2017. Indirect contributions of forests to dietary diversity in Southern Ethiopia. *Ecology and Society* 22(2): 28.

Bhattacharyya, T., Pal, D.K., Mandal, C., et al. 2000. Organic carbon stock in Indian soils and their geographical distribution. *Current Science* 79: 655–660.

Biswas, P.P. and Sharma, P.D. 2008. A new approach for estimating fertilizer response ratio - the Indian Scenario. *Indian Journal of Fertilizers* 4: 59.

Dhyani, S.K. and Tripathi, R.S. 2000. Biomass and production of fine and coarse roots of trees under agrisilvicultural practices in north-east India. *Agroforestry Systems* 50(2): 107–121.

Dhyani, S.K., Ram, A. and Dev, I. 2016. Potential of agroforestry systems in carbon sequestration in India. *Indian Journal of Agricultural Sciences* 86: 1103–1112.

Dhyani, S.K., Ram, A., Newaj, R., et al. 2020. Agroforestry for carbon sequestration in tropical India, pp. 313–331. In: Ghosh, P., Mahanta, S., Mandal, D., et al. (Eds.). *Carbon Management in Tropical and Sub-Tropical Terrestrial Systems*. Springer Nature Singapore.

Dhyani, S.K., Ram, Newaj and Sharma, A.R. 2009. Agroforestry: its relation with agronomy, challenges and opportunities. *Indian Journal of Agronomy* 54(3): 249–266.

FAO. 2009. How to feed the world in 2050. FAO CA website. www.fao.org/wsfs/forum2050

FAO. 2010. An international consultation on integrated crop-livestock systems for development. Integrated Crop Management Vol. 13, 79 p. FAO, Rome.

FAO. 2011. Save and Grow. A policymaker's guide to the sustainable intensification of smallholder crop production, 103 p. FAO Rome.

Garrity, D., Akinnifesi, F., Ajayi, O., et al. 2010. Evergreen Agriculture: a robust approach to sustainable food security in Africa. *Food Security* 2(3): 197–214.

Gathala, M.K., Kumar, V., Sharma, P.C., et al. 2013. Optimizing intensive cereal-based cropping systems addressing current and future drivers of agricultural change in the north-western Indo-Gangetic plains of India. *Agriculture, Ecosystems and Environment* 177: 85–97.

Ghosh, A., Kumar, R.V., Manna, M.C., et al. 2021. Eco-restoration of degraded lands through trees and grasses improves soil carbon sequestration and biological activity in tropical climates. *Ecological Engineering* 162: 106176.

ICAR. 2006. Handbook of Agriculture, 5th Edn. Indian Council of Agricultural Research, New Delhi.

ICRAF. 2009. Creating an evergreen agriculture in Africa for food security and environmental resilience. World Agroforestry Centre, Nairobi, Kenya.

Jat, H.S., Datta, A., Choudhary, M., et al. 2021. Conservation Agriculture: factors and drivers of adoption and scalable innovative practices in Indo-Gangetic plains of India – a review. *International Journal of Agricultural Sustainability* 19(1): 40–55.

Lewis, D., Bell, S., Fay, J., et al. 2011. Community Markets for Conservation (COMACO) links biodiversity conservation with sustainable improvements in livelihoods and food production. *Proceedings of National Academy of Sciences* 108(34): 13957–13962.

Lorenz, K. and R. Lal. 2018. Carbon Sequestration in Agricultural Ecosystems, 393 p. Springer, Nature, Switzerland.

Makumba, W., Janssen, B., Oenema, O., et al. 2006. The long-term effects of a gliricidia-maize intercropping system in southern Malawi, on gliricidia and maize yields, and soil properties. *Agriculture, Ecosystems and Environment* 116: 85–92.

Montagnini, F. and Nair, P.K.R. 2004. Carbon sequestration: An underexploited environmental benefit of agroforestry systems. *Agroforestry Systems* 61: 281–295.

Nair, P.K.R., Kumar, B.M. and Nair, V.D. 2009. Agroforestry as a strategy for carbon sequestration. *Journal of Plant Nutrition and Soil Science* 172:10–23.

NAP (2014). National Agroforestry Policy, 14 pp. Department of Agriculture and Cooperation, Ministry of Agriculture, Government of India.

Newaj, R., Dhyani, S.K., Chavan, S.B., et al. 2014. Methodologies for assessing biomass, carbon stock and carbon sequestration in agroforestry systems. Technical Bulletin 2/2014. ICAR-CAFRI, Jhansi.

Ram, A., Dev, I., Kumar, D., et al. 2016. Effect of tillage and residue management practices on blackgram and greengram under bael (*Aegle marmelos* L.) based agroforestry system, *Indian Journal of Agroforestry* 18(1): 90–95.

Ram, A., Dev, I., Uthappa, A.R., et al. 2017. Reactive Nitrogen in Agroforestry Systems of India. In: The Indian Nitrogen Assessment: Sources of Reactive Nitrogen, Environmental and Climate Effects, Management Options, and Policies. Elsevier Inc.

Rao, A.S. and Saha, R. 2014. Agroforestry for soil quality maintenance, climate change mitigation and ecosystem services. *Indian Farming* 63(11): 26–29.

Saha, S., Nair, P.K.R., Nair, V.D., et al. 2009. Soil carbon stocks in relation to plant diversity of home gardens in Kerala, India. *Agroforestry Systems* 76: 53–65.

Sidhu, H.S., Jat, M.L., Singh, T., et al. 2019. Sub-surface drip fertigation with conservation agriculture in a rice-wheat system: A breakthrough for addressing water and nitrogen use efficiency. *Agricultural Water Management* 216: 273–283.

Srinivasa Reddy, K.M., Kumar, N.G. and Rajagopal, D. 1999. Conservation of soil invertebrates in soybean agro ecosystem. *Journal of Soil Biology and Ecology* 19(2): 81–85.

Swamy, S.L. and Puri, S. 2005. Biomass production and C-sequestration of *Gmelina arborea* in plantation and agroforestry system in India. *Agroforestry Systems* 64: 181–195.

Walia, K.K. and Mathur, S. 1994. Acarine fauna of arable soils and their screening for nematophagy. *Indian Journal of Nematology* 24(1): 69–71.

Zomer, R.J., Trabucco, A., Coe, R., et al. 2009. Trees on farm: Analysis of global extent and geographical patterns of agroforestry. ICRAF Working Paper No. 89. ICRAF, Nairobi, Kenya.

Part IV

Economics, Adoption, and Future of Conservation Agriculture

15 Economic Aspects of Conservation Agriculture

H.S. Dhaliwal and Dharvinder Singh

Introduction

Air pollution, over time, has become a major challenge for governments and communities across the world. Both carrot and stick approaches have been tried to rein in the polluters and alleviate the suffering caused by pollution. There are three main sources of pollution, namely industry, households, and agriculture. It is estimated that around 25% of global greenhouse gases (GHGs) are generated through electricity and heat production, with 24% of GHG emissions coming from agriculture, forestry, and other land uses, 21% from industry, and the remainder being generated through transportation (14%), other energy sources (10%), and buildings (6%) (IPCC 2007). This scenario is no different in India, where air pollution has assumed dangerous proportions, especially during October to December. The northern part of India, especially the states of Punjab, Haryana, and Union Territory of Delhi, wake up to hazy mornings but it soon becomes apparent that this haze is in fact smog and there is little that can be done to assuage the suffering it causes.

So, where does the smog come from? What are its causes? What can be done to stop it from re-occurring? The answer to these questions forms the basis of this chapter in which we explain the extent of the problem and the technologies and machinery to available overcome it.

Open Residue Burning

Rice and wheat are the two main crops grown in India, especially in Punjab, Haryana, Uttar Pradesh, and some parts of Rajasthan. The assured market, meaning that the government guarantees procurement at a specified price, has resulted in the dominance of these crops in the cropping pattern of these states. Rice is generally grown from June to October and wheat from November to April. The residues of both these crops after harvesting are burnt in the field itself, with this practice known as open agricultural burning. The farmers resorting to fires to clear their fields of the standing stubble (rice) or stalks (wheat). The fields are cleared to prepare them for sowing of the next crop.

DOI: 10.4324/9781003292487-19

Table 15.1 Area, production, and percentage of straw burnt in Punjab and Haryana

Particular	Punjab	Haryana
Area under rice (M ha)	3.1	1.3
Production of rice straw (M tonnes)	20	8.5
Burnt crop residue (%)	75	60

Source: PAU (2016–17).

From the farmers' viewpoint and for a short-term benefit, the burning of residues might be the easiest, quickest, cheapest, and most convenient way of clearing the fields. However, the associated costs in terms of the impact on soil, plant and animal health, biodiversity, and the effect on the environment and on air, rail, and road traffic are colossal. Accordingly, society ends up paying heavily for it. Rice straw is burnt on a very large scale in the states of Punjab and Haryana. Table 15.1 depicts the extent of the straw which is burnt.

Why do Farmers Resort to Open Residue Burning?

Some of the reasons as to why farmers practice open burning of crop residue are the following:

a.   After the harvesting of rice in October, the wheat crop has to be sown by the second week of November. Agronomic practices suggest that any delay in sowing of the wheat crop leads to a loss of yield, which results in reduced income for the farmer. The time window between harvesting of the rice crop and sowing of the wheat crop is narrow (15–20 days). Farmers generally resort to open burning to quickly clear the fields.

b.   The economics of using fire seems the best to farmers as otherwise there would be the expenditure incurred of managing the crop residue using machinery. Farmers therefore usually avoid such expenditure and burn the crop residue.

c.   At times, farmers are also unaware of the *in situ* management practices, in which the residue is incorporated into the soil or used as mulch. Such practices can result in lower levels of fertilizer application, thus saving on input costs, enriching soil fertility, and saving at least one irrigation.

d.   At places there is lack of machinery to manage the crop residue. Since the agricultural operations are time-bound and generally carried out during the same time period, the availability of machinery becomes a constraint.

e.   The available machinery is large and heavy and is only used for 25–30 days in a year, and therefore, farmers shy away from buying it. Those who rent out this type of machinery often find it impossible to meet the needs of all farmers in their vicinity.

f.   The age-old practice of open agricultural burning is difficult for farmers to give up as they find it convenient.

Ill Effects of Open Residue Burning

The practice of open residue burning does not affect the quality of air only, but also incinerates the useful elements present in the crop residue. Burning of 1 tonne of rice straw leads to a loss of 5.5 kg N, 2.3 kg P, 25 kg K, 1.2 kg S, and 400 kg organic C (Singh et al. 2008).

Effect on the Environment

Open residue burning also has an impact on the environment. Singh et al. (2008) estimated that one tonne of rice residue releases 1515 kg CO_2, 92 kg CO, 3.83 kg N_2O, 0.4 kg SO_2, and 2.7 kg CH_4. Also, the burning of rice straw leads to the deposition of black carbon on glaciers, which speeds up the greenhouse effect in the cryosphere. This greenhouse effect is directly linked to the prevalent climate changes which manifest themselves in many forms such as severe heat waves, untimely and excessive rains, droughts, etc.

Effect on Human Health

The activity of open residue burning leads to acute asthmatic problems in children, old people, and pregnant women. Also, it can lead to conjunctivitis, leading to teary or watery eyes. Poisonous gases like carbon dioxide and carbon monoxide tend to mix with the red blood cells (RBC) in haemoglobin and reduce the oxygen-carrying capacity of the blood. This can also lead to stunted growth in children, while also limiting their intellectual development.

Effect on Animal Health

Straw burning also affects animals, where severe exposure to the gases emitted can lead to a potential decrease in milk yield of milch animals. There are also instances of eyes and skin diseases in animals, thereby leading to an increase in veterinary expenditures.

Effect on Biodiversity

Field fires cause irreparable damage to plantations and other vegetation on the roadside. These plantations and bushes are homes to some rare species of birds, who build their nest in bushes on trees along roadsides. Crop residue burning leads to the destruction of micro-flora and fauna present in the soil.

Effect on Traffic

Open agricultural burning affects rail, road, and air traffic. Some trains are delayed owing to the dense smog and others are cancelled. Similarly, air traffic is affected where aircraft are often delayed or the flights are diverted as they cannot land in the prevailing conditions. Vehicular traffic takes longer to traverse the same distance when encountering smog-like conditions, thus leading to higher greenhouse gas emissions.

Rice Straw Management Options

Rice straw management options are dealt with in two ways, i.e. *in situ* and *ex situ*. *In situ* management incorporates the straw in the fields or on the field, while *ex situ* management requires lifting the straw and using it for purposes such as fuel, bedding material, composting, etc. (Pathak 2004). Broadly speaking, there are four different methods to manage the straw:

a. *Retaining residue as mulch:* In this method, rice residue is used as mulch, which acts as a buffer in maintaining the soil temperature as well as conserving moisture and suppressing the growth of weeds.
b. *Residue incorporation:* In this method, the rice residue is ploughed back into the soil using machines/equipment such as a rotavator or mouldboard plough. However, this method is energy intensive and takes a lot of time.
c. *Residue collection:* Residue collection aims at collecting the rice straw residue from the fields and then using it as fuel in biomass power generation units, in biogas plants, as compost, in paper/plyboard-making factories, and also as bedding material for animals.
d. *Burning of rice straw:* Burning can be complete or partial. This is the cheapest, easiest, and quickest option to manage straw but the aftereffects are damaging not only to living beings but also to the environment.

Financial and Economic Analysis

Financial evaluation: This involves purely monetary factors. It takes care of direct costs and direct benefits – in order to ascertain the attractiveness of the option from the perspective of farmers. To convince farmers, it is a must to explain the direct monetary benefits of using straw management machinery.

Economic evaluation: This considers the total impacts of the option, from both direct and indirect points of view. The economic evaluation is done in totality, keeping in mind the impact on individuals, the community, and the environment. To carry out an economic evaluation:

a. One needs to determine the direct economic values for all relevant inputs and outputs.
b. Indirect costs and benefits: Health costs (human and animal), biodiversity loss, traffic hazards, environmental degradation etc. are indirect costs, whereas open burning leads to environmental hazards and its indirect risks. The indirect costs include increased veterinary and hospital admissions in the case of animals and humans, and the proportionate expenditure on medicines. In the case of traffic, the impact is on air, road, and rail traffic,

where trains are cancelled or diverted and flights are delayed. These are some of the indirect costs associated with open burning. These may not have a direct impact on the cost of cultivation but society indirectly pays for it in one or other form.

Comparative Financial Assessment of Different Rice Straw Management Options

In this section, the comparative costs of different rice straw management options are presented, and also discussed is the gross marginal analysis of different wheat sowing methods (on contract basis). The custom hiring charges of the machines/ equipment are given in Table 15.2.

In Tables 15.3–15.9 are the comparative costs of different methods of rice straw management.

A cost comparison of sowing wheat with different methods (on contract basis) and using the above tables enables the following analysis to be made:

There are savings of Rs. 3750, 1000, and 500 per ha, respectively, by using HST over CT, ZT, and rotaseeder, respectively. It can be seen that among all the

Table 15.2 Custom-hiring charges of agricultural machinery and equipment in Punjab (2017–18)

S. no.	Machine/equipment	Rate (Rs/ha)
1.	Combine harvester	3000
2.	Combine harvester with SMS	3750
3.	Happy seeder	3500
4.	Zero tillage machine	2250
5.	Rotavator	3500
6.	Rotaseeder	4000
7.	Seed-cum-fertilizer drill (SFD)	1250
8.	Spatial seed drill	2250
9.	Disc harrow	1625
10.	Cultivator	1125
11.	Planker	625
12.	Chopper/ mulcher	3000
13.	Mould board plough	5625
14.	Potato planter	4500
15.	Stubble shaver/cutter	1000
16.	Stubble shaver + rake + baler	6250
17.	Manual removal of loose straw	3000
18.	Manual harvesting and threshing	8750
19.	Grain collection and transportation to market	Rs 1250 for 1 tonne)
20.	Wage rate	40/hr (320 per day)
21.	Tractor	360/hr
22.	Irrigation charges	60/hr

Source: PAU (2016–17).

Table 15.3 When the rice residue is retained as straw mulch

Sr. no.	Method used	Rs/ha)
1	Combine harvester with SMS + Happy seeder (3000+750+3500)	7250
2	Combine harvester with SMS + spatial seed drill (3000+750+2250) (germination gaps, decreases yield)	6000
3	Combine harvester + manual removal of loose straw + zero-till drill (3000+3000+2250)	8250
4	Combine harvester with manual spreading + Happy seeder (3000+800+3500) (immediate spreading takes less time)	7300
5	Combine harvester + PAU straw cutter-cum-spreader + Happy seeder (3000+1000+3500)	7500

Table 15.4 When the rice residue is incorporated into the soil

S. no.	Method used	Rs/ha
1	Combine harvester + disc harrow (2 times) + rotaseeder drill (3000+3250+4000)	10250
2	Combine harvester with SMS + rotaseeder drill (3000+750+4000) (yield, weeds, and other issues)	7750
3	Combine harvester + chopper + MB plough + rotavator+ (SFD) (3000+3000+5625+3500+1250)	16375
4	Combine harvester + chopper + MB plough + rotavator + potato planter (3000+3000+5625+3500+4500)	19625

Table 15.5 When the rice residue is collected

S. no.	Method used	Rs/ha
1	Combine harvester + stubble shaver + rake + baler + no-till drill (3000+6250+2250)	11500
2	Combine harvester + stubble shaver + rake + baler + disc harrow (once) + cultivator (twice) + planker + SFD (3000+6250+1625+2250+625+1250)	15000
3	Manual harvesting and threshing + disc harrow cultivator + planker + SFD (8750+1625+1125+625+1250) (specifically for Basmati)	13375

Note: With residue collection, the soil is deprived of the nutrients which otherwise would have been incorporated in the soil through *in situ* management.

Table 15.6 When the rice residue is burnt

S. no.	Method used	Rs/ha
	Under complete burning	
1	Combine harvester + disc harrow (twice) + cultivator (twice) + planker (twice) +SFD (3000+3250+2250+1250+1250) (most commonly followed practice)	11000
	Under partial burning	
1	Combine harvester + Happy seeder (3000+3500)	6500
2	Combine harvester + spatial seed drill (3000+2250) (germination gaps, decreases yield)	5250
3	Combine harvester + no-till drill (3000+2250)	5250

*Note: These practices are prohibited by law.

Table 15.7 Cost comparison of sowing wheat with different methods

S. no.	Method used	Cost (Rs/ha)	Benefit of HST over other options (Rs/ha)
1	Residue retention as mulch using Happy seeder technology (HST)	7250	–
2	SFD after stubble burning conventional tillage (CT)	11000	3750
3	Zero tillage	8250	1000
4	Rotaseeder	7750	500

Table 15.8 Concepts used in gross marginal analysis

Total product (TP)	Total amount of main output (wheat grain) produced during the production process
By-product (BP)	Total amount of by-product (wheat straw) produced during the production process
Gross return (GR)	(Total product * Price/unit of product) + (Total by-product * Price/unit of by-product)
Fixed cost (FC)	Cost of fixed inputs incurred during the production of the crop (land rent, machinery charges, etc.)
Variable cost (VC)	The cost of using the variable inputs in the production process (seeds, manures, fertilizers, pesticides, insecticides, diesel, etc.)
Returns over variable cost/gross margin (ROVC/GM)	Gross returns (GR)– Variable cost (VC)

Table 15.9 Gross marginal analysis of different wheat sowing methods

Sowing method	Grain yield (t/ha)	Gross returns (×10³ Rs/ha)	Variable cost (×10³ Rs/ha)	Returns over variable cost (×10³ Rs/ha)	Benefit of HST over other methods (×10³ Rs/ha)
HST	5.0	91.25	27.22	64.03	–
CT	5.0	91.25	32.43	58.82	5.21
ZT	4.8	86.69	30.22	56.47	7.56
Rotaseeder	4.8	86.69	31.09	55.59	8.44

Source: PAU (2016–17).

methods of sowing of wheat, HST has a clear advantage as it involves the least cost (Sidhu et al. 2007; Singh et al. 2009).

Gross Marginal Analysis of Different Methods of Sowing of Wheat

It is imperative to know about the different concepts used in gross marginal analysis before setting out to explain the analysis. The concepts used are as follows.

The crop budgeting/estimation is done for the following methods/options:

i.   Happy seeder technology (HST)
ii. Conventional tillage (CT)
iii. Zero tillage (ZT)
iv. Tillage by rotaseeder.

Also, the costs per ha are considered for the following parameters:

i.   Seed and seed treatment
ii. Plant protection
iii. Irrigation
iv. Transportation and marketing cost
v.   Labour and power requirements for various farm operations
vi. Interest components on variable cost.

Also, the returns per ha are calculated, keeping in the mind the quantity of the:

i.   Main product
ii. By-product.

Conclusion

Air pollution caused by open residue burning causes much harm and no one is spared from its adverse effects. The onset of October leads to smog-like conditions in the northern part of India, and most of this is caused by open residue burning where the farmers in Punjab and Haryana burn rice residues on a large scale. The farmers burn the straw for many reasons, such as easiness of operation, small time window between clearing the fields and planting the next crop, lack of information about managing straw, and also at times there is nonavailability of straw management equipment or machines at certain locations. However, to convince farmers to not burn straw, one has to explain to them the costs involved in using straw management machinery. Farmers are usually concerned with the costs that impact them in the short term. It has been shown that using happy seeder technology (HST) is advantageous over using a seed-cum-fertilizer drill (SFD) by around Rs. 3750 per ha, over ZT by around Rs. 1000 per ha, and over rotaseeder by around Rs 500 per ha. As far as the gross marginal analysis of wheat is concerned, it is clear that wheat sowing by HST was advantageous over using CT, ZT, and rotaseeder by Rs. 5212, 7560, and 8437 per ha, respectively.

CA is a good management practice to conserve the elemental properties of soil and to augment production and productivity. However, any such technology or practice which is to be adopted by farmers will always be seen through the financial lens. Unless farmers can be sure that the adopted technique will be profitable, the practice is likely to be shunned. Therefore, the economic considerations become critically important, keeping in view the suitability of adoption of the technology by the farming community. The government should ensure that the requisite number of machines/implements are available to farmers and that youth/entrepreneurs are motivated to adopting the custom-hiring model, where such machines or implements could be rented out on an as-needed basis.

References

IPCC. 2007. Summary for Policymakers. Fourth Assessment Report of the Intergovernmental Panel on Climate Change. Cambridge University Press, New York.

Pathak, B.S. 2004. Crop Residue to Energy, pp. 854–869. In: *Environment and Agriculture*. Eds. K.L. Chadha and M.S. Swaminathan. Malhotra Publishing House, New Delhi.

PAU. 2016–17. Enterprise Budget-Rabi (2016–17), Department of Economics & Sociology, PAU, Ludhiana, Punjab, India.

Sidhu, H.S., Singh, M., Humphreys, E., et al. 2007. The Happy Seeder enables direct drilling of wheat into rice stubble. *Australian Journal of Experimental Agriculture* 47: 844–854.

Singh, R.P., Dhaliwal, H.S., Sidhu, H.S., et al. 2008. Economic assessment of the Happy Seeder for rice-wheat systems in Punjab, India. Conference Paper, AARES 52nd Annual Conference, Canberra. Australia.

Singh, Y., Sidhu, H.S., Singh, M., et al. 2009. Happy Seeder – A conservation agriculture technology for managing rice residues. Technical Bulletin No. 2009/01, Department of Soils, PAU, Ludhiana, India.

16 Adoption of Happy Seeder Technology in North-Western India

Rajbir Singh, Yadvinder-Singh, and A.R. Sharma

Introduction

The rice–wheat system (RWS) is the prevailing production system in the north-western (NW) states of Punjab and Haryana, covering around 4.2 M ha (3.0 M ha in Punjab and 1.2 M ha in Haryana). This is recognized as the heartland of the Green Revolution in India as it contributes the bulk of the foodgrains to the national food basket. The excessive use of irrigation water in the RWS is responsible for exploitation of ground water, with a high energy input cost in pumping ground water. As a result, a decline in water table depth has been observed in the majority of districts in this region. The average rate of decline has been 55 cm per year in Punjab (Aggarwal et al. 2009), and may have further increased over the past decade. The serious decline in the ground water table is threatening the sustainability of the RW production system. This will affect the socio-economic conditions of the small farmers, destroy the eco-logical balance, and adversely affect the sustainable agricultural production of the country.

Since the mid-1980s, the practice of manual harvesting of rice and wheat has been replaced with the advent of combines (Singh and Kaskaoutic 2014) that leave a large amount of residues on the field. Due to the lack of alternative eco-nomic uses of rice (paddy) straw and given the short window for preparing the land for the next wheat crop, the residue is subjected to burning in open fields (Badarinath et al. 2006). The cultivation of high-yielding varieties of rice and wheat produces about 34 million tonnes (M t) of rice residues in this region, of which, Punjab alone contributes 65% of the total residues. With the progres-sive increase in mechanized harvesting coupled with growing labour shortages, rice residue burning has become a serious problem and causes environmental pollution. It is estimated that about 23 M t of rice residues are burnt annually in the NW states. Apart from air pollution, burning causes a huge loss of carbon and plant nutrients, and thus deteriorates the soil health. About 80–90% of N and S, and 15–20% of P and K contained in rice residue are lost during burning (IARI 2012). According to an estimate, burning of 23 M t of rice residue leads to a loss of 1.4×10^5 t of N annually (NAAS 2017).

DOI: 10.4324/9781003292487-20

Crop Residue Burning and its Impact on Ecosystem Services

Residue burning adversely affects our ecosystem which provides a wide range of goods and services for the benefits of humankind. The Millennium Ecosystem Assessment (MEA) grouped ecosystem services into four categories: (i) provisioning services (direct or indirect food for humans, fresh water, fodder, fibre, and fuel); (ii) regulating services (regulation of water and air, climate, floods, erosion, biological processes such as pollination, and diseases); (iii) cultural services (aesthetic, spiritual, therapeutic, educational, and recreational); and (iv) supporting services (nutrient cycling, production, habitat, biodiversity). The impacts of crop residue burning on the four categories of ecosystem services are depicted in Figure 16.1.

Crop residue burning leads to poor air quality and adversely affects the health of millions of people in NW India. WHO consistently reports the poorest air quality ranking related to particulate matter (PM) in several cities in the NW India including New Delhi. A 15-year record (2002–2016) of NASA's satellite measurements have revealed a positive trend in the total fire activity and resulting aerosol loads over the Indo-Gangetic plain. It has been reported that there has been a nearly 43% increase in aerosol loading in NW India owing to post-harvest agricultural fire activities (Jethva et al. 2019). A recent study by the International Food Policy Research Institute (IFPRI) based on the health data of 0.25 million people has shown shocking economic and health implications of agricultural crop-residue burning (ACRB). The study estimated that ACRB

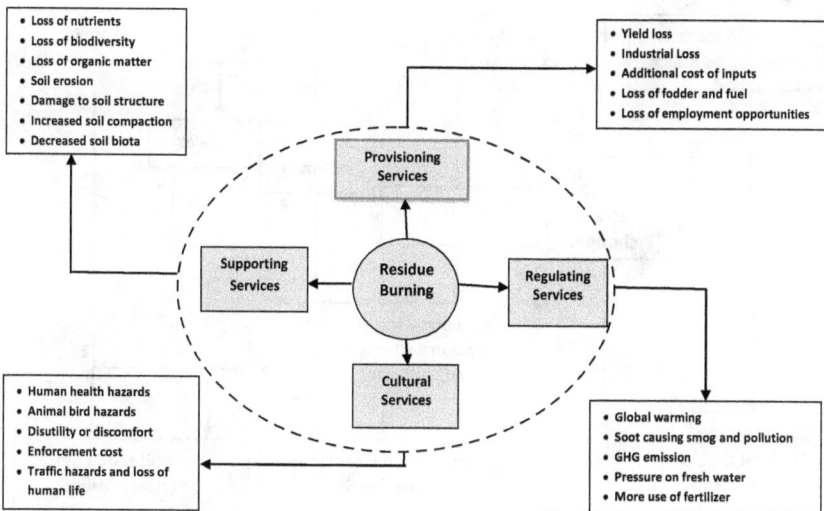

Figure 16.1 Residue burning and its impact on ecosystem services.
Source: Kumar et al. (2019).

causes USD 30 billion economic losses every year to India. Furthermore, it also revealed a threefold increased risk of acute respiratory infection to the exposed general population on account of ACRB (Chakrabarti et al. 2019). Similarly, Kumar et al. (2019) estimated the environmental cost of rice straw burning in NW India at Rs. 8953/ha, and the social cost of burning at around Rs. 31990 M per annum.

Management Options for Rice Residue

The majority of rice and wheat crops in NW India are harvested by combines that leave residues in the field. While about 75% of wheat straw is collected as feed for animals with the help of a special cutting machine known as a straw combine, rice straw is considered poor feed for animals and has no other economic use. Thus, the management of rice straw, rather than wheat straw, is a serious problem, because of the short turnaround time between rice harvesting and wheat sowing. However, there are some alternative options available for farmers for its efficient management either as *in situ* incorporation and mulch, or removal from the field manually or mechanically using a baler (Figure 16.2).

The various options for off-farm uses of rice residue include livestock fodder, livestock bedding, composting, generating electricity, mushroom cultivation, roof

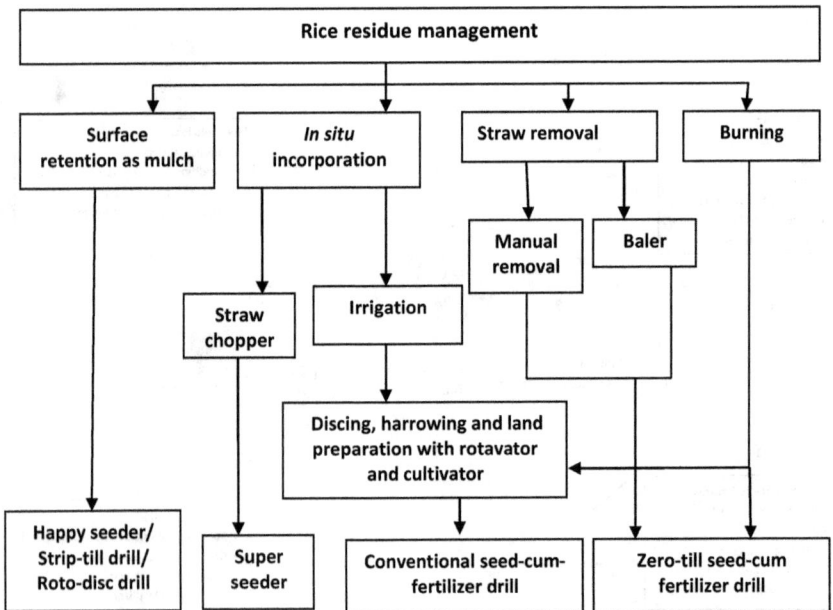

Figure 16.2 Options available for management of rice residue in the field.
Source: Lohan et al. (2018).

thatching, biogas (anaerobic digestion), biofuel, and paper and pulp board manufacturing. Currently, these options together utilize <15% of the total rice residue produced in NW India. Therefore, the need for providing cost-effective and scalable options for the *in situ* management of rice residue is both a major challenge as well as an opportunity for the sustainability of the intensive RWS. Managing 8–10 t/ha of residues efficiently and economically while allowing farmers to plant their wheat crop on time has been a daunting task.

In situ *incorporation of crop residues*

Soil incorporation of crop residues with high C:N ratios is beneficial in recycling nutrients but requires high energy input, cost, and time. This also leads to temporary immobilization of N, resulting in a yield decline over the residue removal or burning, which needs to be corrected by applying extra N fertilizer at the time of residue incorporation (Yadvinder-Singh et al. 2005; Bijay-Singh et al. 2008). Rice straw can be managed successfully *in situ* by allowing sufficient decomposition time (10–20 days) between its incorporation and sowing of the next wheat crop (Yadvinder-Singh et al. 2004). However, incorporating rice residue before wheat planting is challenging for farmers because of the short interval between rice harvest and wheat planting because it delays wheat planting and thereby adversely affects grain yield.

In situ *Retention of Rice Residues*

Science-based evidence suggests that *in situ* retention of crop residues can play an important role in buffering soil moisture and temperature (adaptation to climate risks), replenishing soil nutrients and organic matter, in addition to reducing environmental footprints (mitigation of GHGs). Furthermore, it also helps in checking weed seed germination by preventing light from reaching the soil surface to some extent. Direct seeding of wheat into rice residue allows timely planting of wheat and the elimination of in-field burning, thereby contributing to a sustainable RW production system. Until recently, the availability of suitable machinery was a major constraint for direct drilling of wheat into a heavy load of rice residue. The combination technology of a Super Straw Management System (Super SMS) fitted with a combine harvester and an innovative zero-till seed-cum-fertilizer drill has been recognized as a key technological innovation for direct drilling into a heavy surface residue load, and provides a viable alternative to burning (NAAS 2017).

Happy Seeder Technology for Sowing into Crop Residues

Conventional zero-till seed-cum-fertilizer drills introduced in NW India during the mid-1990s were suitable for sowing of crops in manually harvested fields from which the crop residues are taken out for threshing. Increasing adoption of combine-harvesters led to the crop residues remaining on the field and their widespread burning caused serious environmental problems. Considering this, a variant of the zero-till machine was developed about a decade ago, which could sow seeds in combine-harvested residue-retained fields.

The happy seeder (HS) is a combination of two machines, one having rotating blades in the front, which work on the soil surface, lift the crop residue, and spread it as a mulch after the seed and fertilizer have been placed with the drill behind at the desired depth in the soil. Evidently, the HS requires a heavy-duty tractor and can work in the anchored or loose residue load of 6–8 t/ha in a combine-harvested field. The residue is required to be uniformly spread and should be relatively dry for efficient working of the machine. Also, the soil should have optimum moisture for ensuring proper seed–soil contact for good germination. The seedlings emerge through the residue in rows, while the soil in the inter-row areas with residue mulch remains virtually undisturbed. This machine has undergone several modifications over the last decade and can be used efficiently for sowing most crops including small-seeded mustard to large-seeded maize with proper calibration.

The happy seeder technology was introduced initially for sowing of wheat from the early 2010s. There has been a significant increase in the number of happy seeders over the years and coverage of area under this technology in Punjab, Haryana, and elsewhere in India is estimated at nearly 1 M ha. Keil et al. (2021) reported that HS technology has the potential to eradicate the practice of rice residue burning due to its ability to sow wheat directly into large amounts of anchored and loose residues. This technology leads to significant savings in wheat production costs and benefits in terms of time and water savings, and the societal benefits of reducing air pollution by avoiding burning. Therefore, there is a strong justification for the diffusion of HS technology for *in situ* management of crop residues by farmers.

The happy seeder machine was developed, refined, and validated over several years under diverse farming systems of soil, crops, and weather to establish its significance (Sidhu et al. 2015). This innovative technology takes advantage of residual soil moisture in rice fields, thereby eliminating the need for pre-sowing irrigation. The development and commercialization of the Super SMS attachment for combine harvesters facilitates chopping of straw and spreading of the residue evenly as mulch, and has improved the field efficiency of the HS. Residue mulching in wheat also helps in reducing evaporation losses and buffers soil temperature as well as regulating canopy temperature, thus mitigating terminal heat effects. In addition, residue mulching along with the ZT system facilitates higher microbial and enzyme activities, which lead to better soil quality index and improved soil, plant, and water relations.

Upscaling of HS technology is vital for safeguarding the objectives of sustainable agriculture growth. The technology has been in the market for more than a decade now and is supported by various government interventions, such as financial support in the form of subsidies for purchasing machinery and legislation for completely banning residue burning in the states of Punjab, Haryana, Western Uttar Pradesh, Rajasthan, and the national capital region of Delhi.

Despite the science-based evidence on multiple benefits of this technology, its uptake has been slow due to the lack of proper understanding and knowledge as well as clarity on economic comparisons of various *in situ* management options among different stakeholders. Therefore, instead of 'forceful implementation', 'wilful adoption' of this technology having 'multiple wins' of eliminating residue burning, facilitating timely wheat planting, reducing production costs, and improving soil health should be promoted. Large-scale adoption of HS technology is predicated on a 'scalable business model' that ensures widespread and timely access to the technology through custom-hire services. The well-established business model elucidating its socio-economic benefits also warrants efficiently tapping into the new clientele through strategic expansion plans for concentration and diversification.

Initiatives to Combat Residue Burning

In the NW states of India, crop residue burning, particularly of rice during October–November each year, has assumed serious proportions and become a national issue. In fact, it has international ramifications as the air quality over Delhi becomes so poor that the government has to close traffic, schools, and offices, and enforce restrictions on peoples' movement. In order to address the problem of crop residue burning holistically, the Ministry of Agriculture & Farmers' Welfare, Government of India, initiated the Central Sector Scheme on 'Promotion of agricultural mechanization for *in situ* management of crop residue in the state of Punjab, Haryana, Uttar Pradesh and NCT of Delhi' in 2018 (MoA 2018). This scheme visualizes the promotion of agricultural mechanization for *in situ* management of crop residue with the following objectives:

- Protecting the environment from air pollution and preventing loss of nutrients and soil micro-organisms caused by burning of crop residue.
- Promoting *in situ* management of crop residue by retention and incorporation into the soil through the use of appropriate mechanization inputs.
- Promoting farm machinery banks for custom hiring of *in situ* crop residue management machinery to offset the adverse economics of scale arising due to small landholding and the high cost of individual ownership.
- Creating awareness among stakeholders through demonstration, capacity-building activities, and differentiated information, education, and communication (IEC) strategies for effective utilization and management of crop residues.

The scheme envisages financial assistance to individual farmers, co-operative societies of farmers, FPOs, self-help groups, registered farmers' groups, private entrepreneurs, groups of women farmers or their SHGs, for procurement or for establishment of farm machinery banks for custom hiring of *in situ* crop residue management machinery and equipment. The scheme has three major components: (i) establish farm machinery banks or custom hiring centres (CHCs) of *in situ* crop residue management machinery, (ii) procurement of agricultural machinery and equipment for *in situ* crop residue management, and (iii) information, education, and communication for awareness on *in situ* crop residue management. Under this scheme, financial assistance (subsidies) of 50% of the cost of machinery is provided to individual farmers for the purchase of crop residue management machinery. Further, financial assistance of 80% of the project cost is provided to the Farmers' Cooperative Societies, FPOs, registered farmers societies, and *Panchayats* for establishment of CHCs for renting crop residue management machinery.

With the roll out of this scheme to promote mechanization for *in situ* crop residue management, farmers were made aware and enrolled to procure machines. Over 3 years (2018–21), a total of 1,15,427 machines were provided to farmers for *in situ* management of residue in Punjab (75,372) and Haryana (40,055) (Table 16.1). The major machines procured under this scheme were: happy seeder (HS), super SMS, super seeder, mulchers/chopper, zero-till drill, and rotary slasher, which were used for managing *in situ* paddy straw. In Punjab, the highest number of machines were purchased in Muktsar, followed by Sangrur, Bathinda, Mansa, and Ludhiana. Similarly, in Haryana, most machines were made available in Fatehabad, followed by Kaithal, Sirsa, Jind, and Karnal. Many individual farmers (about 46,000) purchased their own machines by taking advantage of subsidies for the *in situ* management of rice straw.

Establishment of Custom Hiring Centres

This scheme encouraged the establish of CHCs as a business model for providing machines to rent to fellow farmers for scaling out residue management technology. Consequently, a total of 24,392 CHCs were established, with 69,075 machines in Punjab and Haryana over the last 3 years. In Punjab, 20,168 CHCs were established, with 47,839 machines, with Muktsar district having the highest number of CHCs (3,035), followed by Bathinda, Ferozpur and Mansa (Table 16.2). In Haryana, 4,224 CHCs were established with around 21,236 machines, and Fatehbad district had the highest number of CHCs, followed by Kaithal and Karnal.

Table 16.1 Total number of machines distributed on subsidy for *in situ* management of rice straw in Punjab and Haryana during 2018–21

Details	Punjab	Haryana	Total
Number of CHCs established	20,168	4,224	24,392
Number of machines available in CHCs	47,839	21,236	69,075
Number of machines with individual farmers	27,533	18,819	46,352
Total machines available	75,372	40,055	1,15,427

Source: Agriculture Department, Punjab and Haryana.

Table 16.2 Distribution of a number of straw management machines and CHC established in different districts of Punjab and Haryana in 2020–21

Punjab

District	Number of straw management machines	Number of CHCs	Number of machines in CHCs
Barnala	3,983	908	2576
Bathinda	6,105	2,028	4100
Faridkot	3,841	1,155	2873
Fazilka	4,460	1,026	2813
Ferozepur	3,740	1,612	2443
Jalandhar	4,266	1,169	2702
Kapurthala	3,283	920	2217
Ludhiana	4,872	897	2410
Mansa	5,139	1,505	3342
Muktsar	7,881	3,035	6634
Others	27,802	5,913	15729

Haryana

Districts	Number of straw management machines	Number of CHCs	Number of machines in CHCs
Karnal	4,233	522	2,655
Kaithal	4,945	586	2,848
Jind	4,299	515	2,379
Kurukshetra	3,870	503	2,635
Fatehbad	5,492	609	3,083
Sirsa	4,939	394	2,207
Ambala	1,579	183	843
Yamunanagar	1,608	167	881
Hisar	2,159	139	666
Rohtak	1,230	144	651
Others	5,701	462	2,388

Information, Education, and Communication for Mass Awareness

To mobilize the farmers for *in situ* crop residue management, systematic campaigns were initiated by the field staff of state governments. The Department of Agriculture, Cooperation & Farmers' Welfare (DAC&FW), Government of India, also sanctioned a special project to pull in all *Krishi Vigyan Kendras* (KVKs) located in the NW states. The ICAR–ATARIs are implementing the project through 60 KVKs: Punjab (22), Haryana (14), Delhi (1), and UP (23). These KVKs executed the activities under the information, education, and communication (IEC) component. Many workshops and focus group discussions were held with all stakeholders to identify the basic problems and technological gaps. Similarly, farmers' knowledge, attitudes, and practice survey (KAP approach) and benchmark-baseline surveys were conducted, and the results were used as inputs for the synthesis of a robust action plan. The strategic extension campaign plan was developed to provide specific directions in developing and making IEC activities in place. During the campaign, a participatory method was used to stimulate the local community as a whole to create a favourable psychological atmosphere for the adoption of this important technology intervention. A well thought out action plan was then implemented in a strategic manner in collaboration with all stakeholders.

An inter-disciplinary team of KVKs executed a well-designed programme to create awareness against rice residue burning and to promote *in situ* management of residue in Punjab, Haryana, Uttar Pradesh, and NCR of Delhi. Multi-layered innovative and strategic extension methodologies were developed and deployed to accomplish the defined objectives. Mobilizing stakeholders through awareness camps, demonstrating the innovative technology packages at strategic locations, capacity development of farmers and machine/tractor operators, and out-scaling the HS technology were important components of this drive.

KVKs were provided with farm machines to establish machine banks for organizing training, demonstrations, and exhibitions to promote *in situ* residue management. A total of 660 awareness camps were organized to sensitize stakeholders against crop residue burning and more than 72,000 farmers participated. The KVKs also organized 51 *Kisan Melas* (Farmers' Fair) on the theme of crop residue management in which 0.58 million farmers participated. More than 141 special programmes and debates on *in situ* residue management were aired on radio and TV besides coverage of events in mass media and extensive use of ICTs. The KVKs organized 237 training programmes of 5 days each in which 6,500 farmers, machine operators, and CHC owners were trained. More than 12,500 frontline demonstrations (FLDs) on HS-sown wheat have been conducted at farmers' fields at strategic locations in 350 villages over the last 3 years (2018–19, 2019–20 and 2020–21) in Punjab (Singh et al. 2019).

Impact of Initiatives to Stop Rice Residue Burning

The central government initiative in the form of the Central Sector Scheme on the promotion of agricultural mechanization for *in situ* management of crop residues led to the distribution of 1,15,427 machines to farmers for *in situ* management of rice straw and the establishment of 24,392 CHCs in Punjab and Haryana during 2018–2021. In Punjab, a total of 50,815 machines were procured by 2019–20 in which the maximum share was of HS (12,775), followed by ZT drill (10,704).

Impact on Rice Residue Burning

The rice residue burning events were monitored by multiple satellites with thermal sensors during the rice harvest and wheat sowing season from October 1 to November 30 in the states of Punjab and Haryana. High-resolution satellite images at 20 m were acquired for the pre-burning and post-burning periods to map acreage and total area burnt in the districts of Punjab and Haryana in 2018. The area under rice and the area under residue burning was estimated by using moderate-resolution multi-date satellite images. The images were classified using a hierarchical decision rule-based algorithm and ground truth by the ICAR's Consortium for Research on Agro-ecosystem Monitoring and Modelling from Space (CREAMS) Laboratory at the Division of Agricultural Physics, Indian Agricultural Research Institute, New Delhi.

Punjab: In 2019 the total area under rice was estimated to be 2.96 M ha. Residue on about 1.12 M ha area under rice was burnt, which was about 37.4% of the total area compared with 2018 when rice residue on about 1.51 M ha (49.3%) was burnt. The total area under rice in 2019 was reduced by 104,700 ha compared to that in 2018, whereas the area under rice residue burning in 2019 was reduced by 406,220 ha, suggesting about a 12.0% reduction in the estimated area under residue burning compared to 2018. The number of burning events detected in 2016, 2017, 2018, and 2019 were 102,379, 67,079, 59,684, and 50,738, respectively (Figure 16.3). Thus, there were 15%, 24%, and 50% reductions in the numbers of burning events in 2019 as compared to 2018, 2017, and 2016, respectively.

Haryana: In 2019, the total area under rice was estimated to be 1.34 M ha in Haryana. Rice residue on about 237,210 ha was burnt, which was about 17.7% of the total sown area in 2019. There were 31%, 51%, and 60% reductions in the numbers of burning events in 2019 compared to 2018, 2017, and 2016, respectively (Figure 16.4).

Impact of Happy Seeder Technology

Efforts to popularize direct seeding of wheat in anchored and loose stubbles with the help of the HS were started in 2014–15, and demonstrations were laid out at strategic locations in different districts of Punjab and Haryana by the KVKs. The area under wheat sown with HS increased over time in Punjab (Figure 16.5). In 2016, the area under wheat sown with HS was only 15,000 ha, which increased to 35,000 ha in 2017 due to the support provided by the state and central governments. In 2018, with the increased availability of HS with

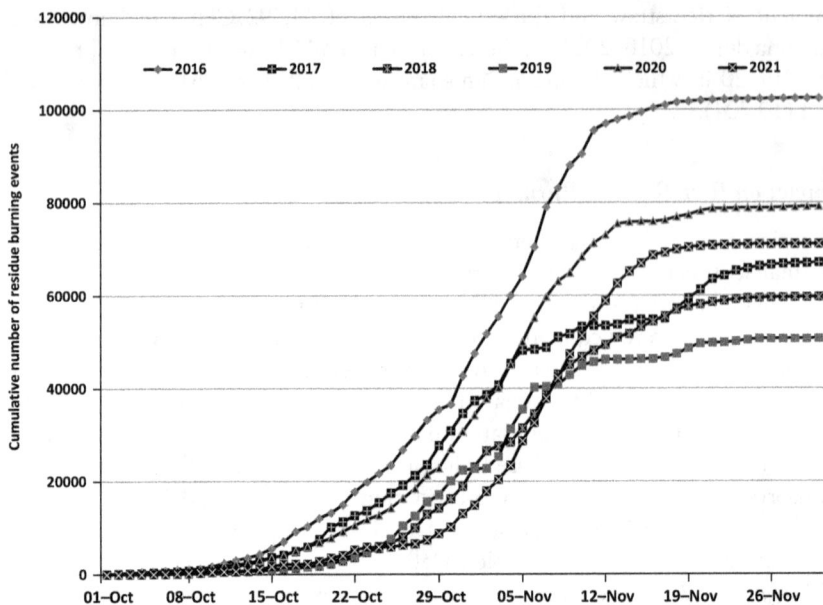

Figure 16.3 Date-wise cumulative number of residue burning events in different years in Punjab.

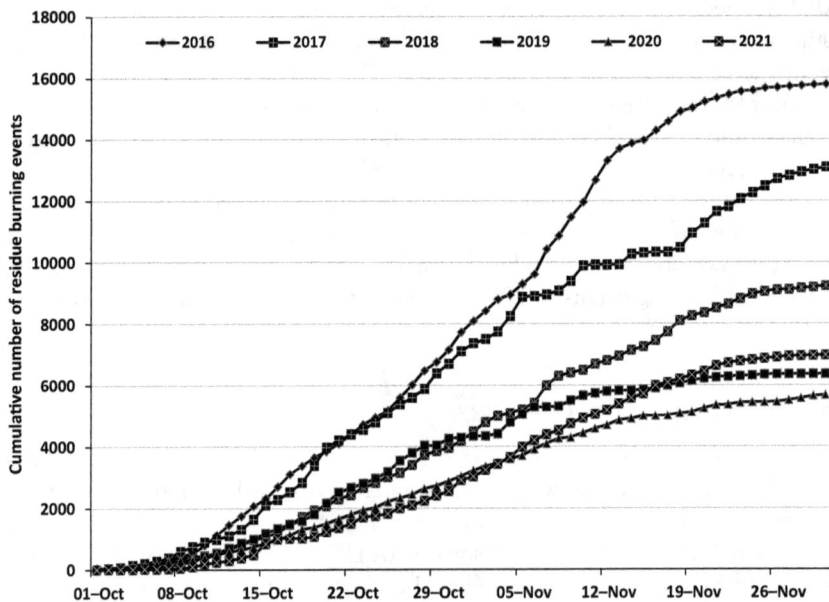

Figure 16.4 Date-wise cumulative number of residue burning events in different years in Haryana.

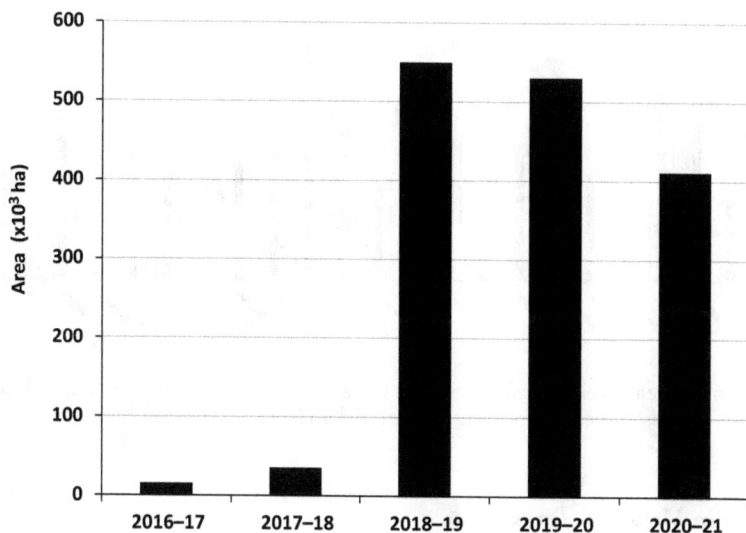

Figure 16.5 Date-wise cumulative number of residue burning events in different years in Punjab.

subsidies, the area under wheat sown with HS reached a new record of 0.55 M ha, which accounted for about 17% of the area under wheat (PRSC, Ludhiana). In 2019, the area under HS-sown wheat was 0.53 M ha, a slight reduction of 3% compared to 2018. However, the area under HS-sown wheat was reduced further to 0.41 M ha mainly due to the introduction of more than 10,000 new super seeder machines on subsidies by the state government which attracted farmers during 2020–21. The district-wise area sown under wheat using HS in Punjab was highest in Ludhiana (57,410 ha), followed by Sangrur (52,500 ha). Print media reported an estimated figure of 0.6 M ha of wheat sown area under paddy residue management (zero paddy straw burning) using various approaches in 2018 (*The Indian Express*, December 21, 2018). Thus, about 0.8 M ha (19%) of area was sown using both HS and ZT drill in Punjab and Haryana during 2018–19.

Remote sensing-based acreage of wheat sown with HS showed a variable trend in different districts of Punjab over the last 3 years (Figure 16.6). In 2018–19, the area was highest in Ludhiana, followed by Sangrur, Hoshiarpur, and Bhatinda. In 2019–20, Sangrur, Mansa, and Bhatinda were the main districts adopting HS-sown wheat technology. In general, the area under HS-sown wheat decreased in most districts in 2020–21, primarily due to the launching of other machines for sowing wheat into rice residues such as the super seeder. It is expected that further improvements in these machines will lead to greater coverage and that they will become popular with farmers in the coming years.

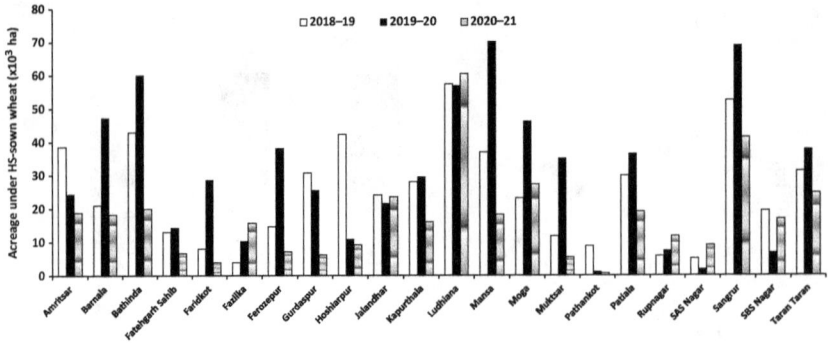

Figure 16.6 Area coverage under HS-sown wheat over the past 5 years in Punjab. Remote sensing-based acreage estimated under HS-sown wheat in Punjab.

Impact in Monetary Terms

Direct seeding of wheat with HS or ZT drill resulted in savings of precious inputs as wheat was directly sown in standing stubble. Our earlier studies based on 4,100 demonstrations laid on 1,640 ha clearly indicated a saving of cost of cultivation to the tune of Rs. 4500 per ha (Singh et al. 2018). This saving is mainly due to a reduction in the costs of tillage (64.8%), weed management (24%), and irrigation (11.1%). If we extrapolate on an area of 0.8 M ha, the saving on cost of cultivation was to the tune of Rs. 3600 M per year in the states of Punjab and Haryana during 2018.

The economic value of the nutrients lost due to burning of paddy straw is estimated to be Rs. 3300 per ha. Thus, the total value of nutrient loss due to burning of rice straw is estimated as Rs. 2640 M/year. Sowing wheat with HS also resulted in the saving of irrigation water of 7 cm or 700 m³/ha and a total water saving of 560 million cubic meters (MCM). Recently, Keil et al. (2021) reported that HS has the potential to eradicate the practice of rice residue burning due to its ability to sow wheat directly into large amounts of residues. It provides significant savings in wheat production costs, amounting to approximately Rs. 8,800 per ha based on average wheat yields in south-eastern Punjab. In addition to monetary savings, the HS technology provided significant benefits in terms of time and water savings.

Impact on Air Quality

The quality of air is measured in terms of the air quality index (AQI), which considers the quantities of eight parameters in the air, namely PM_{10}, $PM_{2.5}$, NO_2, SO_2, CO, O_3, NH_3, and Pb. The Punjab Pollution Control Board (PPCB) monitors AQI through eight continuous ambient air quality monitoring stations (CAAQMSs) across Punjab. In addition to savings in production costs, HS also helps in the overall improvement of air quality due to non-burning of residues and other eco-services. The AQI for Punjab during October 1 to November 15 suggests that there was a significant improvement in the air quality during 2018 as compared to that in 2017. Data for air quality reported by the Punjab government illustrates that the AQI was 273 (poor – breathing discomfort to most people on prolonged exposure) during November 2017, which was significantly reduced to 132 (moderate – breathing discomfort to people with lung, asthma, and heart diseases) during November 2018. The range of AQIs during 2017 was between 168–393 (moderate to very poor), which came down to 69–251 (satisfactory to poor) in 2018. The sharp rise in AQI during October 21, 2017 immediately followed by a sharp decline was attributed to the festival of *Diwali*, which was on October 19, and thereafter, a sharp decline in AQI was also noticed after November 12. Similarly, the sharp rise in AQI on November 8 immediately followed by a sharp decline was also due to the *Diwali* festival on November 7 in 2018. Therefore, the decline in AQI after November 12 in 2018 was not as sharp as that in 2017. This suggests that fluctuations in air quality are also due to factors other than crop residue burning. The government is imposing many restrictions to check pollution during these months, including banning the use of crackers during *Diwali* to improve the air quality.

Residue Burning Villages in Punjab and Haryana

Each KVK has focused on 3–5 adopted villages where intensive efforts have been carried out to convert these villages into residue-burning-free villages. The KVK team worked with farmers and *Panchayats* in a participatory mode. Consequently, in Punjab and Haryana, 132 villages were declared to be 'stubble burning free' in 2018 covering more than 45,000 ha. Villages where residues on 80–85% of the total area were managed without burning were considered as zero-burning villages. Further, 1341 villages across Punjab, constituting about 17% of the total of 8000 villages where rice is cultivated, were declared to be zero-stubble-burning villages in 2018. Similarly, nearly 3200 villages (72% of the total of 4400 rice-growing villages) reported no burning in Haryana during 2018. These figures have improved considerably over the last 3 years due to the intensive efforts of the government, and are likely to achieve the target of zero stubble burning across Punjab and Haryana in the coming years.

Conclusion and Suggestions for Upscaling of HS Technology

Conventional zero-till seed-cum-fertilizer drills introduced in north-western India during the mid-1990s were suitable for sowing of crops in manually harvested fields from which the crop residues are taken out for threshing. The increasing adoption of combine harvesters led to the crop residues remaining on the field and their widespread burning caused a serious environmental problem. Considering this, a variant of the zero-till machine known as the 'happy seeder' was developed about a decade ago, which could sow seeds in combine-harvested residue-retained fields. The happy seeder technology was introduced for sowing of wheat from the early 2010s. There has been a significant increase in the number of happy seeders over the years and the current coverage of area under this technology in Punjab, Haryana, and elsewhere in India is estimated to be more than 1 M ha. This machine is becoming popular for sowing of most crops in many regions where the crops are harvested with combines, and it has the potential to promote the adoption of conservation agriculture in India.

The adoption of HS combination technology is the best remedy to prevent crop residue burning as it provides large financial benefits and ecosystem services. Despite these benefits, the uptake of this technology has been slower than expected. The wider adoption of HS will depend upon the easy availability of the machine at an affordable price. Subsidies have been an important component in making the HS technology affordable during the awareness-raising phase. Therefore, subsidies on the HS by the central and state governments should be continued for some more years. It should also be recognized that there are economic incentives inherent to the HS technology that could be harnessed if existing subsidy policies were revised, notably the provision of free electricity for irrigation. Establishing self-help groups and encouraging unemployed youth to take up custom hiring of HS and other machineries as a profession will help in spreading of the technology over a larger areas. Some farmers have the opportunity to supplement their farm income by the purchase of a HS for use on their own and neighbouring farm. At current hire rates, this business model will provide reasonable returns on the investment with minimum risks. The custom hiring services can also be done through private-sector service providers. Efforts should be made towards developing and promoting use of the HS as a multi-crop planter. Effective policies need to be developed for the engagement of service providers, extension, machinery manufacturers, and farmers in the implementation of HS technology. The private sector should be encouraged to conduct demonstration plots, machinery fairs, and the formation and consolidation of CA farmer mutual support groups. There is a need to equip and train entrepreneurial service providers who often need training in the technical aspects of its correct use, calibration, and maintenance, as well as training on the managerial skills of identifying and running a successful service provision model. There is a need to avoid cross or counter-productive messaging and policies such as promoting super seeders or rotoseeders at the same time and promoting *ex situ* residue management options.

References

Aggarwal, R., Kaushal, M., Kaur, S., et al. 2009. Water resource management for sustainable agriculture in Punjab, India. *Water Science Technology* 60: 2905–2911.

Badarinath, K.V.S., Kiran Chand, T.R. and Krishna Prasad, V. 2006. Agriculture crop residue burning in the Indo-Gangetic Plains: A study using IRS-P6 AWiFS satellite data. *Current Science* 91: 1085–1089.

Bijay-Singh, Shan, Y.H., Johnson-Beeebout, S.E., et al. 2008. Crop residue management for lowland rice-based cropping systems in Asia. *Advances in Agronomy* 98: 118–199.

Chakrabarti, S., Khan, M.T., Kishore, A., et al. 2019. Risk of acute respiratory infection from crop burning in India: estimating disease burden and economic welfare from satellite and national health survey data for 2,50,000 persons. *International Journal of Epidemiology* 48(4): 1113–1124.

IARI. 2012. *Crop Residues Management with Conservation Agriculture: Potential, Constraints and Policy Needs*, vii+32 p. Indian Agricultural Research Institute, New Delhi.

Jethva, H., Torres, O., Field, R.D., et al. 2019. Connecting crop productivity, residue fires, and air quality over Northern India. *Scientific Reports* 9: 16594.

Keil, A., Krishnapriya, P.P., Mitra, A., et al. 2021. Changing agricultural stubble burning practices in the Indo-Gangetic plains: Is the Happy Seeder a profitable alternative? *International Journal of Agricultural Sustainability* 19(2): 128–151.

Kumar, S., Sharma, D.K., Singh, D.R., et al. 2019. Estimating loss of ecosystem services due to paddy straw burning in North-West India. *International Journal of Agricultural Sustainability* 17: 146–157.

Lohan, S.K., Jat, H.S., Yadav, A.K., et al. 2018. Burning issues of paddy residue management in north-west states of India. *Renewable and Sustainable Energy Reviews* 81: 693–706.

MoA. 2018. Central sector scheme on promotion of agricultural mechanization for *in-situ* management of crop residue in the states of Punjab, Haryana, Uttar Pradesh and NCT of Delhi: Operational guidelines. Ministry of Agriculture and Farmers' Welfare, New Delhi.

NAAS. 2017.*Innovative Viable Solution to Rice Residue Burning in Rice-Wheat Cropping System through Concurrent Use of Super Straw Management System-fitted Combines and Turbo Happy Seeder*. Policy Brief No. 2, 16 p. National Academy of Agricultural Sciences, New Delhi.

Sidhu, H.S., Singh, M., Yadvinder-Singh, et al. 2015. Development and evaluation of the Turbo Happy Seeder for sowing wheat into heavy rice residues in NW India. *Field Crops Research* 184: 201–212.

Singh, R.P. and Kaskaoutis, D.G. 2014. Crop residue burning: A threat to South Asian air quality. *EOS Transactions AGU* 95(37): 333–340.

Singh, A.K., Alagusundaram, K., Kumar, A., et al. 2019. *In-situ Crop Residue Management: Key Outcomes and Learning*, 30 p. Division of Agricultural Extension, ICAR, New Delhi.

Singh, R., Kumar, A., Sidhu, R.S. et al. 2018.*Upscaling Happy Seeder Technology: Scientific Evidences from Demonstrations*, 120 p. ICAR-ATARI, Ludhiana, Punjab.

Yadvinder-Singh, Bijay-Singh, Ladha, J.K., et al. 2004. Long-term effects of organic inputs on yield and soil fertility in the rice–wheat rotation. *Soil Science Society of America Journal* 68: 845–853.

Yadvinder-Singh, Bijay-Singh and Timsina, J. 2005. Crop residue management for nutrient cycling and improving soil productivity in rice-based cropping systems in the tropics. *Advances in Agronomy* 85: 269–407.

17 Conservation Agriculture

The Future of Indian Agriculture

Raj Gupta, I.P. Abrol, and A.R. Sharma

Introduction

The global population grew from about 2.5 billion people in 1950 to 6.5 billion in 2005, i.e. by a factor of 2.6. Although the rate of population growth has decreased slightly thereafter, the trends indicate that the global population will reach 8.1 billion by 2025 and 9.6 billion people by 2050. Between 1950 and 2005, the total cultivated area increased slightly, but the acreages of fallow lands was reduced significantly. Global food output as measured in cereal and meat production, in turn, increased even more during the same period due to large increases in fertilizer use, herbicides, plant and animal breeding, and the expansion of irrigated area. Agricultural production systems use public goods – natural resources for the production of private goods (food, feed, fibre, fuel) and production of public services (e.g. mitigating climate change, regulating water, controlling erosion, supporting soil formation, and providing habitats for wildlife). Agricultural activities oriented towards the production of goods change the natural ecosystem with some land degradation. Since 1950, one-third of the soils intensively used for agricultural purposes have been profoundly altered from their natural ecosystem states because of moderate to severe soil degradation. Because of the expansion of agriculture into natural ecosystems and the trend to use more external inputs, the negative impact of agriculture on ecosystem services supply will require increasing attention. For these reasons, enhancing total food production without due diligence on soil degradation processes and ecosystem services has been a fundamental flaw of the Indian strategy for food security under the new realities of climate change.

Globally, research outputs have indicated that the inclusion of M^3 research, namely soil organic Matter, soil Microbes, and soil Moisture retention, is critical in arresting and reversing soil degradation processes. Most researchers are in agreement that M^3 soil attributes enhance soil productivity, improve nutrient- and water-use efficiency, reduce production costs, and significantly benefit the environment. There is an urgent need to move away from traditional agriculture (which has many conflicting and unsustainable practices) to production management systems that are more sustainable and that benefit the environment.

DOI: 10.4324/9781003292487-21

Innovative Soil Management Approaches

In recent years, two production systems, namely organic agriculture and conservation agriculture, have been in general circulation. Organic agriculture is a holistic production management system that promotes and enhances agro-ecosystem health including biodiversity, biological cycles, and soil biological activity (CAC 2001). It emphasizes the use of locally adapted management practices in preference to the use of off-farm inputs. This is usually accomplished using cultural, biological, and mechanical methods instead of synthetic material inputs in crop production. The sustainability of organic agriculture is quite often a subject of intense debate, with divergent views regarding its feasibility and productivity potential in resource-poor areas. Soil-borne diseases and pests are also known to play a role in productivity dips in the transition from conventional to organic farming. This calls for additional research on the subject of root health. It is also worth mentioning here that most of the information about organic farming is from temperate countries, with little science-based relevant information from the tropical region.

The second innovative production system is known as conservation agriculture. As part of sustainable land management approaches, conservation or zero-tillage agriculture is one of the most important technological innovations. Conservation agriculture (CA) consists of three broad intertwined management practices: (i) drastic reduction in soil disturbance, (ii) maintenance of continuous vegetative soil cover, and (iii) sound crop rotations. The CA production system mimics natural agro-ecosystems, and hence results in numerous environmental benefits such as decreased soil erosion and water loss due to runoff, decreased CO_2 emissions and higher C sequestration, SOM build-up, efficient nutrient cycling, reduced fuel consumption, increased water productivity, less flooding, and recharging of underground aquifers (World Bank 2005; IAASTD 2009). This also reduces compaction in the subsoil and cracking in vertisols. A zero-till agriculture with residue retention on the soil surface has been found to be more carbon efficient and helpful in producing more at less cost while also improve soil health in the process (Dubey and Lal 2009). Together with direct dry seeding, CA has huge potential for reducing the acreage of fallow lands in the vertisols of the central plateau region of India. It is our strong belief that CA can produce more food, sustainably, at less cost, while improving environmental quality and preserving natural resources.

The new technological innovations in CA are perceived as fundamental shifts from the age-old practice of excessive ploughing. Even though the term 'zero-till' refers to tillage practice, it is a complete soil and crop management system, sensitive to local situations and resource endowments of the farmers. The ZT system is 'divisible' in application and 'flexible in application' under diverse situations. It is a holistic agricultural system that incorporates crop rotations, use of cover crops, and maintenance of plant cover throughout the year, with positive economic, environmental, and social impacts. In the context of M^3 research, the role of soil microbes vis-a-vis soil organic matter needs additional discussion here.

Role of Soil Organic Matter

Long-term experiments, from all over the world, have pointed out that adequate and balanced use of mineral fertilizers generally result in an increase in SOM and better biological life in fertilized than non-fertilized plots (Ladha et al. 2011; Korschens et al. 2013). Given the fundamental coupling of microbial C and N cycling, the dominant occurrence of both elements in SOM and the close correlation between soil C and N, mineralization practices that lead to loss of soil organic C also have serious implications on soil N. Considerable evidence from ^{15}N-tracer investigations indicates that plant uptake is generally greater from native soil N than from N applied via fertilizers (Stevens et al. 2005). Thus, native soil N dictates the efficiency of applied fertilizer N as well as the quantity of N lost from the soil–plant system. This has implications for soil functioning and crop productivity. While it is considered that the availability of carbon substrate is normally the primary limiting factor for microbial activity in soils, this is not necessarily the case, and there is accumulating evidence that soil microbes may frequently be N limited (Schimel et al. 2005; Kibblewhite et al. 2008). For Indian agriculture, there is evidence that unrealized yield gaps due to N nutrient limitations (1 – Ya/Ymax) are less than 0.3, suggesting that there will be little or no expected yield benefit from a simple increase of N fertilization without paradigm improvements in soil–water and crop management practices for cropping systems (Lassaletta et al. 2014; Gupta et al. 2021). Hence, Indian agriculture needs a paradigm shift in land management practices leading to sustainability of agriculture during the summer monsoon season.

The foregoing discussion suggests that nutrients, land use, and management practices act as controlling inputs for the processes within the soil system governing soil health. If there are no additions of nutrients to replace those lost through crop off-take and other processes, the capacity of the soil ecosystem to deliver production and other services declines together with the health of the soil. The impact of nutrient additions on the assemblages of minerals, soil organic matter constituents, and soil organisms is complex. This is because the organisms involved in organic matter decomposition, nutrient cycling, and soil structure formation ultimately themselves become the primary or secondary constituents of soil organic matter (Rao 2017). This soil carbon sponge is of great significance for its influence on soil aggregation, water infiltration, water retention, and access to essential nutrients, and it supports a diverse range of microbial processes. Soil organisms respond sensitively to land management practices and climate, and correlate well with beneficial soil and ecosystem functions.

Dependence of Conventional Production Systems on Fertilizers

Conventional agricultural production systems have become increasingly dependent on chemical fertilizers to meet the nutritional requirement of crops. Moreover, the use of major nutrients (N:P:K) is rather in an unbalanced form

(7:3:1) compared with the generalized recommended ratio of 4:2:1. This has had consequences on soil fertility and the productivity capacity of soils.

The Prime Minister of India has urged that our dependence on chemical nutrient fertilizers be reduced to half in the coming years to protect the environ-ment and improve soil health. Despite the fact that this is a laudable objective and a challenge for agricultural researchers, developmental agencies and farmers, it is yet to ignite a debate on how this objective can be achieved sooner rather than later. This challenge would call for a major shift away from the singular focus on chemical fertilizers to more of the biological approaches to sustain and enhance the current levels of crop production. To save on fertilizers, we need to: (i) iden-tify potential agronomic practices that reduce the use of synthetic nutrients, and (ii) identify production management systems that have the targeted effect. The enunciated policy statement implies that we immediately promote the adoption of agronomic practices, such as the following:

i.   inclusion of legumes in the cropping systems
ii.   conjunctive use of chemical fertilizer and organics
iii. rely on nutrient recycling with cropping pattern differing in rooting pattern
iv.   application of beneficial symbiotic microbial associations
v.   deploy *in situ/ex situ* composting techniques to improve biotic activity in the soils
vi.   increase biological N_2 fixation
vii. raise green manuring crops
viii. use microbial inoculants to improve nutrient access in soils (arbuscular mycorrhiza and P-solubilizing bacteria), and
ix.   promote rational use of nutrients in cropping systems.

A production management system strategy requires promoting the adoption of soil, water, and crop management strategies that improve resource-use effi-ciency and build soil organic carbon. CA is one such innovative production management system, which is closer to organic farming. CA allows the use of agrochemicals and its yield potential is hardly debatable, unlike organic farming. CA as a production management system has the targeted effect of reducing the use of synthetic fertilizers through attributes such as the following:

i.   tillage practices that reduce the rate of SOM decomposition, runoff, and soil erosion (avoidance of summer deep ploughing), and conserve soil moisture to improve soil health
ii.   enable seeding in excessively moist (surface seeding) or dry soils (dry seeding) and inclusion of high biomass-producing crops
iii. residue retention and brown manuring
iv.   switch from monoculture to rotation cropping, and annual to perennial crops
v.   reduce fallowing during the rainy season
vi.   avoid sudden land use changes, and
vii. adoption of agroforestry systems.

SOM content and carbon levels are central to the ability of soil to provide essential services to society. Since soils have the potential to help mitigate climate change, they should be part of the solution, not part of the problem. On balance, it appears that to be able to move in the 'implied direction', it would be even more important that we reorient our production management systems such that they begin to promote and enhance agro-ecosystem health under new realities of climate change.

Meeting Water Requirements

According to the Water Resources Group (WRG) Report, India will need 1,498 billion m^3 water by 2030. The break-up of the water demand was projected at 80% for agriculture, 13% for industry, and 7% for domestic needs. Against this demand, India's current water supply is approximately 740 billion m^3. As a result, most of India's river basins could face severe deficits (758 billion m^3) by 2030 unless concerted action is taken in the Ganga, Krishna, and the Indian portion of the Indus – facing the biggest absolute gap (WRG 2009). Agricultural water productivity measures contribute towards closing the water gap, increasing the 'crop per drop' through a mix of improved water application and improvement in crop yield. Given the fact that water application through surface flooding of basins is popular with most farmers in India, there is an urgent need to switch over to ridge-furrow systems of irrigation using gated pipes and pressurized drip and sprinkler systems wherever possible to save on water. Molden et al. (2007) suggested that in areas with low water productivity, such as South Asia, reducing evaporation and improving soil quality remain important options for increasing water productivity. Practices that improve soil health, mulching, breeding for early crop vigour to shade the ground as rapidly as possible, and longer superficial roots can reduce evaporation and increase productive transpiration. Improvement of soil fertility can significantly improve transpiration efficiency, and improving soil physical properties including infiltration and water storage capacity can reduce evaporation. Besides this, ZT, residue management, and dry seeding practices can also result in saving of water. Together these methods can result in large increases in crop water productivity. We can further improve water productivity through a change-over from basin flooding to ridge-furrow irrigation methods and the use of gated lay-pipes for water conveyance. The World Water Report suggested that India's base case 2030 supply–demand gap could be solved with agricultural measures provided there is a strong shift in favour of CA.

Need for Improving Root Health

Skills are needed to be developed to address yield stagnation and declining factor productivity in long-term cropping systems such as the rice–wheat system. The productivity of this system seems to depend on plant-parasitic nematodes and plant-pathogenic fungi. Soil-borne pests and diseases are often difficult to control because symptoms can be hard to diagnose and management options are limited. Nematodes prevent good root system establishment and function, and their damage can diminish crop tolerance to abiotic stresses such as seasonal dry spells and heat waves and competitiveness with weeds. Several reports suggest that the cereal systems of South and Central Asia grown in areas with high nematode pressure will become increasingly susceptible to yield losses from nematodes under the temperature increase with climate change (Padgham et al. 2004; McDonald and Nicol 2005). Bio-fumigation of soils is achieved by the generation of isothiocynate compounds, which are secondary metabolites released from the degradation of fresh *Brassica* residues in soil. They have a similar mode of action to metam sodium, a common synthetic replacement for methyl bromide, and have been used to control a range of soil-borne fungal pathogens including *Rhizoctonia*, *Sclerotinia*, and *Verticillium*. For many plant parasitic nematodes, significant control is often achieved when solarization is combined with bio-fumigation.

Potential of CA in India

Indian agriculture has made some rapid strides during the Green Revolution era but food gains and their growth rates have subsided since the 2000s due partly to the fatigue of natural resources and poor R&D investments (Abrol and Gupta 2019). Additional food now has to increasingly come through improved productivity routes and use of lands lying fallow (>15 M ha) for various reasons. Therefore, harnessing the untapped potential of rainy and winter season fallow lands and improving the productivity of external inputs, particularly in the rainy season, in both irrigated and rainfed agriculture is of great importance (Gupta et al. 2021). In Indian agriculture, N fertilizers are used extensively, however their productivity has decreased over the years. Recent evidence suggests that when yield gaps are less than 0.3, there is little or no expected yield benefit from a simple increase of N fertilization without paradigm improvements in soil–water and crop management practices for cropping systems (Lassaletta et al. 2014; Gupta et al. 2021). Hence, Indian agriculture is in great need of a paradigm shift from tilled to no-till land management practices for sustainability of agriculture. Besides the shift in tillage practices, plant breeding efforts in rainfed/dryland systems would have to focus on increasing plant biomass, developing crop stress tolerance, and harnessing genotype × tillage × environment to improve crop yields and reduce production costs. This can be achieved through a paradigm shift from crop-based to resource-based research and developmental planning.

Indian farmers are generally small-holder farmers who grow two or more crops in a year depending on the availability of surface and ground waters. Irrigation water supplies primarily depend on the monsoon patterns, wherein >85% of the total annual precipitation is received through the southwest summer monsoons. Summer rains are highly variable in space and time, and are received in several high-intensity storms following the prolonged hot rainless summer periods. Monsoon rains provide relief from the sweltering summer heat conditions, and replenish depleted profile moisture to breathe new life into soils. Pre-monsoon rains in summer deep-ploughed bare fields promote slaking and break down of soil aggregates such as to facilitate erosion of fertile soil with runoff rain water. In the SAT region, land degradation by erosion is largely a monsoon phenomenon spread over a period of 2 months involving the loss of some or all of the following: soil, soil productivity, vegetation cover, biomass, biodiversity, ecosystem services, and environmental resilience (UNCCD 2017; Abrol and Gupta 2019). This requires that we make rainy season crop planting independent of the onset of monsoon rains so as to reduce crop losses due to land fallowing, late planting, soil moisture, and terminal heat stresses, besides providing surface cover against monsoon rain-enhanced soil erosion to prevent loss of fertile surface soil. The only solution, available to Indian farmers for all the above maladies, rests in the adoption of CA practices such as direct dry seeding, minimal soil disturbance, retention of crop residues, use of cover crops, and use of farmyard manure. In the Indian SAT region, CA has significant potential for enhancing the productivity of both irrigated and rainfed agriculture, besides harnessing the untapped potential of >15 M ha of rainy and winter season fallow lands. CA practices are also known to improve the resilience of saline and alkali soils (totalling 7 M ha in India) as well as offering significant potential for sequestration of carbon to offset climate change effects (Gupta and Rao 1994). In light of the above, it is our strong belief that CA principles can and must form an important component of the national strategy to produce more food sustainably at lower cost, while improving environmental quality and preserving natural resources.

The adoption of CA by a large number of farmers requires that quality planting machinery prototypes are available to them. This requires that private entrepreneurs be encouraged to set up new machinery manufacturing units in different agro-eco-regions to develop appropriately designed new prototypes suitable for the local farming conditions. Local government should train local blacksmiths and Industrial Training Institute (ITI) students in the setting up and repair of new agriculture machinery. Rural banks and line departments should be encouraged to provide subsidies and loans to farmers, custom service providers, and the farmers' producer organizations (FPOs) for the purchase of new prototypes, and training and setting up of repair workshops in villages/blocks. Thus, a shift from tilled to no-till agriculture requires institutional innovations and changes to the way research and extension activities are conducted in the country.

CA – The Future of Agriculture in India

It is widely perceived that conventional tillage practices are ecologically unsustainable as they degrade land by destroying soil structure and biodiversity, reduce soil organic matter content, cause soil compaction, increase runoff and soil erosion, and contaminate water bodies with pollutants and sediments, threatening land productivity, the environment, and human health. In addition, they produce unacceptable levels of greenhouse gas emissions, speeding up climate change. On the other hand, CA avoids many of the negative consequences of conventionally tilled agriculture by enhancing natural processes through the continuous avoidance of soil tillage, and permanent maintenance of a mulch cover through which diverse crops are directly seeded or planted and through which rainfall is allowed to enter the soil and be retained, thus reducing/preventing erosion. CA enhances the root environment (soil structure, carbon, nutrients, and moisture) and reduces the build-up of pests and diseases. Thus, CA leads to productive agriculture for enhanced food security and improved rural livelihoods. Many economic, social, and environmental benefits described by a number of researchers justify a fundamental reappraisal of the conventional farming methods.

Soils are a 'finite non-renewable' natural resource, considered to be the core component of agricultural development and ecological sustainability. These soils are visualized to provide almost 95% of the total food sources for sustainability of innumerable lives and provide a basis for many critical ecosystem services. It is rightly said that the survival of humankind depends on a handful of the surface soil, and that a nation which destroys its soil, also destroys itself. CA and many of its locally adapted variants offer alternative avenues to Indian farmers for practicing productive farming methods.

Accumulated positive experiences and scientific knowledge about CA are leading to its rapid adoption worldwide. Farmers now apply CA on over 200 M ha in more than 100 countries across a diverse range of agro-ecological zones and farm sizes in all continents except Asia, Africa, and Europe. In regions of North and South America, and Australia, CA has enhanced farm production and reduced farming costs, while conserving and enhancing the natural resources of land, water, biodiversity, and climate. The World Congress on Conservation Agriculture (WCCA) in its 8th meeting held in Switzerland in 2021 gave a clarion call to all concerned stakeholders to recognize that the conservation of natural resources is the co-responsibility, past, present, and future, of all sections of society (WCCA 2021). Further, the WCCA has called for the enacting of appropriate long-term strategies embracing the concepts of CA as a fundamental element in achieving the agriculture-related SDGs. Importantly, the WCCA has also suggested that CA practitioners not only nurture soil health but also contribute to environmental services needed by society at large and must be better appreciated and recognized as such. The CA community should aim at bringing at least 50% of the global cropped area or 700 M ha under good-quality CA systems by 2050.

India is a large country with diverse agro-ecological conditions in both irrigated and rainfed regions. With a continuously increasing population, there is enormous stress on natural resources. Available estimates suggest that about half the geographical land area in India has been degraded due to erosion and other land degradation processes. Undesirable practices such as stubble burning following combine harvesting in most parts of the country are not only causing environmental pollution but also depriving the soil of much-needed organic matter. The problem has assumed serious proportions in the north-western parts of the country during harvesting/sowing of the major rice–wheat system followed in the region. Furthermore, the increased use of chemicals and indiscriminate exploitation of ground water are taking a heavy toll on soil and water resources.

The real adverse effects of global warming are being witnessed now not only in terms of stagnant or decreasing productivity but also manifested in major disasters in the environmentally sensitive hilly and coastal zones of the country. In view of these challenges, the Government of India is envisaging promotion of alternative CA variant systems such as nature or organic farming, which are considered harmonious with nature and that address the concerns of climate change also.

Conservation agriculture is a potential strategy for mitigating climate change, and ensuring food and nutritional security in the tropical and sub-tropical environments of India. Despite the enormous amount of data generated through intensive research over the last 25–30 years showing largely positive effects on crop productivity, profitability, and soil health, there remains limited adoption of CA in farmers' fields. Some estimates suggest that CA principles are being followed, albeit in a partial manner, on some 5 M ha of irrigated lands in the north-western, central, and southern parts of the country. CA practices also have been partially adopted by many rainfed farmers in the country. These farmers now have recognized the benefits of reduced-till systems. These successes indicate that there is vast scope for replicating such success stories in similar environments elsewhere, for which concerted efforts at each level – research, extension, development, and policy – are needed. Moreover, it requires significant willpower to shift from conventional tilled agriculture to no-till innovative farming practices like CA and provide the much needed policy support for its accelerated adoption. We believe that CA has come to stay and 'co-evolving with agents' for change in South Asia. We must make all-out efforts to cover the targeted 20 M ha under CA by 2030 as per the call given by the leading policy makers in the country. CA has been the fastest adopted technology globally, and undoubtedly it has the potential to transform and become truly the agriculture of the future in India.

Conclusion

CA has emerged as a new approach to addressing problems of production agriculture in the wake of the second-generation challenge which accompanied the adoption of the Green Revolution over the past 5–6 decades. The CA approach also recognizes new challenges in the form of accelerated climate-related aberrations facing the world. More explicitly, the approach addresses: (i) the need for continued productivity improvements over shrinking land mass to meet increasing food demands, (ii) urgency of reversing processes contributing to resource and environmental degradation and improving soil health, (iii) urgency of enhancing the use efficiency of water in agriculture with increasing competitive demands from other development sectors, and (iv) reducing production costs and making agriculture more relevant to frequent climate aberrations becoming the new norm.

CA practices hold the potential to contribute to the sustainability of agricultural developments and their adaptability to changing climate. CA recognizes the need for a paradigm shift from tilled agriculture to minimal soil disturbance and maintaining surface cover to achieve the goals of food security and reversal of land degradation processes for achieving sustainability goals. This approach aims at achieving enhanced use efficiency of external agricultural inputs, optimal use of available land and water resources, and reversing resource degradation processes. The CA approach has to be operationalized in an eco-regional or resource management domain framework. Furthermore, it has to integrate the focus of research to address both the short- and longer-term contexts of resource use problems. CA is a problem-solving approach that calls for institutional innovations for organizing multi-disciplinary farmer participatory research teams for working out integrated solutions towards resolution of regionally defined problems.

References

Abrol, I.P. and Gupta, Raj. 2019. Climate change-land degradation-food security nexus: addressing India's challenge. *Journal of Agronomy Research* 2(2): 17–35.

CAC. 2001. Codex Alementarius Commission. *Organically Produced Foods*. FAO and WHO, Rome.

Dubey, A. and Lal, R. 2009. Carbon footprint and sustainability of agricultural production systems in Punjab, India, and Ohio, USA. *Journal of Crop Improvement* 23(4): 332–350.

Gupta, R.K. and Rao, D.L.N. 1994. Potential of wastelands for sequestering carbon by reforestation. *Current Science* 66: 378–380.

Gupta, Raj, Benbi, D.K. and Abrol, I.P. 2021. Indian agriculture needs a strategic shift for improving fertilizer response and overcome sluggish growth in foodgrains production. *Journal of Agronomy Research* 4(3): 1–16.

IAASTD. 2009. International Assessment of Agricultural Knowledge, Science, and Technology for Development. Global report: Agriculture at a crossroads. Washington, DC: Island Press; 2009. 590 p.

Kibblewhite, M., Ritz, K. and Swift, M. 2008. Soil health in agricultural systems. *Philosophical Transactions of Royal Society B Biological Sciences* 27: 685–701.

Körschens, M., Albert, E., Armbruster, M., et al. 2013. Effect of mineral and organic fertilization on crop yield, nitrogen uptake, carbon and nitrogen balances, as well as soil organic carbon content and dynamics: results from 20 European long-term field experiments of the twenty-first century. *Archives of Agronomy and Soil Science* 59(8): 1017–1040.

Ladha, J.K., Reddy, C.K., Padre, A.T., et al. 2011. Role of nitrogen fertilization in sustaining organic matter in cultivated soils. *Journal of Environmental Quality* 40(6): 1756–1766.

Lassaletta, L., Billen, G., Grizzetti, B., et al. 2014. 50 year trends in nitrogen use efficiency of world cropping systems: the relationship between yield and nitrogen input to cropland. *Environmental Research Letters* 9: 105011. doi:10.1088/1748-9326/9/10/105011

McDonald, A.H. and Nicol, J.M. 2005. Nematode parasite of cereals, pp. 131–191. In: M. Luc et al. (Ed). *Plant Parasitic Nematodesin Subtropical and Tropical Agriculture*. 2ndEd. Wallingford: CABI.

Molden, D., Oweis, T., Steduto, P., et al. 2007. Pathways for increasing agricultural water productivity. In: D. Molden (Ed). *Water for Food, Water for Life: A comprehensive Assessment of Water Management in Agriculture*. Earthscan, London and IWMI, Colombo.

Padgham, J.L., Abawi, G.S., Duxbury, J.M., et al. 2004. Impact of wheat on *Meloidogynegraminicola* populations in the rice–wheat system of Bangladesh. *Nematropica* 34(2): 183–190.

Rao, D.L.N. 2017. Microbial and biochemical origins of soil organic matter: Insights from history and recent ecological and bio-molecular advances, pp. 77–89. In: S.K. Sanyal (Ed.). Souvenir 82ndAnnual Convention and National Seminar of Indian Society of Soil Science, Kolkata.

Schimel, J.P., Bennett, J. and Fierer, N. 2005. Microbial community composition and soil nitrogen cycling: Is there really a connection? In: Bardgett R. et al. (Eds.). *Biological Diversity and Function in Soils*. Cambridge: Cambridge University Press.

Stevens, W.B., Hoeft, R.G. and Mulvaney, R.L. 2005. Fate of nitrogen-15 in a long-term nitrogen rate study: II. Nitrogen uptake efficiency. *Agronomy Journal* 97(4): 1046–1053.

UNCCD. 2017. Global Land Outlook, 336 p. First Edition. Secretariat of the United Nations Convention to Combat Desertification. Bonn, Germany.

WCCA. 2021. Declaration. 8th World Congress on Conservation Agriculture – The Future of Farming, 4 p. Bern, Switzerland, 21–23 June, 2021.

WRG. 2009. Charting our water future: Economic frameworks to inform decision-making. Water Resources Group.

World Bank. 2005. Agriculture investment sourcebook [Internet]. Report No.: 34392. Available from: http://documents.worldbank.org.

Index

For Product Safety Concerns and Information please contact our EU
representative GPSR@taylorandfrancis.com
Taylor & Francis Verlag GmbH, Kaufingerstraße 24, 80331 München, Germany

www.ingramcontent.com/pod-product-compliance
Lightning Source LLC
Chambersburg PA
CBHW052118230326
41598CB00080B/3812